Wireless Networks

Series Editor

Xuemin Sherman Shen, University of Waterloo, Waterloo, ON, Canada

The purpose of Springer's Wireless Networks book series is to establish the state of the art and set the course for future research and development in wireless communication networks. The scope of this series includes not only all aspects of wireless networks (including cellular networks, WiFi, sensor networks, and vehicular networks), but related areas such as cloud computing and big data. The series serves as a central source of references for wireless networks research and development. It aims to publish thorough and cohesive overviews on specific topics in wireless networks, as well as works that are larger in scope than survey articles and that contain more detailed background information. The series also provides coverage of advanced and timely topics worthy of monographs, contributed volumes, textbooks and handbooks.

** Indexing: Wireless Networks is indexed in EBSCO databases and DPLB **

More information about this series at http://www.springer.com/series/14180

Jing Yan • Haiyan Zhao • Yuan Meng •
Xinping Guan

Localization in Underwater Sensor Networks

Jing Yan
Institute of Electrical Engineering
Yanshan University
Qinhuangdao
Hebei, China

Haiyan Zhao
Institute of Electrical Engineering
Yanshan University
Qinhuangdao
Hebei, China

Yuan Meng
Institute of Electrical Engineering
Yanshan University
Qinhuangdao
Hebei, China

Xinping Guan
Department of Automation
Shanghai Jiao Tong University
Shanghai, China

ISSN 2366-1186 ISSN 2366-1445 (electronic)
Wireless Networks
ISBN 978-981-16-4833-5 ISBN 978-981-16-4831-1 (eBook)
https://doi.org/10.1007/978-981-16-4831-1

This Springer imprint is published by the registered company Springer Nature Singapore Pte Ltd.
The registered company address is: 152 Beach Road, #21-01/04 Gateway East, Singapore 189721, Singapore

Preface

The ocean covers 70.8% of the Earth's surface, and it plays a significant role in supporting the system of life on Earth. Nevertheless, more than 80% of the ocean's volume remains unmapped, unobserved, and unexplored. With regard to this, underwater sensor networks (USNs), which incorporate ubiquitous computation, efficient communication, and reliable control, have emerged as a promising solution to understand and explore the ocean. The deployment of USNs can enhance the monitoring capacity for various applications such as intrusion surveillance, marine resource protection, navigation, geographic mapping, and petroleum exploration. In order to support these applications, accurate location information of sensor nodes is required for correctly analyzing and interpreting the sampled data. However, the openness and weak communication characteristics of USNs make underwater localization much more challenging as compared with the terrestrial sensor networks.

In this book, we focus on the localization problem of USNs with consideration of the unique characteristics of USNs. The localization problem is of great necessary and importance since fundamental guidance on design and analysis on the localization of USNs is very limited at present. We first introduce the network architecture and briefly review the prior arts in localization of USNs. Then, we consider the asynchronous clock and node mobility during the localization procedure, through which a mobility prediction-based least squares estimator is developed to seek the locations of sensor nodes. Note that the first-order linearization is required for least squares estimator to calculate the Jacobian matrix; however, it can introduce large model errors. In view of this, we further design a consensus-based unscented Kalman filtering (UKF) localization estimator to relax the linearization requirement and improve the localization accuracy. In addition, we also employ the reinforcement learning (RL) to relax the linearization requirement and avoid the local minimum during the least squares-based localization procedure, such that an RL-based localization algorithm is provided. Besides that, we investigate the privacy-preserving localization issue for USNs. To this end, three privacy-preserving localization protocols are designed to hide the position information of reference nodes. Accordingly, the least squares and deep reinforcement learning (DRL) based localization estimators are developed, respectively, to jointly achieve privacy preservation, asynchronous localization, and stratification compensation for USNs. Finally, rich implications from the book provide guidance on the design for future localization schemes on USNs.

The results in this book reveal from the system perspective that the underwater localization accuracy is closely related to the communication protocol and optimization estimator. Researchers, scientists, and engineers in the field of USNs can benefit a lot from this book. As such, the valuable knowledge, useful methods, and practical algorithms can provide the guidance on understanding and exploring the ocean. To use this book for underwater applications, knowledge of wireless communication and signal processing is needed.

The book is organized as follows.

Chapter 1 provides the network architecture of USNs and briefly reviews the prior arts in localization of USNs. Besides that, the weak communication characteristics of USNs, including asynchronous clock, stratification effect, and node mobility, are also summarized in this chapter.

Chapter 2 considers the asynchronous clock and node mobility. A hybrid network architecture including autonomous underwater vehicles as well as active and passive sensor nodes is constructed. Then, an asynchronous localization solution with mobility prediction is developed for USNs, where iterative least squares estimators are conducted to seek the position information.

Chapter 3 presents a consensus-based UKF localization algorithm. Compared with the results in Chap. 2, the stratification effect is incorporated into the developed localization protocol in this chapter, and more importantly, the model error can be significantly reduced since the first-order linearization is not required by the localization algorithm in Chap. 3.

Chapter 4 covers an RL-based localization algorithm for USNs in weak communication channel. Note that the least squares-related localization estimators are adopted in Chaps. 2 and 3. However, the least squares-related estimators can easily fall into local minimum. In view of this, Chap. 4 employs the RL to seek the global optimization localization solution.

Chapters 2–4 assume the monitoring area is safe and the position privacy is ignored. However, USNs are usually deployed in open environment, and it is necessary to utilize the information-hiding technology to develop a privacy-preserving localization protocol for USNs. In view of this, Chap. 5 presents a privacy-preserving solution for the asynchronous localization of USNs, in which the asynchronous clock and node mobility are also considered.

Chapter 6 further considers the stratification effect and the forging attack in underwater environment, through which a privacy-preserving localization estimator is developed for USNs. It is worth mentioning that the malicious attacks can be detected and the straight-line localization bias can be compensated in this chapter, which are not available in Chap. 5.

Chapters 5–6 employ the least squares-based estimators to seek the position information of sensor nodes, which can easily fall into local minimum. In order to solve this issue, Chap. 7 develops DRL-based privacy-preserving localization protocol and estimator for USNs. Per knowledge of the authors, this is the first work that incorporates DRL and stratification compensation into the privacy-preservation localization of USNs.

Chapter 8 provides the future research direction on the localization of USNs. We have tried to provide complete instructions for the underwater localization and meanwhile share insights into the underwater localization from the system perspective. We hope this book has reached our goal.

This book was supported in part by the National Natural Science Foundation of China under Grants 62033011, 61873345, and 61973263, by Youth Talent Support Program of Hebei under Grants BJ2018050 and BJ2020031, and by Outstanding Young Foundation of Hebei under Grant 2020203002. We would like to acknowledge Prof. Cailian Chen for her contributions on preparing material and Prof. Xiaoyuan Luo for valuable discussions.

Qinhuangdao, China — Jing Yan
Qinhuangdao, China — Haiyan Zhao
Qinhuangdao, China — Yuan Meng
Shanghai, China — Xinping Guan
July 2021

Contents

About the Authors

Jing Yan received his B.Eng. degree in automation from Henan University, Kaifeng, China, in 2008, and his Ph.D. degree in control theory and control engineering from Yanshan University, Qinhuangdao, China, in 2014. In 2014, he was a research assistant with the Key Laboratory of System Control and Information Processing, Ministry of Education, Shanghai Jiao Tong University, Shanghai, China. From January 2016 to September 2016, he was a postdoc at University of North Texas, Denton, USA. From October 2016 to January 2017, he was a research associate at University of Texas at Arlington, Arlington, USA. Currently, he is a full professor at Yanshan University, Qinhuangdao, China. Meanwhile, he is also an associate editor for IEEE Access. His research interests cover underwater acoustic sensor networks, networked teleoperation systems, and cyber-physical systems. He has published more than 80 peer-reviewed papers in leading academic journals and conferences. He has also received numerous awards, including the Excellence Paper Award from the National Doctoral Academic Forum of System Control and Information Processing in 2012, the Outstanding Doctorate Dissertation of Hebei Province in 2015, the Excellence Paper Award from the National Doctoral Academic Forum of System Control and Information Processing in 2012, the Youth Talent Support Program of Hebei Province in 2019, the Outstanding Young Foundation of Hebei Province in 2020, and the Excellence Adviser from Oceanology International Underwater Robot Competition in 2017.

Haiyan Zhao received her B.S. degree in automation from Yanshan University in 2017. Currently, she is pursuing her Ph.D. degree in control theory and control engineering at Yanshan University, Qinhuangdao, China. Her research interests cover in underwater acoustic sensor networks and autonomous underwater vehicle. She won the national scholarship in 2019 and presided over Postgraduate Innovation Fund Project of Hebei in 2019.

Yuan Meng received her B.S. degree in measurement and control technology and Instruments from Liaoning Technical University, Huludao, China, in 2019.

Currently, she is pursuing her Ph.D. degree in control theory and control engineering at Yanshan University, Qinhuangdao, China. Her research interests include localization of underwater sensor networks and networked underwater robot control. Besides that, she won the national scholarship in 2021 and presided over Postgraduate Innovation Fund Project of Hebei in 2021.

Xinping Guan received his B.S. degree in applied mathematics from Harbin Normal University, Harbin, China, in 1986, and his M.S. degree in applied mathematics as well as Ph.D. degree in electrical engineering from the Harbin Institute of Technology, Harbin, in 1991 and 1999, respectively. He is currently a chair professor at Shanghai Jiao Tong University, Shanghai, China. He has authored and/or co-authored four research monographs, more than 270 papers in IEEE and other peer-reviewed journals, and numerous conference papers. His current research interests include industrial cyber-physical systems, wireless networking and applications in smart city and smart factory, and underwater sensor networks. Dr. Guan was a recipient of the National Outstanding Youth Honored by the NSF of China, the Changjiang Scholar by the Ministry of Education of China, and the State-Level Scholar of New Century Bai Qianwan Talent Program of China. He is an executive committee member of the Chinese Automation Association Council and the Chinese Artificial Intelligence Association Council.

Acronyms

AOA	Angle of arrival
AUVs	Autonomous underwater vehicles
CRLB	Cramér-Rao lower bound
DRL	Deep reinforcement learning
DVL	Doppler velocity log
FOG	Fiber optic gyroscope
GPS	Global positioning system
HAPs	High-altitude platforms
LBL	Long-baseline
LS	Least squares
ML	Maximum likelihood
PPDP	Privacy-preserving diagonal product
PPS	Privacy-preserving summation
RF	Radio frequency
RL	Reinforcement learning
RMSE	Root mean square error
RSS	Received signal strength
SBL	Short-baseline
TDOA	Time difference of arrival
TOA	Time of arrival
TOF	Time of flight
UAVs	Unmanned aerial vehicles
UKF	Unscented Kalman filtering
USNs	Underwater sensor networks
UWB	Ultra-wideband

Symbols

$\mathcal{R}$	Field of real numbers
$\mathcal{R}^n$	n-Dimensional real Euclidean space
$\mathcal{R}^{n\times m}$	Space of $n \times m$ real matrices
$\mathbf{I}$	Identity matrix
$\mathbf{A}$	System matrix
$\mathbf{A}^{-1}$	Inverse of matrix $\mathbf{A}$
$\mathbf{A}^{\mathrm{T}}$	Transpose of matrix $\mathbf{A}$
$\operatorname{argmin} f$	Value of the variable that minimizes function f
$\operatorname{argmax} f$	Value of the variable that maximizes function f
$\mathrm{tr}(\mathbf{A})$	Trace of matrix $\mathbf{A}$
$\mathrm{rank}(\mathbf{A})$	Rank of matrix $\mathbf{A}$
$\lVert\cdot\rVert$	Euclidean norm
∇f	The gradient of function f
$\forall$	For all
$\in$	Belong to
$\sum$	Sum
$\mathbf{E}\{\cdot\}$	Mathematical expectation operator
$\mathbf{var}\{\cdot\}$	Mathematical variance operator
$(\mathbf{A})_{\mathrm{s}}$	The sth column of matrix $\mathbf{A}$
$L_f\mathbf{A}$	Lie derivative of $\mathbf{A}$ to f
$\lceil\cdot\rceil$	Ceiling function
$\lfloor\cdot\rfloor$	Floor function
ϕ	Empty set
$\mathrm{diag}\{\cdot\}$	Diagonal matrix

Chapter 1
Introduction

Abstract This chapter presents the network architecture of underwater sensor networks (USNs). According to the different measurement ways, the localization schemes for wireless sensor networks are briefly reviewed. Based on this, the weak communication characteristics of USNs are summarized, through which the problems studied in this book are provided.

Keywords Underwater sensor networks (USNs) · Localization · Underwater · Weak communication

1.1 Underwater On-Line Monitoring System

The twenty-first century is the ocean century, and the development of ocean enterprise will become a main theme in the twenty-first century. In order to understand and explore the ocean, strong ocean observation ability is required to offer technical support. As for this issue, many instruments, e.g., acoustic Doppler current profiler, sonar array and multi-beam swath bathymeter, have been deployed to provide ocean monitoring services. These instruments acquire and store the measurement data, such that the data recovery and analysis can be implemented later based on salvaging the instruments manually. However, the spatial and temporal coverage ability of the above instruments is very limited, because they are bulky and expensive which are not conductive to large scale deployment. Besides that, they are off-line systems, which cannot meet the real-time requirement for some specific scenarios. Therefore, it is necessary to connect underwater instruments by means of wireless networks to enable real-time monitoring of given underwater region.

With the rapid development of communication and network technology, Underwater Sensor Networks (USNs) have made on-line ocean monitoring a reality. To be specific, USNs are consist of a large number of static sensors and mobile vehicles, which are deployed to perform cooperative monitoring tasks (Akyildiz et al. 2005). Compared with the off-line systems, USNs have the advantages of increased data source, improved real-time characteristic and reduced failure rate. Figure 1.1 illustrates a concept of underwater monitoring system via USNs. In

J. Yan et al., *Localization in Underwater Sensor Networks*, Wireless Networks,
https://doi.org/10.1007/978-981-16-4831-1_1

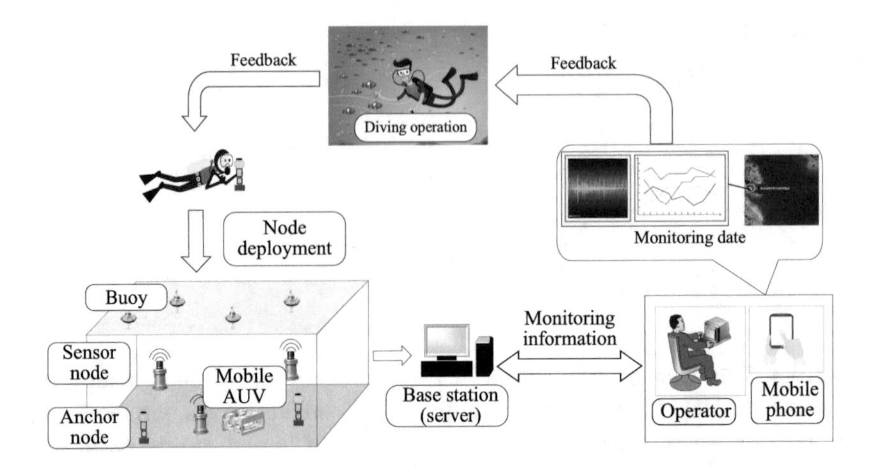

Fig. 1.1 Conceptual sketch of underwater on-line monitoring system via USNs

this system, static sensor nodes deployed in monitoring area periodically collect marine environment information, while the mobile autonomous underwater vehicles (AUVs) collect marine environment information by approaching or passing areas that cannot be monitored by static sensor nodes. Through acoustic and electromagnetic communication links, an operator on coast (or in a vessel on water offshore) collects all available data and then makes mission decision for the monitoring task. Accordingly, the seamless spatial observation and on-line data acquisition can be realized through the cooperation networking of sensor nodes and AUVs.

In such a system, the accurate locations of sensor nodes and AUVs are required for the routing, networking and control of USNs, since these services are valid only when the location information is accurately acquired. Although there exist many recently designed localization schemes for wireless sensor networks, the unique characteristics of USNs, such as asynchronous clock, node mobility, stratification effect and limited bandwidth, require very reliable and efficient new localization solutions for USNs. In view of this, we first present an overview of the localization schemes for wireless sensor networks. Based on this, the unique characteristics of USNs are provided. To address these characteristics, we put forward the problems studied in this book.

1.2 Localization Schemes for Wireless Sensor Networks

According to the different measurement ways, the localization schemes for wireless sensor networks can be classified into the following three categories: (1) angle of arrival (AOA) based localization; (2) distance-related localization; (3) received signal strength (RSS) profiling-based localization.

1.2.1 Localization with AOA Measurements

The localization with AOA measurements is to measure the azimuth of arrival of radio beam signal from the transmitter by using receiver antenna, through which the position of transmitter can be estimated by using optimization algorithm. In general, beamforming is one of the basic types of AOA measurement technology, whose measurement unit can be of small size as compared with the signal wavelength. Suppose that the beam of receiver antenna is rotating, and then the direction of the maximum received signal strength can be considered as the direction of the transmitter. In view of this, an improved polar localization scheme was developed in Wang and Ho (2018), which can make AOA localization unified in the near and far position. Besides that, an AOA-based weighted localization scheme was provided in Zheng et al. (2019) to reduce the median localization error by evaluating the multipath effect. Also of relevance, some other AOA-based localization schemes can be found in Al-Sadoon et al. (2020) and Huang and Zheng (2018).

In the absence of noise and interference, the azimuth lines from two or more receivers can intersect at a certain unique position, that is, the estimated position of the transmitter, as shown in Fig. 1.2a. However, the noise and interference cannot be ignored in practice, i.e., two or more azimuth lines cannot intersect at a unique position if the noise and interference are considered, as depicted by Fig. 1.2b. Thereby, it is necessary to design localization estimator to reduce the influences of noise and interference.

In the following, the details of two dimensional localization by using AOA measurement are presented. Let $\mathbf{x}_\mathrm{t} = [x_\mathrm{t}, y_\mathrm{t}]^\mathrm{T}$ represent the real position vector of the transmitter to be estimated from bearing measurements $\boldsymbol{\beta} = [\beta_1, \beta_2, \beta_3]^\mathrm{T}$, and one takes three receivers as the example. Denote $\boldsymbol{\theta}(\mathbf{x}_\mathrm{t}) = [\theta_1(\mathbf{x}_\mathrm{t}), \theta_2(\mathbf{x}_\mathrm{t}), \theta_3(\mathbf{x}_\mathrm{t})]^\mathrm{T}$ as the bearing vector of transmitter at position $\mathbf{x}_\mathrm{t}$. Particularly, $\theta_i(\mathbf{x}_\mathrm{t}) = \arctan \frac{y_\mathrm{t}-y_i}{x_\mathrm{t}-x_i}$ for $1 \le i \le 3$, and $\hat{\mathbf{x}}_\mathrm{t}$ is the estimated position of the transmitter. Besides, $\mathbf{x}_i = [x_i, y_i]^\mathrm{T}$ is the real position vector of the receiver i.

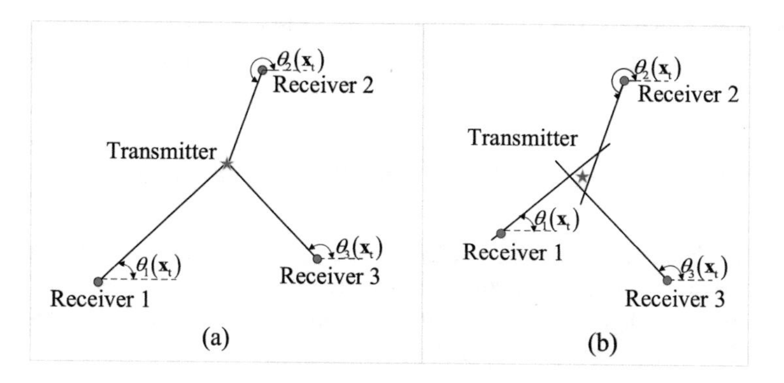

Fig. 1.2 Description of the azimuth lines from three receivers

The measurement bearing of transmitter is composed of true bearing and additional noise $\boldsymbol{\varepsilon} = [\varepsilon_1, \varepsilon_2, \varepsilon_3]^{\mathrm{T}}$ with zero mean and covariance matrices $\mathbf{S} = \mathrm{diag}\{\sigma_1^2, \sigma_2^2, \sigma_3^2\}$, i.e.,

$$\boldsymbol{\beta} = \boldsymbol{\theta}(\mathbf{x}_t) + \boldsymbol{\varepsilon}. \tag{1.1}$$

Based on this, the maximum likelihood (ML) estimation of the transmitter position is designed as

$$\begin{aligned}\hat{\mathbf{x}}_{\mathrm{t}} &= \arg\min \frac{1}{2}[\boldsymbol{\theta}(\mathbf{x}_{\mathrm{t}}) - \boldsymbol{\beta}]^{\mathrm{T}} \mathbf{S}^{-1} [\boldsymbol{\theta}(\mathbf{x}_{\mathrm{t}}) - \boldsymbol{\beta}] \\ &= \arg\min \frac{1}{2}\sum_{i=1}^{3} \frac{(\theta_i(\mathbf{x}_{\mathrm{t}}) - \beta_i)^2}{\sigma_i^2}.\end{aligned} \tag{1.2}$$

Stanfield's method assumes that the measurement error is small enough, i.e., $\varepsilon_i \approx \sin\varepsilon_i$, and hence, Eq. (1.2) can be arranged as

$$\hat{\mathbf{x}}_{\mathrm{t}} = \arg\min \frac{1}{2}\sum_{i=1}^{3} \frac{\sin^2(\theta_i(\mathbf{x}_{\mathrm{t}}) - \beta_i)}{\sigma_i^2}. \tag{1.3}$$

After trigonometric transformation, Eq. (1.3) is sorted into

$$\hat{\mathbf{x}}_{\mathrm{t}} = \arg\min \frac{1}{2}\sum_{i=1}^{3} \frac{[(y_{\mathrm{t}} - y_i)\cos\beta_i - (x_{\mathrm{t}} - x_i)\sin\beta_i]^2}{\sigma_i^2 r_i^2}, \tag{1.4}$$

where $r_i = \sqrt{(x_{\mathrm{t}} - x_i)^2 + (y_{\mathrm{t}} - y_i)^2}$.

Therefore, one can get an estimation of the position $\mathbf{x}_{\mathrm{t}}$, i.e.,

$$\hat{\mathbf{x}}_{\mathrm{t}} = (\mathbf{A}^{\mathrm{T}}\mathbf{R}^{-1}\mathbf{S}^{-1}\mathbf{A})^{-1}\mathbf{A}^{\mathrm{T}}\mathbf{R}^{-1}\mathbf{S}^{-1}\mathbf{b}, \tag{1.5}$$

where $\mathbf{A} = [\sin\beta_1, -\cos\beta_1; \sin\beta_2, -\cos\beta_2; \sin\beta_3, -\cos\beta_3]$, $\mathbf{R} = \mathrm{diag}\{r_1^2, r_2^2, r_3^2\}$ and $\mathbf{b} = [x_1\sin\beta_1 - y_1\cos\beta_1, x_2\sin\beta_2 - y_2\cos\beta_2, x_3\sin\beta_3 - y_3\cos\beta_3]^{\mathrm{T}}$.

1.2.2 Localization with Distance-Related Measurements

Distance-related measurements can be classified into four types: (1) time-of-arrival (TOA) measurement; (2) time-difference-of-arrival (TDOA) measurement; (3) RSS measurement; (4) lighthouse measurement. In the following, the details of these four localization schemes are presented.

1.2.2.1 Localization with TOA Measurement

The main idea of TOA-based localization is to employ the propagation time measurement to obtain the relative distance between transmitter and receiver, through which the position of transmitter can be estimated by trilateral localization or maximum likelihood estimation. In view of this, a channel-aware localization scheme with quantized ToA measurement and transmission uncertainty was developed in Yan et al. (2021). In Yuan et al. (2019), factor graphs were employed to construct a TOA-based passive localization scheme. Also of relevance, some other TOA-based localization schemes were developed in Wu et al. (2019) and Chen et al. (2020).

As shown in Fig. 1.3, the TOA strategy can be divided into one-way propagation time measurement and roundtrip propagation time measurement. For one-way propagation time measurement, the difference between the sending time of transmitter and the receiving time of receiver can be measured, i.e., the difference between t_{t} and t_i for $i \in \{1, 2, 3\}$ as shown in Fig. 1.3a. Accordingly, the TOA between transmitter and receiver under one-way propagation time measurement can be given as

$$\begin{aligned} \Delta T_i &= t_i - t_{\mathrm{t}}, \\ &= \frac{1}{\mathrm{c}} \|\mathbf{x}_i - \mathbf{x}_{\mathrm{t}}\|, i \in \{1, 2, 3\}, \end{aligned} \tag{1.6}$$

where t_{t} is the sending time by transmitter, t_i is the receiving time by receiver i for $i \in \{1, 2, 3\}$, c is the propagation speed of the signal, $\mathbf{x}_i$ is the position vector of receiver i, and $\mathbf{x}_{\mathrm{t}}$ is the position vector of transmitter.

Of note, the main drawback of one-way propagation time measurement is that it requires the local time of transmitter and receiver to be accurately synchronized. This requirement may increase the cost and complexity of sensor nodes by requiring a sophisticated synchronization mechanism. In order to relax the

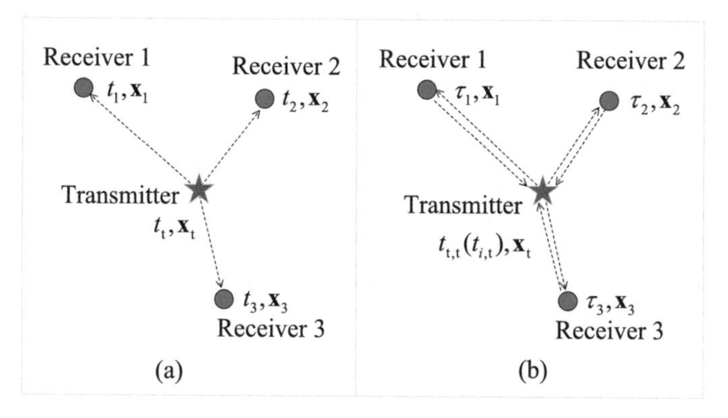

Fig. 1.3 An illustration of the TOA measurement. (**a**) One-way measurement. (**b**) Roundtrip measurement

clock synchronization requirement, the roundtrip propagation time measurement is appreciated. As shown in Fig. 1.3b, the difference between the time when a signal is sent by transmitter and the time when transmitter receives a signal from receiver is measured, i.e., the difference between $t_{t,t}$ and $t_{i,t}$ for $i \in \{1, 2, 3\}$. Accordingly, the TOA between transmitter and receiver under roundtrip propagation time measurement can be given as

$$\begin{aligned} \Delta T_i &= (t_{i,t} - t_{t,t}) - \tau_i, \\ &= \frac{2}{c} \|\mathbf{x}_i - \mathbf{x}_t\|, i \in \{1, 2, 3\}, \end{aligned} \tag{1.7}$$

where $t_{t,t}$ is the sending time by transmitter, $t_{i,t}$ is the time when a signal is received by transmitter from receiver i, and τ_i is the delay for the receiver i to handle the signal. It is noted that the delay is either known by a priori calibration, or measured to be subtracted. The delay measurement is a relatively mature filed, where the most popular method for delay measurement is the generalized cross-correlation (Chen et al. 2011; Zhou et al. 2017).

By applying TOA measurement, the 2D localization problem can be formulated as follows. Define $\Delta\tilde{\mathbf{T}} = [c\Delta T_1, c\Delta T_2, c\Delta T_3]^T$ (or $\Delta\tilde{\mathbf{T}} = [\frac{c}{2}\Delta T_1, \frac{c}{2}\Delta T_2, \frac{c}{2}\Delta T_3]^T$), $h(\mathbf{x}_t) = [\|\mathbf{x}_1 - \mathbf{x}_t\|, \|\mathbf{x}_2 - \mathbf{x}_t\|, \|\mathbf{x}_3 - \mathbf{x}_t\|]^T$, and $\mathbf{w} = [\varepsilon_1, \varepsilon_2, \varepsilon_3]^T$, where $\mathbf{w}$ is measurement noise with zero mean and covariance matrices $\bar{\mathbf{R}}$. Thus, the TOA measurement with noise measurement can be rewritten as $\Delta\tilde{\mathbf{T}} = h(\mathbf{x}_t) + \mathbf{w}$. Based on this, the ML estimator of the transmitter location $\mathbf{x}_t$ can be given as

$$\hat{\mathbf{x}}_t = \arg\min\{[\Delta\tilde{\mathbf{T}} - h(\mathbf{x}_t)]^T \bar{\mathbf{R}}^{-1} [\Delta\tilde{\mathbf{T}} - h(\mathbf{x}_t)]\}. \tag{1.8}$$

According to the principle of trilateration (Zhang et al. 2012), the nonlinear minimization problem is solved, i.e., the position of transmitter can be estimated.

1.2.2.2 Localization with TDOA Measurement

The localization with TDOA measurement is to apply the propagation time measurement to obtain the difference, such that optimization algorithms can be conducted to estimate the position of transmitter. Particularly, some TDOA-based localization schemes can be found in Ge et al. (2020) and Dai et al. (2020).

As shown in Fig. 1.4a, t_t denotes the time when a signal is sent by transmitter, and $\mathbf{x}_t$ denotes the position of transmitter. Similarly, t_i denotes the time when a signal is received by receiver $i \in \{1, 2, 3\}$ from transmitter, while $\mathbf{x}_i$ denotes the position of receiver i. Based on this, the TDOA between receivers i and j can be given as

$$\begin{aligned} \Delta t_{i,j} &= t_i - t_j, \\ &= \frac{1}{c}(\|\mathbf{x}_i - \mathbf{x}_t\| - \|\mathbf{x}_j - \mathbf{x}_t\|), i \neq j. \end{aligned} \tag{1.9}$$

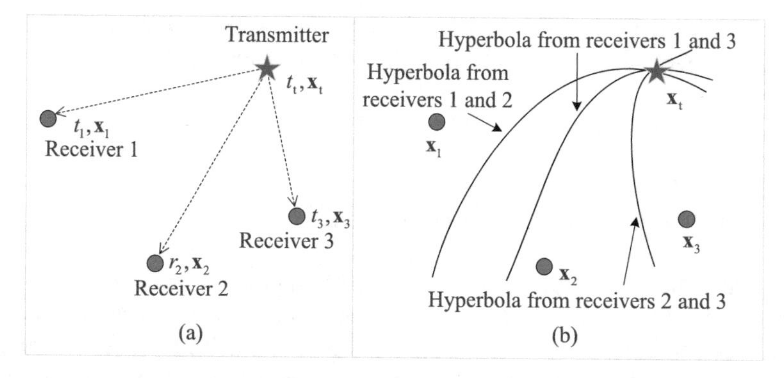

Fig. 1.4 An illustration of (**a**) TDOA and (**b**) intersecting hyperbolas from three receivers

From (1.9), a branch of hyperbola whose foci is at the position of receiver can be defined. Meanwhile, the transmitter must be on the branch, as illustrated in Fig. 1.4b. Because $\Delta t_{i,j}$ is not available, the noisy TDOA measurement of $\Delta \tilde{t}_{i,j}$ can be given as

$$\Delta \tilde{t}_{i,j} = \Delta t_{i,j} + n_{i,j}, \tag{1.10}$$

where $n_{i,j}$ is measurement noise, and it is usually assumed to be an independent zero-mean Gaussian distributed random variable.

Accordingly, TDOA measurements with three receivers can be stacked into

$$\underbrace{\begin{bmatrix} \Delta \tilde{t}_{1,2} \\ \Delta \tilde{t}_{1,3} \\ \Delta \tilde{t}_{2,3} \end{bmatrix}}_{:= \Delta \tilde{\mathbf{t}}} = \underbrace{\begin{bmatrix} \frac{(\|\mathbf{x}_1 - \mathbf{x}_t\| - \|\mathbf{x}_2 - \mathbf{x}_t\|)}{c} \\ \frac{(\|\mathbf{x}_1 - \mathbf{x}_t\| - \|\mathbf{x}_3 - \mathbf{x}_t\|)}{c} \\ \frac{(\|\mathbf{x}_2 - \mathbf{x}_t\| - \|\mathbf{x}_3 - \mathbf{x}_t\|)}{c} \end{bmatrix}}_{:= \varphi(\mathbf{x}_t)} + \underbrace{\begin{bmatrix} \varepsilon_{1,2} \\ \varepsilon_{1,3} \\ \varepsilon_{2,3} \end{bmatrix}}_{:= \bar{\mathbf{w}}} \tag{1.11}$$

where $\Delta \tilde{\mathbf{t}}$ is the TDOA measurement vector, $\varphi(\mathbf{x}_t)$ is the nonlinear vector function about transmitter, and $\bar{\mathbf{w}}$ is measurement noise with zero mean and covariance matrices $\bar{\mathbf{S}}$.

Based on this, the 2D localization problem of TDOA measurement can be formulated as $\Delta \tilde{\mathbf{t}} = \varphi(\mathbf{x}_t) + \bar{\mathbf{w}}$. Thus, the ML estimator of $\mathbf{x}_t$ can be give as

$$\hat{\mathbf{x}}_t = \arg\min\{[\Delta \tilde{\mathbf{t}} - \varphi(\mathbf{x}_t)]^{\mathrm{T}} \bar{\mathbf{S}}^{-1} [\Delta \tilde{\mathbf{t}} - \varphi(\mathbf{x}_t)]\}. \tag{1.12}$$

Clearly, there is no closed solution to (1.12) because of the nonlinear function of $\varphi(\mathbf{x}_t)$. According to the knowledge of Taylor series, one can linearize $\varphi(\mathbf{x}_t)$ around a reference point $\mathbf{x}_0$, i.e.,

$$\varphi(\mathbf{x}_t) \approx \varphi(\mathbf{x}_0) + \nabla \varphi(\mathbf{x}_0)(\mathbf{x}_t - \mathbf{x}_0), \tag{1.13}$$

where $\nabla\varphi(\mathbf{x}_0)$ is the partial derivative of $\varphi(\mathbf{x}_t)$ at the value $\mathbf{x}_0$. Referring to Wang et al. (2020), a recursive solution of ML estimator can be obtained, i.e.,

$$\hat{\mathbf{x}}_{t,k+1} = \hat{\mathbf{x}}_{t,k} + (\nabla^T\varphi(\mathbf{x}_{t,k})\bar{\mathbf{S}}^{-1}\nabla\varphi(\mathbf{x}_{t,k}))^{-1}\nabla^T\varphi(\mathbf{x}_{t,k})\bar{\mathbf{S}}^{-1}[\Delta\tilde{\mathbf{t}} - \varphi(\mathbf{x}_{t,k})]. \quad (1.14)$$

1.2.2.3 Localization with RSS Measurement

The localization with RSS measurement is to employ the RSS to estimate the relative distance between transmitter and receiver, through which optimization algorithms can be conducted to estimate the position of transmitter. In Zhang et al. (2016a), a RSS-based localization solution was developed, where a Newton-Raphson algorithm was provided to seek the position of target. In our previous works (Sebastian and Petros 2018; Yan et al. 2017), RSS was employed to estimate the position of remote unmanned aerial vehicle, thorough which a robust long-range aerial communication channel was established with the assistance of directional antennas. An advantage of this strategy is that the time synchronization is not required. However, RSS is sensitive to man-made and ambient noises.

Next, a brief introduction on RSS-based localization is presented. Particularly, the RSS in dBm on the receiver can be modelled as

$$P_i(d_i) = P_0(d_0) - 10n_p \log_{10}(\frac{d_i}{d_0}) + n_{i,\sigma}, \quad (1.15)$$

where $P_i(d)$ is the received power of receiver i for $i \in \{1, 2, 3\}$, $P_0(d_0)$ is the known reference power value at a reference distance d_0 from the transmitter, d_i is the distance between the transmitter and receiver i, and n_p is the path loss exponent. Besides, $n_{i,\sigma}$ is the zero mean Gaussian measurement noise with variance σ^2, which explains the random effect of shadow (Rappaport 2001).

By using RSS measurement, the 2D localization problem can be formulated as follows. Let $\mathbf{P} = [P_1(d_1), P_2(d_2), P_3(d_3)]^T$ be the measurement vector. Meanwhile, $\mathbf{g}(\mathbf{x}_t) = [P_0(d_0) - 10n_p \log_{10}(\frac{d_1}{d_0}), P_0(d_0) - 10n_p \log_{10}(\frac{d_2}{d_0}), P_0(d_0) - 10n_p \log_{10}(\frac{d_3}{d_0})]^T$ is the nonlinear vector function about transmitter x_t, and $\mathbf{n} = [n_{1,\sigma}, n_{2,\sigma}, n_{3,\sigma}]^T$ is the measurement noise vector, Based on the RSS measurement in (1.15), the likelihood function can be defined as

$$p(\mathbf{P}; \mathbf{x}_t) = \frac{1}{\sqrt{2\pi}\sigma} \exp[-\frac{1}{2\sigma^2} \|\mathbf{P} - \mathbf{g}(\mathbf{x}_t)\|^2]. \quad (1.16)$$

Thus, the ML solution of x_t is given as

$$\hat{\mathbf{x}}_t = \arg\max p(\mathbf{P}; \mathbf{x}_t). \quad (1.17)$$

Clearly, the optimization problem in (1.17) is highly nonconvex, and it is difficult to solve analytically. In view of this, the Newton-Raphson method (Gnetchejo et al. 2021) can be introduced to solve the optimization problem, i.e.,

$$\hat{\mathbf{x}}_{t,k+1} = \hat{\mathbf{x}}_{t,k} - \left[\frac{\partial^2 \ln p(\mathbf{P}; \mathbf{x}_t)}{\partial \mathbf{x}_t \partial \mathbf{x}_t^T} \right]^{-1} \left. \frac{\partial \ln p(\mathbf{P}; \mathbf{x}_t)}{\partial \mathbf{x}_t} \right|_{\mathbf{x}_t = \hat{\mathbf{x}}_{t,k}}, \tag{1.18}$$

where $\frac{\partial \ln p(\mathbf{P}; \mathbf{x}_t)}{\partial \mathbf{x}_t}$ is the gradient of the log-likelihood function with respect to x_t, and $\frac{\partial^2 \ln p(\mathbf{P}; \mathbf{x}_t)}{\partial \mathbf{x}_t \partial \mathbf{x}_t^T}$ is the second derivation with respect to x_t.

1.2.2.4 Localization with Lighthouse Approach

This solution is to obtain the relative distance by measuring the time duration that the receiver dwells in the beam (Gibson et al. 2016). In view of this, a lighthouse localization system was developed in Campos et al. (2020) to allow sensor nodes to localize themselves for high-reliability, latency-bounded, industrial automation tasks. In Kilberg et al. (2020), a quadrotor-based lighthouse localization algorithm with the bearing-only measurements was provided. In the following, a brief introduction on lighthouse approach based localization method is presented.

Step 1: Distance Measurement of Lighthouse Approach As shown in Fig. 1.5a, the transmitter is equipped with a parallel optical beam of width b, and the optical beam rotates around Z axis at an angular velocity ω. According to the triangle relationship, one has

$$\begin{aligned} d_1 &= \frac{b}{2\sin(\alpha_1/2)} = \frac{b}{2\sin(\omega t_1/2)}, \\ d_2 &= \frac{b}{2\sin(\alpha_2/2)} = \frac{b}{2\sin(\omega t_2/2)}, \end{aligned} \tag{1.19}$$

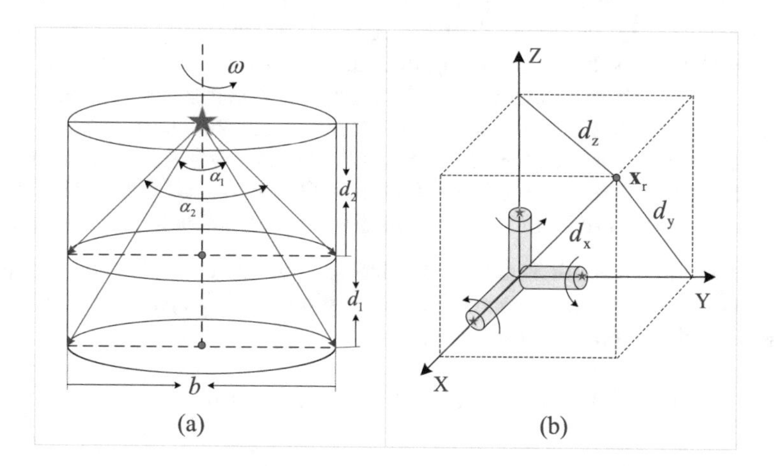

Fig. 1.5 An illustration of the lighthouse approach (**a**) measurement and (**b**) localization

where t_1 is the time duration of the beam detected by the receiver 1, t_2 is the time duration of the beam detected by the receiver 2, d_1 is the distance of receiver 1 between the X-Y plane and the Z plane, and d_2 is the distance of receiver 2 between the X-Y plane and the Z plane. Note that ω can be derived from the difference between the time when the beam is detected by the optical receiver at the first time and the time when the beam is detected by the optical receiver at the second time.

Step 2: Position Estimation of Receiver To locate the optical receiver autonomously, three transmitters with mutually perpendicular axes are used, as shown in Fig. 1.5b. Assumed that measurements d_{x}, d_{y} and d_{z} can be obtained by Step 1. Then, the following relationship can be built

$$\begin{aligned} d_{\mathrm{x}}^2 &= y_{\mathrm{r}}^2 + z_{\mathrm{r}}^2, \\ d_{\mathrm{y}}^2 &= x_{\mathrm{r}}^2 + z_{\mathrm{r}}^2, \\ d_{\mathrm{z}}^2 &= x_{\mathrm{r}}^2 + y_{\mathrm{r}}^2, \end{aligned} \tag{1.20}$$

where $\mathbf{x}_{\mathrm{r}} = [x_{\mathrm{r}}, y_{\mathrm{r}}, z_{\mathrm{r}}]^{\mathrm{T}}$ is the position vector of receiver.

Thus, the solution to (1.20) can be obtained by using optimization algorithm. Note that there are eight solutions by solving (1.20), due to the eight quadrants. Only one solution can be selected by using prior knowledge of which quadrant the receiver is located.

1.2.3 Localization with RSS Profiling Measurements

The main idea of RSS profiling based localization is to construct a form of signal behavior map, such that the positions of sensor nodes can be estimated by map matching. Particularly, map construction can be divided into the following two ways: offline and online. Offline way is mainly realized by a priori measurement, while online way needs to deploy devices in known locations to collect signal strength vectors in real time (Krishnan et al. 2004). In view of this, an accurate WiFi localization solution by fusing multiple fingerprints was developed in Guo et al. (2018a), where multiple classifiers were designed to exploit the intrinsic complementarity. In Li (2009), a simple generic localization scheme without prior profiling information was presented. Also of relevance, some other RSS profiling based localization schemes can be found in Guo et al. (2018b, 2020). In the following, a brief introduction of RSS profiling based localization is presented.

Step 1: Signal Strength Behavior Modeling Initially, control center constructs the signal behavior map by associating the position of each deployment device with the RSS by the deployment device, i.e.,

$$\{(x_1, y_1), \ldots, (x_i, y_i), P_1(d_1), \ldots, P_i(d_i)\}, \tag{1.21}$$

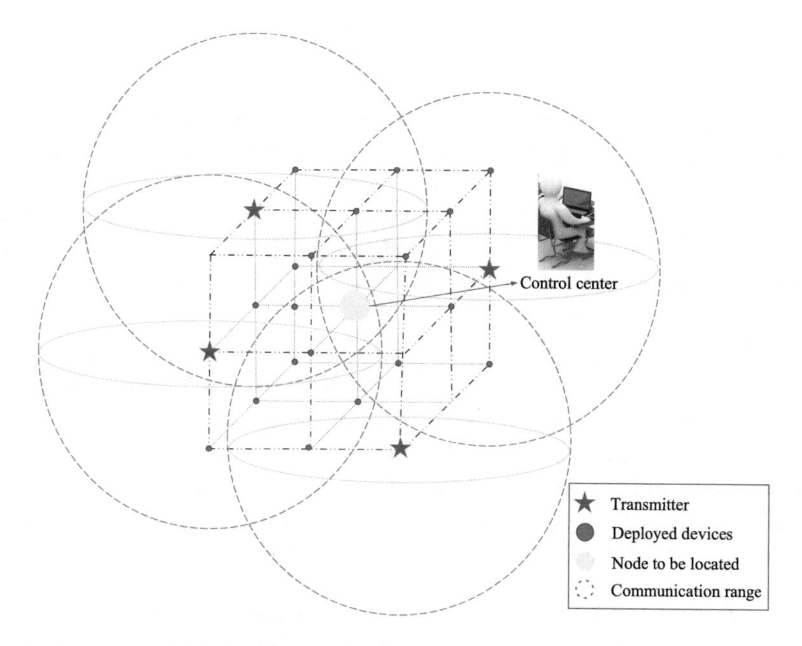

Fig. 1.6 Deployment of devices and nodes in coverage area

where (x_i, y_i) represents the position of deployment device for $1 \leq i \leq n$. In addition, n represents the total number of deployed devices, and $P_i(d_i)$ is the signal strength received by the deployment devices.

Step 2: Design of Localization Algorithm As shown in Fig. 1.6, control center selects a position of the deployment device that is closest to the node to be located. Similarly, the control center transmits the estimated position to the node to be located, or sends the full RSS model to the node to be located for its own position estimation. Particularly, sensor node measures the signal strength online, and hence, the measured signal is matched with the RSS behavior map. In general, the localization procedure can be based on deterministic method, probabilistic method, k-nearest neighbor (KNN), artificial neural network or their combination.

Step 3: Position Estimation for the Node to be Located Based on the weighted version of the RSS-profiling localization algorithm, the estimated position of the node to be located is given as

$$\hat{\mathbf{x}}_{\mathrm{t}} = \sum_{j=1}^{n} \frac{\frac{1}{\|\gamma-\beta_i\|^2}}{\sum_{j=1}^{n}\frac{1}{\|\gamma-\beta_j\|^2}}\mathbf{x}_j, \tag{1.22}$$

where $\|\gamma - \beta_j\|$ represents the Euclidean distance between γ and β_j for $1 \leq j \leq n$, and the total number of deploy devices is n.

1.3 Unique Characteristics of USNs

In underwater environment, the radio and optical waves are strongly absorbed, which can only travel to short distances as illustrated by Table 1.1. Alternatively, the acoustic waves can transmit a signal over distances greater than 100 m, and hence, acoustic communication emerges as the most efficient communication way for the localization of USNs (Stojanovic and Preisig 2009). However, the following unique characteristics of USNs make underwater localization more difficult as compared with the terrestrial sensor networks.

- **Asynchronous Clock.** Propagation delay of underwater acoustic wave ($\approx$1500 m/s) is five orders of magnitude higher than the radio wave ($\approx 3 \times 10^8$ m/s). Meanwhile, the global positioning system (GPS) is not available in underwater environment. Because of the long propagation delay and the unavailability of GPS, the time clock of node in USNs is always asynchronous, i.e., the time clock is affected by unknown time *skew* and *offset.* As a result, the clock synchronization-based range estimation strategy (e.g. the results in Shu et al. 2015; Cheng et al. 2021) is invalid for USNs.
- **Stratification Effect.** Water medium is inhomogeneous, and hence, the propagation delay of acoustic wave varies in different depth levels, which is called as stratification effect. In view of the stratification effect, some researchers point out that the isogradient sound speed profile can be employed to depict the acoustic propagation speed, e.g., Zhang et al. (2016b) and Ramezani et al. (2013). It should be noted that, the ignorance of stratification effect in USNs can increase the ranging biases, and thereby reduce the localization accuracy.
- **Node Mobility**. Notice that the node positions in terrestrial sensor networks are usually fixed, however the nodes in USNs often have passive motions derived by water current. In view of this, the round-trip propagation delays between any two sensor nodes are not equal, which make it hard to estimate the actual relative distance between two sensor nodes.
- **Energy Constraint**. Underwater sensor nodes are usually battery-powered, which are deployed in harsh underwater environment. So it is difficult or

Table 1.1 Property comparison for the underwater communication methods

Method	Range	Rate	Delay	Disadvantage	Advantage
Acoustic	~km	~kbps	667 ms/km	Low bandwidth, highdelay, high energy consumption	Long communication range
Optical	~100 m	~Mbps	0.03 ms/km	Line-of-sight, short communication range	High bandwidth, low delay, low energy consumption
Radio	~10 m	~Mbps	0.03 ms/km	Short communication range	High bandwidth, low delay, low energy consumption

impossible to replenish the batteries mounted on sensor nodes. Besides that, the communication energy consumption in water is much higher than the one in terrestrial environment. Although the energy can be extended with added battery, the extension is at a cost of extra space occupation, which will significantly increase the cost of sensor nodes.

- **Privacy Violation**. USNs are usually deployed in harsh environments, which make sensor nodes and AUVs extremely vulnerable to many security attacks (Li et al. 2015). For example, an attacker can wreck the entire localization system when the position information of anchor nodes is harvested. Another example is that the attacker can infer the position information of sensor nodes by correlating the history measurements.

1.4 Problems Studied in This Book

With consideration of the above characteristics, this book is concerned with the localization issue in USNs. The structural relationship of chapters can be depicted by Fig. 1.7. Specifically, Chap. 2 investigates the localization problem for USNs in the presence of clock asynchronization and node mobility constraints. We first present a hybrid architecture for USNs, which includes three types of nodes, i.e., AUVs, active sensor nodes, and passive sensor nodes. Of note, AUVs provide auxiliary localization information for sensor nodes. Meanwhile, active sensor nodes

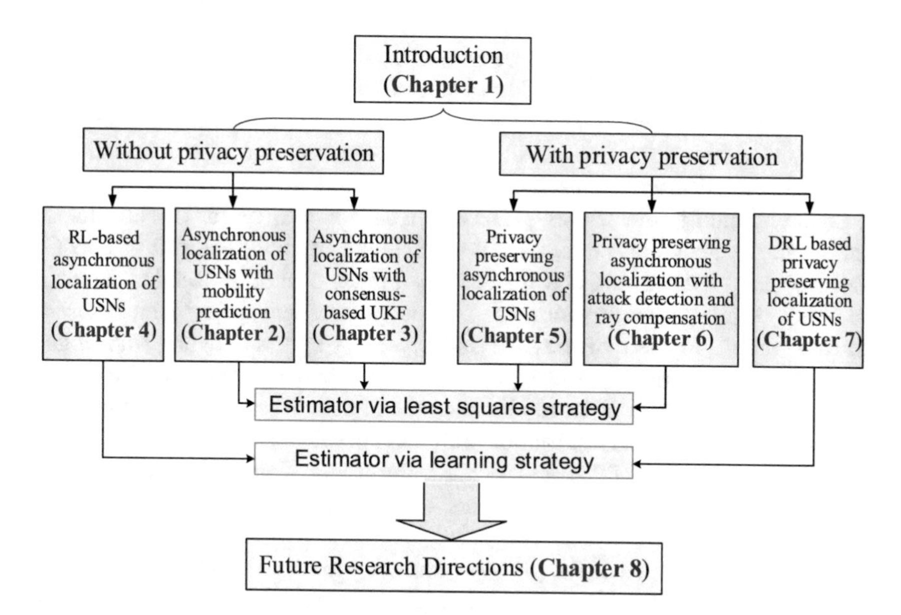

Fig. 1.7 Structural relationship of chapters in this book

initiate the whole localization process by broadcasting initialization localization messages to AUVs and passive sensor nodes. Besides, the passive sensor nodes listen to the messages and localize themselves without broadcasting any messages. To this end, an asynchronous localization algorithm with mobility prediction is developed to estimate the positions of sensor nodes, where the iterative least squares estimators are employed to solve the optimization problems.

Chapter 3 studies the issue of underwater target localization, in the presence of the asynchronous clock, the stratification effect and the strong-noise measurements. We first construct the relationship between the propagation delay and the position. Particularly, the ray tracing approach is adopted to model the stratification effect. Of note, the first-order linearization is required in Chap. 2 to compute the Jacobian matrix, such that linear measurements can be obtained. However, the calculation of the Jacobian matrix is not an easy task. Besides that, we note that most of the existing works rely on a single measurement, yet the fusion of all available measurements can yield improved estimator accuracy and robustness. In view of this, Chap. 3 develops a consensus-based unscented Kalman filtering (UKF) localization algorithm in USNs to reduce the influences of measurement and model errors.

Chapter 4 presents an AUV-aided localization solution for USNs. Specifically, we first construct a hybrid network architecture which includes four types of nodes, i.e., surface buoys, AUVs, active sensor nodes and passive sensor nodes. To be specific, AUVs are employed to play the role of mobile anchor nodes. Active sensor nodes initiate the whole localization process by periodically broadcasting initialization messages to the networks. At the same time, the passive sensor nodes monitor the messages and locate themselves without transmitting any messages to the networks. Note that the least squares-related localization estimators are employed in Chaps. 2 and 3. However, the least squares-related localization estimators can easily fall into local minimum. Chapter 4 develops a reinforcement learning (RL) based localization algorithm for USNs, which can seek the global optimization localization solution with value iteration.

Chapter 5 studies the privacy-preserving localization issue for USNs, which is ignored in the above chapters. Of note, USNs are usually deployed in harsh or even insecure environment, and the security threats cannot be avoided. Ignoring the effect of privacy preservation can lead to privacy leakage or even failure of the localization. In view of this, we first present a hybrid network architecture, which is consist of four types of nodes, i.e., surface buoys, anchors, active sensors and ordinary sensors. In order to eliminate the effects of asynchronous clock and mobility, an asynchronous localization protocol is developed, through which privacy preserving summation (PPS) and privacy-preserving diagonal product (PPDP) based localization algorithms are designed to hide privacy information and achieve localization jointly.

Chapter 6 further studies the privacy-preserving localization issue for USNs, in the presence of stratification effect and forging attack. We first design a privacy-preserving asynchronous transmission protocol to collect timestamp measurements. Note that part of the anchor nodes may be captured as malicious nodes, and hence, the localization accuracy can be sharply degraded, since the malicious anchor nodes would forge the signal characteristics, e.g., the false timestamps. In view of this, the RSS is incorporated into the transmission protocol, whose aim is to detect the malicious anchor nodes. Based on this, a privacy-preserving difference strategy is designed to hide the private position information of anchor nodes. With the measured and collected timestamp information, a least squares localization estimator is finally designed to estimate the position of underwater target.

Chapters 5–6 employ the least squares-based estimators to acquire the position information. As mentioned above, the least squares-based localization estimators can easily fall into local minimum. Inspired by this, Chap. 7 presents a DRL-based privacy-preserving localization solution for USNs. Instead of the encryption technique, we adopt the information-hiding technology to develop a privacy-preserving underwater localization protocol. An honest-but-curious model is regarded for USNs, such that a ray compensation strategy is designed to remove the effect of inhomogeneous water medium. Based on this, deep reinforcement learning (DRL) based localization estimators are conducted to seek the positions of sensor nodes, wherein the unsupervised, supervised and semisupervised scenarios are considered, respectively.

At last, Chap. 8 presents several research directions that depict future investigation on the underwater localization, including the network architecture, communication protocol and optimization estimator.

References

Akyildiz I, Pompili D, Melodia T (2005) Underwater acoustic sensor networks: research challenges. Ad Hoc Networks 3(3):257–279

Al-Sadoon M, Asif R, Al-Yasir Y, Abd-Alhameed R, Excell P (2020) AOA localization for vehicle-tracking systems using a dual-band sensor array. IEEE Trans Antennas Propag 68(8):6330–6345

Campos F, Schindler C, Kilberg B, Pister K (2020) Lighthouse localization of wireless sensor networks for latency-bounded, high-reliability industrial automation tasks. In: Proc. IEEE WFCS, Porto

Chen L, Liu Y, Kong F, He N, Acoustic source localization based on generalized cross-correlation time-delay estimation. Procedia Eng 15(1):4912–4919 (2011)

Chen H, Wang G, Wu X (2020) Cooperative multiple target nodes localization using TOA in mixed LOS/NLOS environments. IEEE Sensors J 20(3):1473–1484

Cheng L, Li Y, Xue M, Wang Y (2021) An indoor localization algorithm based on modified joint probabilistic data association for wireless sensor network. IEEE Trans Ind Inf 17(1):63–72

Dai Z, Wang G, Jin X, Lou X (2020) Nearly optimal sensor selection for TDOA-based source localization in wireless sensor networks. IEEE Trans Veh Technol 69(10):12031–12042

Ge T, Tharmarasa R, Lebel B, Florea M, Kirubarajan T (2020) A multidimensional TDOA association algorithm for joint multitarget localization and multisensor synchronization. IEEE Trans Aerosp Electron Syst 56(3):2083–2100

Gibson J, Haseler C, Lassiter H, Liu R, Lewin G (2016) Lighthouse localization for unmanned applications. In: Proc. IEEE SIEDS, Charlottesville, pp. 187–192

Gnetchejo P, Essiane S, Dadjé A, Ele P, Chen Z (2021) A self-adaptive algorithm with Newton Raphson method for parameters identification of photovoltaic modules and array. Trans Electr Electron Mater. https://doi.org/10.1007/s42341-021-00312-5

Guo X, Li L, Ansari N, Liao B (2018a) Accurate WiFi localization by fusing a group of fingerprints via global fusion profile. IEEE Trans Veh Technol 67(8):7314–7325
Guo X, Li L, Ansari N, Liao B (2018b) Knowledge aided adaptive localization via global fusion profile. IEEE Internet Things J 5(2):1081–1089
Guo X, Elikplim N, Ansari N, Li L, Wang L (2020) Robust WiFi localization by fusing derivative fingerprints of RSS and multiple classifiers. IEEE Trans Ind Inf 16(5):3177–3186
Huang H, Zheng Y (2018) Node localization with AOA assistance in multi-hop underwater sensor networks. Ad Hoc Networks 78(1):32–41
Kilberg B, Campos F, Schindler C, Pister K (2020) Quadrotor-based lighthouse localization with time-synchronized wireless sensor nodes and bearing-only measurements. Sensors 20(14):3888–3904
Krishnan P, Krishnakumar A, Ju W, Mallows C, Gamt S (2004) A system for LEASE: location estimation assisted by stationary emitters for indoor RF wireless networks. In: Proc. IEEE INFOCOM, Hong Kong, pp. 1001–1011
Li X (2009) Ratio-based zero-profiling indoor localization. In: Proc. IEEE 6th international conference on mobile adhoc and sensor systems, Macau, pp 40–49
Li H, He Y, Cheng X, Zhu H, Sun L (2015) Security and privacy in localization for underwater sensor networks. IEEE Commun Mag 53(11):56–62
Ramezani H, Jamali-Rad H, Leus G (2013) Target localization and tracking for an isogradient sound speed profile. IEEE Trans Signal Process 61(6):1434–1446
Rappaport T (2001) Wireless communications: principles and practice. Prentice Hall PTR, London
Sebastian S, Petros S (2018) RSSI-based indoor localization with the internet of things. IEEE Access 6(1):30149–30161
Shu T, Chen Y, Yang J (2015) Protecting multi-lateral localization privacy in pervasive environments. IEEE/ACM Trans Networking 23(5):1688–1701
Stojanovic M, Preisig J (2009) Underwater acoustic communication channels: propagation models and statistical characterization. IEEE Commun Mag 47(1):84–89
Wang Y, Ho K (2018) Unified near-field and far-field localization for AOA and hybrid AOA-TDOA positionings. IEEE Trans Wireless Commun 17(2):1242–1254
Wang T, Xiong H, Ding H, Zheng L (2020) TDOA-based joint synchronization and localization algorithm for asynchronous wireless sensor networks. IEEE Trans Commun 68(5):3107–3124
Wu S, Zhang S, Xu K, Huang D (2019) A TOA-based localization algorithm with simultaneous NLOS mitigation and synchronization error elimination. IEEE Sensors Lett 3(3):1–4
Yan J, Wan Y, Fu S, Xie J, Li S, Lu K (2017) Received signal strength indicator-based decentralised control for robust long-range aerial networking using directional antennas. IET Control Theory Appl 11(11):1838–1847
Yan Y, Yang G, Wang H, Shen X (2021) Semidefinite relaxation for source localization with quantized TOAmeasurements and transmission uncertainty in sensor networks. IEEE Trans Commun 69(2):1201–1213
Yuan W, Wu N, Guo Q, Huang X, Li Y, Hanzo L (2019) TOA-based passive localization constructed over factor graphs: a unified framework. IEEE Trans Commun 67(10):6952–6965
Zhang A, Ye X, Hu H (2012) Point in triangle testing based trilateration localization algorithm in wireless sensor networks. KSII Trans Internet Inf Syst 6(10):2567–2586
Zhang B, Wang H, Tao X, Zheng L, Yang Q (2016a) Received signal strength-based underwater acoustic localization considering stratification effect. In: Proc. IEEE OCEANS, Shanghai, China, pp 1–8
Zhang M, Xu W, Xu Y (2016b) Inversion of the sound speed with radiated noise of an autonomous underwater vehicle in shallow water waveguides. IEEE J Oceanic Eng 41(1):204–216
Zheng Y, Sheng M, Liu J, Li J (2019) Exploiting AOA estimation accuracy for indoor localization: a weighted AOA-based approach. IEEE Wireless Commun Lett 8(1):65–68
Zhou Y, Zhong L, Cai H, Tian J, Li D, Lu X (2017) White light scanning interferometry based on generalized cross-correlation time delay estimation. IEEE Photonics J 9(5):1–11

Chapter 2
Asynchronous Localization of Underwater Sensor Networks with Mobility Prediction

Abstract This chapter considers the clock asynchronization and node mobility, and then an asynchronous localization solution with mobility prediction is developed for USNs. Particularly, an asynchronous localization scheme with mobility prediction is designed to eliminate the effect of asynchronous clocks and compensate the mobility. Based on this, iterative least squares estimators are conducted to estimate the positions of sensor nodes. Besides that, the Cramér-Rao lower bound and convergence analysis are also presented. Finally, simulation results represent that the developed localization solution in this chapter can reduce the localization time as compared with the exhaustive search-based localization method. Meanwhile, it can effectively eliminate the influences of clock asynchronization and node mobility.

Keywords Asynchronous clock · Localization · Mobility · Underwater Sensor Networks (USNs) · Prediction

2.1 Introduction

Compared with the terrestrial localization issue, underwater localization has some fundamental differences. Firstly, GPS is unavailable in water, due to the strong attenuation of electromagnetic waves. Although some approaches have been developed to achieve the underwater localization such as the inertial measurement unit (Wang et al. 2014) and Doppler velocity log (Wang and Xie 2015), these methods suffer from the error accumulation problem. In addition, the propagation delay in water is five orders of magnitude higher than the one in radio frequency (RF) terrestrial channels, which has been pointed out in Chap. 1. By adopting TOA or TDOA, some range-based approaches are developed to determine distances between sensor nodes, such as long-baseline (LBL) and short-baseline (SBL) methods as well as their many improvements (Luo et al. 2016; Liu et al. 2014; Yan et al. 2017; Isbitiren and Akan 2011). Nevertheless, these localization approaches rely on the clock synchronization assumption, i.e., the clocks for transmitter and receiver are assumed to be synchronized. Since the high propagation delay can lead to synchronization errors (i.e., the clocks for sensor nodes are asynchronous), it is not easy to apply

J. Yan et al., *Localization in Underwater Sensor Networks*, Wireless Networks,
https://doi.org/10.1007/978-981-16-4831-1_2

these localization approaches to the USNs. Also of relevance, some asynchronous localization algorithms were proposed in Carroll et al. (2014), Cheng et al. (2008), Xia et al. (2015), Luo and Fan (2017), and Mortazavi et al. (2017), however the sensor nodes are assumed to be static and they only work in static underwater environment.

In underwater environment, mobility is an important factor that requires be considered, as sensor nodes often have passive mobility caused by water current or tides (Novikov and Bagtzoglou 2006). It is difficult to estimate the real distance between two nodes, which in turn leads to the decrease of localization accuracy. In Caruso et al. (2008), a meandering current mobility model was presented to describe the movement of sensor nodes, where the effects of meandering sub-surface currents and vortices are both considered. Based on this, some mobility prediction strategies were applied to the underwater localization problem (Zhou et al. 2011; Zhang et al. 2016). In Luo et al. (2016), a floating model for double-head node was constructed to design a drifting restricted localization algorithm. Nevertheless, these mobility-based localization algorithms face the clock synchronization assumption. How to relax the clock synchronization assumption and design a mobility-based asynchronous localization algorithm for USNs are largely unexplored.

To address the above challenges, this chapter studies the localization problem for USNs, subject to clock asynchronization and node mobility constraints. An asynchronous localization algorithm with mobility prediction is provided to estimate the locations of active and passive sensor nodes, where iterative least squares estimators are carried out to solve the optimization problems. The main contributions of this chapter are threefold.

1. Unlike other algorithms assuming the clock synchronization, we propose an asynchronous localization algorithm for USNs. Meanwhile, the propagation delays on the way to and from nodes are not necessarily equal.
2. Considering the node mobility, a mobility prediction strategy is provided to estimate the future locations of sensor nodes. The localization accuracy can be improved comparing with the previous works in Carroll et al. (2014) and Cheng et al. (2008).
3. Iterative least squares estimators are designed to solve the optimization problems, where convergence analysis and Cramér-Rao lower bound are also given. Compared with the exhaustive search-based localization method in Carroll et al. (2014), the localization time in this chapter can be reduced.

2.2 Network Architecture and Overview of the Localization

2.2.1 Network Architecture

In order to achieve the localization mission, a network architecture is required to be provided. In Isbitiren and Akan (2011), anchor nodes float at the water surface to provide auxiliary localization services, where sensor nodes send messages to anchor nodes frequently for localization calculation. Nevertheless, due to the energy constraint and passive mobility of sensor nodes, the architecture in Isbitiren and Akan (2011) can cause localization errors while reducing the energy effectiveness for sensor nodes, especially for the nodes located at the bottom of USNs. In Carroll et al. (2014), anchor nodes are deployed under the water surface to achieve the localization. Nevertheless, this architecture suffers from the error accumulation problem. Inspired by the above considerations, we develop a network architecture that comprises of three different types of nodes, as depicted by Fig. 2.1.

- **AUVs**: AUVs are installed with inertial navigation systems, and they act as anchor nodes. Upon floating at the water surface, AUVs acquire their positions through GPS. After receiving localization messages from sensor nodes, AUVs submerge into the water and move around actively to provide localization information for sensor nodes. Different from Liu et al. (2014), AUVs are not assumed to be clock synchronized. With the submerging of AUVs, long-distance transmissions can be replaced as short-distance ones, which means there exists less influence of noise and energy attenuation, such that sensors can obtain more accurate localization performance.
- **Active sensor nodes**: Active sensor nodes are capable of sensing, transmission and computation. They initiate the localization procedure by broadcasting the localization messages to AUVs and passive sensor nodes. Because of the effect of current, active sensor nodes move passively, and their task is to locate themselves with the developed localization method.
- **Passive sensor nodes**: Passive sensor nodes have the ability of sensing and computation. Without broadcasting any messages, passive sensor nodes listen

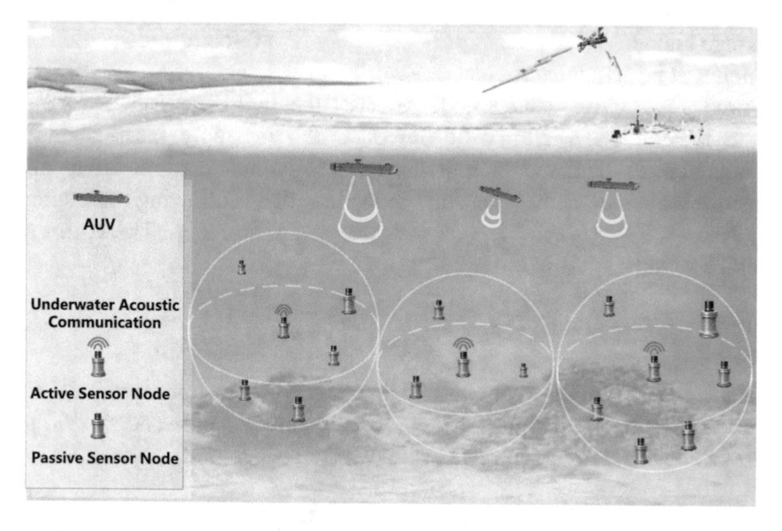

Fig. 2.1 Network architecture of the USNs in this chapter

to the messages from AUVs and active sensor nodes. Similar to active sensor nodes, they can move passively, and their task is to localize themselves with the proposed localization method.

2.2.2 *Overview of the Localization*

The localization procedure can be divided into two subprocesses: active sensor node localization and passive sensor node localization. We assume that each sensor node is outfitted with a depth sensor. Hence, only the positions on X axis and Y axis need to be calculated by the range-based localization schemes. In view of this, three AUVs can be deployed to provide auxiliary localization services for active and passive sensor nodes.

At the beginning, AUVs float at the water surface and obtain their accurate locations via GPS. Besides that, the velocity vector can be acquired and stored as historical information. Upon the localization procedure begins, AUVs submerge and broadcast 'HELLO' message to the whole sensor network. After the 'HELLO' message is received by an active sensor node, the localization procedure for a sensor node can be given as follows:

1. At time $T_{\mathrm{s,s}}$, active sensor node sends out an initialization message, containing the sending order for AUVs, i.e., $n = 1, 2, 3$. After that, active sensor node switches into the waiting mode for the replies from AUVs.
2. At time $t_{\mathrm{s},n}$, AUV n receives the initialization message. After that, AUV n switches into the listening mode and receives messages from AUV 1 to $n-1$, where the arrival times are denoted by $t_{k,n}$ ($k = 1, ..., n-1$). Subsequently, at time $t_{n,n}$, AUV n sends out its localization message. Particularly, the localization message includes AUV ID n, $t_{\mathrm{s},n}$, $\{t_{k,n}\}_{\forall k}$, $t_{n,n}$, the velocity and position of AUV (which will be estimated in Sect. 2.3.2).
3. At time $T_{n,\mathrm{s}}$, active sensor node receives the reply message from AUV n, where $n = 1, 2, 3$. Similarly, the other AUVs' reply messages can also be obtained, as depicted in Fig. 2.2. After a complete round of localization transmission, the active sensor node has the following measurements: $t_{\mathrm{s},1}, T_{1,\mathrm{s}}, t_{\mathrm{s},2}, T_{2,\mathrm{s}}, t_{\mathrm{s},3}, T_{3,\mathrm{s}}, t_{1,1}, t_{1,2}, t_{1,3}, t_{2,2}, t_{2,3}$ and $t_{3,3}$. These timestamps can be shown as

$$\left\{t_{\mathrm{s},n}, t_{n,n}, T_{n,\mathrm{s}}\right\}_{n=1}^{3}, \quad \left\{t_{k,n}\right\}_{n=2,k=1}^{3,n-1}, \tag{2.1}$$

   where $t_{k,n}$ is the time instant when AUV n receives message from AUV k.
4. By receiving the localization messages from active sensor and AUVs, passive sensor node has the following measurements

$$T_{\mathrm{s,p}}, \left\{t_{\mathrm{s},n}, T_{n,\mathrm{p}}, t_{n,n}\right\}_{n=1}^{3}, \left\{t_{k,n}\right\}_{n=2,k=1}^{3,n-1}, \tag{2.2}$$

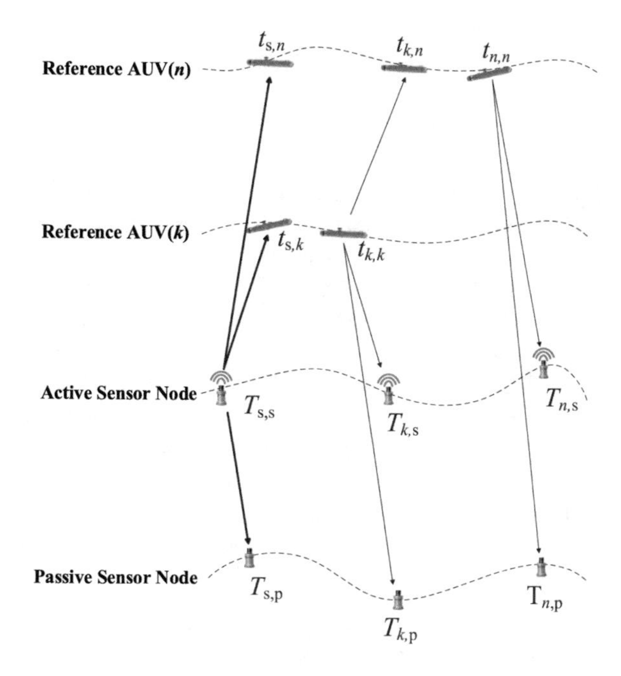

Fig. 2.2 Description of the localization strategy

where $T_{s,p}$ is the time instant when passive sensor node receives the initialization message from active sensor node. $T_{k,p}$ and $T_{n,p}$ are the time instants when passive sensor receives messages from AUVs k and n, respectively.

5. With the collected localization messages, an asynchronous localization algorithm with mobility prediction can be proposed to localize active and passive sensor nodes, in which the message broadcasting is not required for passive sensor node, as presented in Sect. 2.3.3.

Similar to Carroll et al. (2014), we ignore the clock skew and only deal with the clock offset. Then the asynchronization clock model is considered as

$$T = t + \beta, \tag{2.3}$$

where T is the measured time, t is the real time, and β is the clock offset.

Localization Objective Considering the clock asynchronization and node mobility constraints, we attempt to develop a mobility prediction strategy to estimate the future locations of active and passive sensor nodes. With the predicted locations, we aim to provide an asynchronous localization scheme to improve the localization accuracy.

2.3 Asynchronous Localization Approach Design

2.3.1 Relationship Between Delay and Position

For AUV n, the initial position information can be accurately acquired by the assistance of GPS, where $n = 1, 2, 3$. Meanwhile, the real-time position of AUV n is denoted by $(\mathcal{X}_n, \mathcal{Y}_n, \mathcal{Z}_n)$. The initial and real-time positions of an active sensor node are denoted by $(x_{\rm s}, y_{\rm s}, z_{\rm s})$ and $(x'_{\rm s}, y'_{\rm s}, z'_{\rm s})$, respectively. Besides, the initial and real-time positions of a passive sensor node are defined by $(x_{\rm p}, y_{\rm p}, z_{\rm p})$ and $(x'_{\rm p}, y'_{\rm p}, z'_{\rm p})$, respectively.

With the aforementioned definitions, the initial relative distance between active sensor node and AUV n can be given as $d_{{\rm s},n}$. The initial relative distance between AUV k and AUV n is given as $d_{k,n}$, while the initial relative distance between active and passive sensor node is denoted by $d_{\rm s,p}$. Meanwhile, the real-time relative distance between AUV n and active sensor node is set as $d'_{n,{\rm s}}$, and the real-time relative distance between AUV n and passive sensor node is set as $d'_{n,{\rm p}}$. Particularly, they can be shown as

$$\begin{cases} d_{{\rm s},n}=\sqrt{(x_{\rm s}-\mathcal{X}_n)^2+(y_{\rm s}-\mathcal{Y}_n)^2+(z_{\rm s}-\mathcal{Z}_n)^2} \\ d_{k,n}=\sqrt{(\mathcal{X}_k-\mathcal{X}_n)^2+(\mathcal{Y}_k-\mathcal{Y}_n)^2+(\mathcal{Z}_k-\mathcal{Z}_n)^2} \\ d'_{n,{\rm s}}=\sqrt{(\mathcal{X}_n-x'_{\rm s})^2+(\mathcal{Y}_n-y'_{\rm s})^2+(\mathcal{Z}_n-z'_{\rm s})^2} \\ d_{\rm s,p}=\sqrt{(x_{\rm s}-x_{\rm p})^2+(y_{\rm s}-y_{\rm p})^2+(z_{\rm s}-z_{\rm p})^2} \\ d'_{n,{\rm p}}=\sqrt{(\mathcal{X}_n-x'_{\rm p})^2+(\mathcal{Y}_n-y'_{\rm p})^2+(\mathcal{Z}_n-z'_{\rm p})^2} \end{cases}, \tag{2.4}$$

where the positions of sensor nodes on the Z axis (i.e., $z_{\rm s}$, $z_{\rm p}$, $z'_{\rm s}$ and $z'_{\rm p}$) are assumed to be known by pressure sensor. Meanwhile, the positions of sensor nodes on the X axis and Y axis are unknown, which requires to be computed.

Based on (2.4), the relationship between propagation delay and position can be constructed as

$$\begin{aligned} &\tau_{{\rm s},n}=\frac{d_{{\rm s},n}}{\rm c},\ \tau_{k,n}=\frac{d_{k,n}}{\rm c}, \\ &\tau'_{n,{\rm s}}=\frac{d'_{n,{\rm s}}}{\rm c},\ \tau_{\rm s,p}=\frac{d_{\rm s,p}}{\rm c},\ \tau'_{n,{\rm p}}=\frac{d'_{n,{\rm p}}}{\rm c}, \end{aligned} \tag{2.5}$$

where $\tau_{{\rm s},n}$ denotes the relative propagation delay of active sensor node and AUV n during the time interval $[T_{\rm s,s}, t_{{\rm s},n}]$, $\tau_{k,n}$ denotes the relative propagation delay of AUV k and AUV n during the time interval $[t_{k,k}, t_{k,n}]$, $\tau'_{n,{\rm s}}$ denotes the relative propagation delay of AUV n and active sensor node during the time interval $[t_{n,n}, T_{n,{\rm s}}]$, $\tau_{\rm s,p}$ denotes the relative propagation delay of active sensor node and passive sensor node during the time interval $[T_{\rm s,s}, T_{\rm s,p}]$, $\tau'_{n,{\rm p}}$ denotes the relative propagation delay of AUV n and passive sensor node during the time interval $[t_{n,n}, T_{n,{\rm p}}]$, and c $\approx$ 1500 m/s denotes the propagation speed of sound in water.

2.3.2 Mobility Prediction for AUVs and Sensor Nodes

Note that AUVs act as anchor nodes to provide auxiliary localization services, and the position $(\mathcal{X}_n, \mathcal{Y}_n, \mathcal{Z}_n)$ is required to be known. Based on this, a linear prediction algorithm is first provided to predict the positions of AUVs. Notice that $\mathcal{Z}_n$ can be obtained by pressure sensor installed on AUV. We only predict the vertical and horizontal positions, i.e., $\mathcal{X}_n$ and $\mathcal{Y}_n$. For AUV n, the vertical and horizontal velocities at time instant t are given as $\mu_{n,\mathrm{x}}(t)$ and $\mu_{n,\mathrm{y}}(t)$, respectively. Accordingly, the predicted velocity is given as

$$\boldsymbol{\mu}_n(t) = \sum\nolimits_{m=1}^{l} a_m \boldsymbol{\mu}_n(t - m\delta), \tag{2.6}$$

where $\boldsymbol{\mu}_n(t - m\delta) = (\mu_{n,\mathrm{x}}(t - m\delta), \mu_{n,\mathrm{y}}(t - m\delta))$ is the historical velocity of AUV n at time instant $t - m\delta$. It can be acquired through the different methods. Besides that, l denotes the length of prediction step, and δ denotes the sampling period. a_m denotes the prediction coefficient, which satisfies the conditions of $\sum_{m=1}^{l} a_m = 1$ and $a_m \in [0, 1]$. a_m can be acquired by Durbin algorithm in Rowden (1992).

With (2.6), the predicted position at time instant t can be calculated as

$$(\mathcal{X}_n, \mathcal{Y}_n) = (\mathcal{X}_{n,\mathrm{start}}, \mathcal{Y}_{n,\mathrm{start}}) + \delta \sum\nolimits_{m_{1,n}=0}^{l_{1,n}} \boldsymbol{\mu}_n(t_{\mathrm{start}} + m_{1,n}\delta), \tag{2.7}$$

where $(\mathcal{X}_{n,\mathrm{start}}, \mathcal{Y}_{n,\mathrm{start}})$ denotes the initial position of AUV n, and this position information can be accurately obtained via the assistance of through GPS, since AUV n floats on water surface at the initial time instant. $l_{1,n} = (t - t_{\mathrm{start}})/\delta$ denotes the total updating step till time instant t, and t_{start} denotes the initial time instant.

For sensor nodes, their initial positions are unknown, so the prediction algorithm of AUVs in (2.6) cannot be directly applied to sensor nodes. In Bagtzoglou and Andrei (2007) and Novikov and Bagtzoglou (2006), it was pointed out that the movement of one object was closely related to its neighboring objects according to the spatial correlations existed in USNs. In view of this, we employ a common spatial correlation-based model (Zhou et al. 2010, 2011) to predict the mobility of sensor nodes. Let $\mathbf{v}_{\mathrm{s}} = (v_{\mathrm{s,x}}, v_{\mathrm{s,y}})$ is the vertical and horizontal velocities of active sensor node, and $\mathbf{v}_{\mathrm{p}} = (v_{\mathrm{p,x}}, v_{\mathrm{p,y}})$ is the velocities of passive sensor node. Thus, the predicted velocities of active and passive sensor nodes are denoted as

$$\begin{aligned} v_{\mathrm{s,x}} &= \sum\nolimits_{n=1}^{3} \varsigma_{n,\mathrm{s}} \mu_{n,\mathrm{x}},\ v_{\mathrm{s,y}} = \sum\nolimits_{n=1}^{3} \varsigma_{n,\mathrm{s}} \mu_{n,\mathrm{y}}, \\ v_{\mathrm{p,x}} &= \sum\nolimits_{n=1}^{3} \varsigma_{n,\mathrm{p}} \mu_{n,\mathrm{x}},\ v_{\mathrm{p,y}} = \sum\nolimits_{n=1}^{3} \varsigma_{n,\mathrm{p}} \mu_{n,\mathrm{y}}, \end{aligned} \tag{2.8}$$

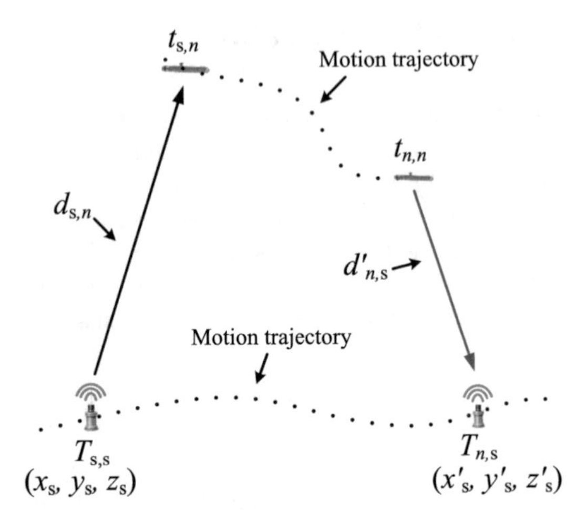

Fig. 2.3 The mobility of active sensor node

with

$$\varsigma_{n,\mathrm{s}} = \frac{\frac{1}{d'_{n,\mathrm{s}}}}{\sum_{n=1}^{3} \frac{1}{d'_{n,\mathrm{s}}}} \text{ and } \varsigma_{n,\mathrm{p}} = \frac{\frac{1}{d'_{n,\mathrm{p}}}}{\sum_{n=1}^{3} \frac{1}{d'_{n,\mathrm{p}}}}, \tag{2.9}$$

where $\varsigma_{n,\mathrm{s}}$ and $\varsigma_{n,\mathrm{p}}$ are the interpolation coefficients.

As presented in (2.9), the distances (i.e., $d'_{n,\mathrm{s}}$ and $d'_{n,\mathrm{p}}$) are required to be known. Nevertheless, due to the impact of asynchronous clocks, sensor nodes cannot directly obtain the distances. In order to solve this issue, we attempt to establish the relationship between distance and localization information, such that $d'_{n,\mathrm{s}}$ (or $d'_{n,\mathrm{p}}$) can be indirectly acquired. Next, we take the active sensor node as the example (see Fig. 2.3). With the timestamp information in (2.1) and the measurement noise $w_{n,\mathrm{s}}$, we have

$$\left[(T_{n,\mathrm{s}} - T_{\mathrm{s,s}}) - (t_{n,n} - t_{\mathrm{s},n})\right]\mathrm{c} = d_{\mathrm{s},n} + d'_{n,\mathrm{s}} + w_{n,\mathrm{s}}. \tag{2.10}$$

According to the definitions in (2.4) and (2.5), it is obtained that $d'_{n,\mathrm{s}}$ is related to the unknown initial position $(x_\mathrm{s}, y_\mathrm{s}, z_\mathrm{s})$. Based on this, $d'_{n,\mathrm{s}}$ can be expressed as a function of $(x_\mathrm{s}, y_\mathrm{s}, z_\mathrm{s})$, and then $\mathbf{v}_\mathrm{s}$ can also be expressed as a function of $(x_\mathrm{s}, y_\mathrm{s}, z_\mathrm{s})$. In order to acquire the unknown variable $(x_\mathrm{s}, y_\mathrm{s}, z_\mathrm{s})$, we define the asynchronous localization problem (see Sect. 2.3.3), whose solution is given in Sect. 2.4.1. With the solved variable $(x_\mathrm{s}, y_\mathrm{s}, z_\mathrm{s})$, $\mathbf{v}_\mathrm{s}$ is obtained. Then, the relationship of $(x'_\mathrm{s}, y'_\mathrm{s})$ and $(x_\mathrm{s}, y_\mathrm{s})$ is constructed as

$$(x'_\mathrm{s}, y'_\mathrm{s}) = (x_\mathrm{s}, y_\mathrm{s}) + \mathbf{v}_\mathrm{s}(T_{n,\mathrm{s}} - T_{\mathrm{s,s}}). \tag{2.11}$$

Similarly, the relationship of $(x'_{\mathrm{p}}, y'_{\mathrm{p}})$ and $(x_{\mathrm{p}}, y_{\mathrm{p}})$ is

$$(x'_{\mathrm{p}}, y'_{\mathrm{p}}) = (x_{\mathrm{p}}, y_{\mathrm{p}}) + \mathbf{v}_{\mathrm{p}}(T_{n,\mathrm{p}} - T_{\mathrm{s,p}}). \tag{2.12}$$

2.3.3 Asynchronous Localization Optimization Problem

In order to remove the effect of clock offsets, we define the following time differences for active sensor node, i.e.,

$$\begin{aligned} \Delta\mathcal{T}_{n,\mathrm{s}} &= (T_{n,\mathrm{s}} - T_{\mathrm{s,s}}) - (t_{n,n} - t_{\mathrm{s},n}), \\ \Delta\mathcal{T}_{k,n} &= (t_{k,n} - t_{\mathrm{s},n}) - (t_{k,k} - t_{\mathrm{s},k}), \end{aligned} \tag{2.13}$$

where $n = 1, 2, 3$ and $k = 1, ..., n - 1$.

With the result in (2.13), the relationship of time differences and propagation delays is established, i.e.,

$$\begin{aligned} \Delta\mathcal{T}_{n,\mathrm{s}} &= \tau_{\mathrm{s},n} + \tau'_{n,\mathrm{s}} + w_{n,\mathrm{s}}, \\ \Delta\mathcal{T}_{k,n} &= \tau_{\mathrm{s},k} + \tau_{k,n} - \tau_{\mathrm{s},n} + w_{k,n}, \end{aligned} \tag{2.14}$$

where $w_{k,n}$ is the measurement noise, and measurement is assumed to have variance δ^2_{mea}. Noting with the measurement properties of $\tau_{\mathrm{s},n}$, $\tau'_{n,\mathrm{s}}$, $\tau_{\mathrm{s},k}$, and $\tau_{k,n}$, it is obtained that $w_{n,\mathrm{s}}$ and $w_{k,n}$ satisfy the following distributions: $w_{n,\mathrm{s}} \sim \mathcal{N}(0, 2\delta^2_{\mathrm{mea}})$ and $w_{k,n} \sim \mathcal{N}(0, 3\delta^2_{\mathrm{mea}})$. Correspondingly, the variance on distance measurement is defined as $\delta^2_{\mathrm{d}} = \mathrm{c}^2\delta^2_{\mathrm{mea}}$.

Problem 2.1 Localization problem for active sensor node is formulated as

$$\begin{aligned} &(\tilde{x}_{\mathrm{s}}, \tilde{y}_{\mathrm{s}}) \\ &= \underset{(x_{\mathrm{s}}, y_{\mathrm{s}})}{\operatorname{argmin}} \left\{ \frac{\mathrm{c}^2}{4\delta^2_{\mathrm{d}}} \sum_{n=1}^{3} (\Delta\mathcal{T}_{n,\mathrm{s}} - \tau_{\mathrm{s},n} - \tau'_{n,\mathrm{s}})^2 \right. \\ &\left. + \frac{\mathrm{c}^2}{6\delta^2_{\mathrm{d}}} \sum_{n=2}^{3} \sum_{k=1}^{n-1} \left[\Delta\mathcal{T}_{k,n} - (\tau_{\mathrm{s},k} + \tau_{k,n} - \tau_{\mathrm{s},n})\right]^2 \right\} \\ &= \underset{(x_{\mathrm{s}}, y_{\mathrm{s}})}{\operatorname{argmin}} \left\{ \frac{1}{4\delta^2_{\mathrm{d}}} \sum_{n=1}^{3} (\mathrm{c}\Delta\mathcal{T}_{n,\mathrm{s}} - d_{\mathrm{s},n} - d'_{n,\mathrm{s}})^2 \right. \\ &\left. + \frac{1}{6\delta^2_{\mathrm{d}}} \sum_{n=2}^{3} \sum_{k=1}^{n-1} \left[\mathrm{c}\Delta\mathcal{T}_{k,n} - (d_{\mathrm{s},k} + d_{k,n} - d_{\mathrm{s},n})\right]^2 \right\}. \end{aligned} \tag{2.15}$$

For passive sensor node, we define the following time differences, i.e.,

$$\begin{aligned} \Delta\mathcal{T}_{\mathrm{s},n,\mathrm{p}} &= (T_{n,\mathrm{p}} - T_{\mathrm{s,p}}) - (t_{n,n} - t_{\mathrm{s},n}), \\ \Delta\mathcal{T}_{k,n,\mathrm{p}} &= (T_{n,\mathrm{p}} - T_{k,\mathrm{p}}) - (t_{n,n} - t_{k,n}), \end{aligned} \tag{2.16}$$

where $n = 1, ..., 3$ and $k = 1, ..., n-1$.

Similar to (2.14), the relationship of time differences and propagation delays can be established, i.e.,

$$\begin{aligned}\Delta\mathcal{T}_{\mathrm{s},n,\mathrm{p}} &= \tau_{\mathrm{s},n} + \tau'_{n,\mathrm{p}} - \tau_{\mathrm{s,p}} + w_{\mathrm{s},n,\mathrm{p}},\\ \Delta\mathcal{T}_{k,n,\mathrm{p}} &= \tau_{k,n} + \tau'_{n,\mathrm{p}} - \tau'_{k,\mathrm{p}} + w_{k,n,\mathrm{p}},\end{aligned} \tag{2.17}$$

where $w_{\mathrm{s},n,\mathrm{p}}$ and $w_{k,n,\mathrm{p}}$ are the measurement noises. $w_{\mathrm{s},n,\mathrm{p}}$ and $w_{k,n,\mathrm{p}}$ satisfy the following distributions: $w_{\mathrm{s},n,\mathrm{p}} \sim \mathcal{N}(0, 3\delta^2_{\mathrm{mea}})$ and $w_{k,n,\mathrm{p}} \sim \mathcal{N}(0, 3\delta^2_{\mathrm{mea}})$.

Problem 2.2 Localization problem for passive sensor node is formulated as

$$\begin{aligned}&(\tilde{x}_\mathrm{p}, \tilde{y}_\mathrm{p})\\ &= \underset{(x_\mathrm{p}, y_\mathrm{p})}{\mathrm{argmin}} \frac{\mathrm{c}^2}{6\delta_\mathrm{d}^2}\Big\{\sum_{n=2}^{3}\sum_{k=1}^{n-1}[\Delta\mathcal{T}_{k,n,\mathrm{p}} - (\tau_{k,n} + \tau'_{n,\mathrm{p}}\\ &-\tau'_{k,\mathrm{p}})]^2 + \sum_{n=1}^{3}[\Delta\mathcal{T}_{\mathrm{s},n,\mathrm{p}} - (\tau_{\mathrm{s},n} + \tau'_{n,\mathrm{p}} - \tau_{\mathrm{s,p}})]^2\Big\}\\ &= \underset{(x_\mathrm{p}, y_\mathrm{p})}{\mathrm{argmin}}\Big\{\frac{1}{6\delta_\mathrm{d}^2}\sum_{n=2}^{3}\sum_{k=1}^{n-1}[\mathrm{c}\Delta\mathcal{T}_{k,n,\mathrm{p}} - (d_{k,n} + d'_{n,\mathrm{p}}\\ &-d'_{k,\mathrm{p}})]^2 + \frac{1}{6\delta_\mathrm{d}^2}\sum_{n=1}^{3}[\mathrm{c}\Delta\mathcal{T}_{\mathrm{s},n,\mathrm{p}} - (d_{\mathrm{s},n} + d'_{n,\mathrm{p}} - d_{\mathrm{s,p}})]^2\Big\}.\end{aligned} \tag{2.18}$$

2.4 Position Solving and Performance Analysis

2.4.1 Position Solving for Sensor Nodes

With (2.5), we rearrange (2.15) and have the following results

$$\underbrace{\begin{bmatrix}\frac{\mathrm{c}\Delta T_{1,\mathrm{s}}}{2}\\ \frac{\mathrm{c}\Delta T_{2,\mathrm{s}}}{2}\\ \frac{\mathrm{c}\Delta T_{3,\mathrm{s}}}{2}\\ \frac{\mathrm{c}\Delta T_{1,2}}{\sqrt{6}}\\ \frac{\mathrm{c}\Delta T_{1,3}}{\sqrt{6}}\\ \frac{\mathrm{c}\Delta T_{2,3}}{\sqrt{6}}\end{bmatrix}}_{:=\boldsymbol{\varphi}_\mathrm{s}} - \underbrace{\begin{bmatrix}\frac{d_{\mathrm{s},1}+d'_{1,\mathrm{s}}}{2}\\ \frac{d_{\mathrm{s},2}+d'_{2,\mathrm{s}}}{2}\\ \frac{d_{\mathrm{s},3}+d'_{3,\mathrm{s}}}{2}\\ \frac{d_{\mathrm{s},1}+d_{1,2}-d_{\mathrm{s},2}}{\sqrt{6}}\\ \frac{d_{\mathrm{s},1}+d_{1,3}-d_{\mathrm{s},3}}{\sqrt{6}}\\ \frac{d_{\mathrm{s},2}+d_{2,3}-d_{\mathrm{s},3}}{\sqrt{6}}\end{bmatrix}}_{:=\mathbf{h}_\mathrm{s}} = \underbrace{\begin{bmatrix}\frac{\mathrm{c}w_{1,\mathrm{s}}}{2}\\ \frac{\mathrm{c}w_{2,\mathrm{s}}}{2}\\ \frac{\mathrm{c}w_{3,\mathrm{s}}}{2}\\ \frac{\mathrm{c}w_{1,2}}{\sqrt{6}}\\ \frac{\mathrm{c}w_{1,3}}{\sqrt{6}}\\ \frac{\mathrm{c}w_{2,3}}{\sqrt{6}}\end{bmatrix}}_{:=\mathbf{W}_\mathrm{s}}. \tag{2.19}$$

From (2.19), the following nonlinear relation is constructed

$$\boldsymbol{\varphi}_{\mathrm{s}} - \mathbf{h}_{\mathrm{s}} = \mathbf{W}_{\mathrm{s}}, \tag{2.20}$$

where $\mathbf{h}_{\mathrm{s}} = \mathbf{h}_{\mathrm{s}}(\boldsymbol{\eta}_{\mathrm{s}}, \boldsymbol{\xi})$, $\boldsymbol{\eta}_{\mathrm{s}} = (x_{\mathrm{s}}, y_{\mathrm{s}})$ denotes position of active sensor node, and $\boldsymbol{\xi} = [\mathcal{X}_1, \mathcal{Y}_1; \mathcal{X}_2, \mathcal{Y}_2; \mathcal{X}_3, \mathcal{Y}_3]$ is the real-time position matrix of AUVs.

To solve the nonlinear estimation in (2.20), we carry out the Taylor series expansion. Then, we have

$$\boldsymbol{\varphi}_{\mathrm{s}} - \mathbf{h}_{\mathrm{s}}(\hat{\boldsymbol{\eta}}_{\mathrm{s}}^{i}, \boldsymbol{\xi}) = \mathbf{J}_{\mathrm{s}}\boldsymbol{\zeta}_{\mathrm{s}} + \mathbf{W}_{\mathrm{s}}, \tag{2.21}$$

where $\hat{\boldsymbol{\eta}}_{\mathrm{s}}^{i}$ is the estimation in the ith iteration. $\mathbf{J}_{\mathrm{s}}$ is the Jacobian matrix, i.e.,

$$\mathbf{J}_{\mathrm{s}} = \frac{\partial[\boldsymbol{\varphi}_{\mathrm{s}} - \mathbf{h}_{\mathrm{s}}(\boldsymbol{\eta}_{\mathrm{s}}, \boldsymbol{\xi})]}{\partial \boldsymbol{\eta}_{\mathrm{s}}} \Big|_{\boldsymbol{\eta}_{\mathrm{s}} = \hat{\boldsymbol{\eta}}_{\mathrm{s}}^{i}},$$

which can be expressed as

$$\mathbf{J}_{\mathrm{s}} = - \begin{bmatrix}
\frac{1}{2}(\frac{x_{\mathrm{s}} - \mathcal{X}_1}{d_{\mathrm{s},1}} + \frac{x'_{\mathrm{s}} - \mathcal{X}_1}{d'_{1,\mathrm{s}}}) & \frac{1}{2}(\frac{y_{\mathrm{s}} - \mathcal{Y}_1}{d_{\mathrm{s},1}} + \frac{y'_{\mathrm{s}} - \mathcal{Y}_1}{d'_{1,\mathrm{s}}}) \\
\frac{1}{2}(\frac{x_{\mathrm{s}} - \mathcal{X}_2}{d_{\mathrm{s},2}} + \frac{x'_{\mathrm{s}} - \mathcal{X}_2}{d'_{2,\mathrm{s}}}) & \frac{1}{2}(\frac{y_{\mathrm{s}} - \mathcal{Y}_2}{d_{\mathrm{s},2}} + \frac{y'_{\mathrm{s}} - \mathcal{Y}_2}{d'_{2,\mathrm{s}}}) \\
\frac{1}{2}(\frac{x_{\mathrm{s}} - \mathcal{X}_3}{d_{\mathrm{s},3}} + \frac{x'_{\mathrm{s}} - \mathcal{X}_3}{d'_{3,\mathrm{s}}}) & \frac{1}{2}(\frac{y_{\mathrm{s}} - \mathcal{Y}_3}{d_{\mathrm{s},3}} + \frac{y'_{\mathrm{s}} - \mathcal{Y}_3}{d'_{3,\mathrm{s}}}) \\
\frac{1}{\sqrt{6}}(\frac{x_{\mathrm{s}} - \mathcal{X}_1}{d_{\mathrm{s},1}} - \frac{x_{\mathrm{s}} - \mathcal{X}_2}{d_{\mathrm{s},2}}) & \frac{1}{\sqrt{6}}(\frac{y_{\mathrm{s}} - \mathcal{Y}_1}{d_{\mathrm{s},1}} - \frac{y_{\mathrm{s}} - \mathcal{Y}_2}{d_{\mathrm{s},2}}) \\
\frac{1}{\sqrt{6}}(\frac{x_{\mathrm{s}} - \mathcal{X}_1}{d_{\mathrm{s},1}} - \frac{x_{\mathrm{s}} - \mathcal{X}_3}{d_{\mathrm{s},3}}) & \frac{1}{\sqrt{6}}(\frac{y_{\mathrm{s}} - \mathcal{Y}_1}{d_{\mathrm{s},1}} - \frac{y_{\mathrm{s}} - \mathcal{Y}_3}{d_{\mathrm{s},3}}) \\
\frac{1}{\sqrt{6}}(\frac{x_{\mathrm{s}} - \mathcal{X}_2}{d_{\mathrm{s},2}} - \frac{x_{\mathrm{s}} - \mathcal{X}_3}{d_{\mathrm{s},3}}) & \frac{1}{\sqrt{6}}(\frac{y_{\mathrm{s}} - \mathcal{Y}_2}{d_{\mathrm{s},2}} - \frac{y_{\mathrm{s}} - \mathcal{Y}_3}{d_{\mathrm{s},3}})
\end{bmatrix}.$$

So the iterative least squares increment $\boldsymbol{\zeta}_{\mathrm{s}} = \left[\zeta_{\mathrm{s,x}}, \zeta_{\mathrm{s,y}}\right]^{\mathrm{T}}$ is

$$\boldsymbol{\zeta}_{\mathrm{s}} = -\left[\mathbf{J}_{\mathrm{s}}^{\mathrm{T}}\mathbf{J}_{\mathrm{s}}\right]^{-1}\mathbf{J}_{\mathrm{s}}^{\mathrm{T}}\mathbf{D}_{\mathrm{s}}, \tag{2.22}$$

where $\mathbf{D}_{\mathrm{s}} = [\boldsymbol{\varphi}_{\mathrm{s}}[1] - \mathbf{h}_{\mathrm{s}}(\hat{\boldsymbol{\eta}}_{\mathrm{s}}^{i}, \boldsymbol{\xi})[1], \boldsymbol{\varphi}_{\mathrm{s}}[2] - \mathbf{h}_{\mathrm{s}}(\hat{\boldsymbol{\eta}}_{\mathrm{s}}^{i}, \boldsymbol{\xi})[2], \ldots, \boldsymbol{\varphi}_{\mathrm{s}}[6] - \mathbf{h}_{\mathrm{s}}(\hat{\boldsymbol{\eta}}_{\mathrm{s}}^{i}, \boldsymbol{\xi})[6]]^{\mathrm{T}}$.

Based on this, the estimation in the $(i+1)$st iteration is updated as

$$\hat{\boldsymbol{\eta}}_{\mathrm{s}}^{i+1} = \hat{\boldsymbol{\eta}}_{\mathrm{s}}^{i} + \boldsymbol{\zeta}_{\mathrm{s}}. \tag{2.23}$$

When $\left\|\hat{\boldsymbol{\eta}}_{\mathrm{s}}^{i+1} - \hat{\boldsymbol{\eta}}_{\mathrm{s}}^{i}\right\| < \varepsilon_{\mathrm{s}}$, the localization task can be considered to be achieved, and $\varepsilon_{\mathrm{s}} > 0$ denotes limit of the iterative termination criteria.

We use iterative least squares method to solve the optimization problem (2.18). According to the relationship in (2.5) and (2.18), we have

$$\underbrace{\begin{bmatrix} c\Delta T_{1,2,\mathrm{p}} \\ c\Delta T_{1,3,\mathrm{p}} \\ c\Delta T_{2,3,\mathrm{p}} \\ c\Delta T_{\mathrm{s},1,\mathrm{p}} \\ c\Delta T_{\mathrm{s},2,\mathrm{p}} \\ c\Delta T_{\mathrm{s},3,\mathrm{p}} \end{bmatrix}}_{:=\boldsymbol{\varphi}_{\mathrm{p}}} = \underbrace{\begin{bmatrix} (d_{1,2} + d'_{2,\mathrm{p}} - d'_{1,\mathrm{p}}) \\ (d_{1,3} + d'_{3,\mathrm{p}} - d'_{1,\mathrm{p}}) \\ (d_{2,3} + d'_{3,\mathrm{p}} - d'_{2,\mathrm{p}}) \\ (d_{\mathrm{s},1} + d'_{1,\mathrm{p}} - d_{\mathrm{s},\mathrm{p}}) \\ (d_{\mathrm{s},2} + d'_{2,\mathrm{p}} - d_{\mathrm{s},\mathrm{p}}) \\ (d_{\mathrm{s},3} + d'_{3,\mathrm{p}} - d_{\mathrm{s},\mathrm{p}}) \end{bmatrix}}_{:=\mathbf{h}_{\mathrm{p}}} + \underbrace{\begin{bmatrix} cw_{1,2,\mathrm{p}} \\ cw_{1,3,\mathrm{p}} \\ cw_{2,3,\mathrm{p}} \\ cw_{\mathrm{s},1,\mathrm{p}} \\ cw_{\mathrm{s},2,\mathrm{p}} \\ cw_{\mathrm{s},3,\mathrm{p}} \end{bmatrix}}_{:=\mathbf{W}_{\mathrm{p}}}. \tag{2.24}$$

From (2.24), the following nonlinear relation is constructed

$$\boldsymbol{\varphi}_{\mathrm{p}} - \mathbf{h}_{\mathrm{p}} = \mathbf{W}_{\mathrm{p}}, \tag{2.25}$$

where $\mathbf{h}_{\mathrm{p}} = \mathbf{h}_{\mathrm{p}}(\boldsymbol{\eta}_{\mathrm{p}}, \boldsymbol{\xi})$, and $\boldsymbol{\eta}_{\mathrm{p}} = (x_{\mathrm{p}}, y_{\mathrm{p}})$ denotes the position of passive sensor node.

Similar to (2.21), one has

$$\boldsymbol{\varphi}_{\mathrm{p}} - \mathbf{h}_{\mathrm{p}}(\hat{\boldsymbol{\eta}}_{\mathrm{p}}^{i}, \boldsymbol{\xi}) = \mathbf{J}_{\mathrm{p}}\boldsymbol{\zeta}_{\mathrm{p}} + \mathbf{W}_{\mathrm{p}}, \tag{2.26}$$

where $\hat{\boldsymbol{\eta}}_{\mathrm{p}}^{i}$ is estimation in the ith iteration, and $\mathbf{J}_{\mathrm{p}}$ is Jacobian matrix, i.e.,

$$\mathbf{J}_{\mathrm{p}} = \frac{\partial[\boldsymbol{\varphi}_{\mathrm{p}} - \mathbf{h}_{\mathrm{p}}(\boldsymbol{\eta}_{\mathrm{p}}, \boldsymbol{\xi})]}{\partial \boldsymbol{\eta}_{\mathrm{p}}} \Big|_{\boldsymbol{\eta}_{\mathrm{p}}=\hat{\boldsymbol{\eta}}_{\mathrm{p}}^{i}}, \tag{2.27}$$

which is expressed as

$$\mathbf{J}_{\mathrm{p}} = -\begin{bmatrix} \frac{\mathcal{X}_1 - x'_{\mathrm{p}}}{d'_{1,\mathrm{p}}} - \frac{\mathcal{X}_2 - x'_{\mathrm{p}}}{d'_{2,\mathrm{p}}} & \frac{\mathcal{Y}_1 - y'_{\mathrm{p}}}{d'_{1,\mathrm{p}}} - \frac{\mathcal{Y}_2 - y'_{\mathrm{p}}}{d'_{2,\mathrm{p}}} \\ \frac{\mathcal{X}_1 - x'_{\mathrm{p}}}{d'_{1,\mathrm{p}}} - \frac{\mathcal{X}_3 - x'_{\mathrm{p}}}{d'_{3,\mathrm{p}}} & \frac{\mathcal{Y}_1 - y'_{\mathrm{p}}}{d'_{1,\mathrm{p}}} - \frac{\mathcal{Y}_3 - y'_{\mathrm{p}}}{d'_{3,\mathrm{p}}} \\ \frac{\mathcal{X}_2 - x'_{\mathrm{p}}}{d'_{2,\mathrm{p}}} - \frac{\mathcal{X}_3 - x'_{\mathrm{p}}}{d'_{3,\mathrm{p}}} & \frac{\mathcal{Y}_2 - y'_{\mathrm{p}}}{d'_{2,\mathrm{p}}} - \frac{\mathcal{Y}_3 - y'_{\mathrm{p}}}{d'_{3,\mathrm{p}}} \\ \frac{x_{\mathrm{s}} - x_{\mathrm{p}}}{d_{\mathrm{s},\mathrm{p}}} - \frac{\mathcal{X}_1 - x'_{\mathrm{p}}}{d'_{1,\mathrm{p}}} & \frac{y_{\mathrm{s}} - x_{\mathrm{p}}}{d_{\mathrm{s},\mathrm{p}}} - \frac{\mathcal{Y}_1 - y'_{\mathrm{p}}}{d'_{1,\mathrm{p}}} \\ \frac{x_{\mathrm{s}} - x_{\mathrm{p}}}{d_{\mathrm{s},\mathrm{p}}} - \frac{\mathcal{X}_2 - x'_{\mathrm{p}}}{d'_{2,\mathrm{p}}} & \frac{y_{\mathrm{s}} - y_{\mathrm{p}}}{d_{\mathrm{s},\mathrm{p}}} - \frac{\mathcal{Y}_2 - y'_{\mathrm{p}}}{d'_{2,\mathrm{p}}} \\ \frac{x_s - x_{\mathrm{p}}}{d_{\mathrm{s},\mathrm{p}}} - \frac{\mathcal{X}_3 - x'_p}{d'_{3,\mathrm{p}}} & \frac{y_s - y_{\mathrm{p}}}{d_{\mathrm{s},\mathrm{p}}} - \frac{\mathcal{Y}_3 - y'_{\mathrm{p}}}{d'_{3,\mathrm{p}}} \end{bmatrix}. \tag{2.28}$$

Thus, the iterative least squares increment $\boldsymbol{\zeta}_{\mathrm{p}} = \left[\zeta_{\mathrm{p,x}}, \zeta_{\mathrm{p,y}}\right]^{\mathrm{T}}$ is

$$\boldsymbol{\zeta}_{\mathrm{p}} = -\left[\mathbf{J}_{\mathrm{p}}^{\mathrm{T}}\mathbf{J}_{\mathrm{p}}\right]^{-1}\mathbf{J}_{\mathrm{p}}^{\mathrm{T}}\mathbf{D}_{\mathrm{p}}, \tag{2.29}$$

where $\mathbf{D}_{\mathrm{p}} = [\boldsymbol{\varphi}_{\mathrm{p}}[1]-\mathbf{h}_{\mathrm{p}}(\hat{\boldsymbol{\eta}}_{\mathrm{p}}^{i}, \boldsymbol{\xi})[1], \boldsymbol{\varphi}_{\mathrm{p}}[2]-\mathbf{h}_{\mathrm{p}}(\hat{\boldsymbol{\eta}}_{\mathrm{p}}^{i}, \boldsymbol{\xi})[2], \ldots, \boldsymbol{\varphi}_{\mathrm{p}}[6]-\mathbf{h}_{\mathrm{p}}(\hat{\boldsymbol{\eta}}_{\mathrm{p}}^{i}, \boldsymbol{\xi})[6]]^{\mathrm{T}}$.

Based on the estimation $\hat{\boldsymbol{\eta}}_{\mathrm{p}}^{i}$ in the ith iteration, the estimation in the $(i+1)$st iteration is updated as

$$\hat{\boldsymbol{\eta}}_{\mathrm{p}}^{i+1} = \hat{\boldsymbol{\eta}}_{\mathrm{p}}^{i} + \boldsymbol{\zeta}_{\mathrm{p}}. \tag{2.30}$$

When $\left\|\hat{\boldsymbol{\eta}}_{\mathrm{p}}^{i+1} - \hat{\boldsymbol{\eta}}_{\mathrm{p}}^{i}\right\| < \varepsilon_{\mathrm{p}}$, the localization task can be considered to be achieved, and $\varepsilon_{\mathrm{p}} > 0$ denotes limit of the iterative termination criteria.

2.4.2 Convergence of the Iterative Squares Estimators

In this section, we investigate the convergence rate of the developed iterative squares localization estimators (2.23) and (2.30). For the analysis of convergence rate, we have the following assumption.

Assumption 2.1 *Assume that Jacobian matrices $\mathbf{J}_s$ and $\mathbf{J}_p$ are rank-full. Besides, $\mathbf{J}_s$ is Lipschitz continuous in a neighborhood of an optimal solution $\tilde{\boldsymbol{\eta}}_s$, while $\mathbf{J}_p$ is Lipschitz continuous in a neighborhood of an optimal solution $\tilde{\boldsymbol{\eta}}_p$.*

Next, the convergence rate of iterative squares estimator (2.23) is provided.

Theorem 2.1 *Consider the iterative least squares estimator (2.23) with Assumption 2.1 , if the starting iterative vector $\hat{\boldsymbol{\eta}}_s^0$ sufficiently closes to $\tilde{\boldsymbol{\eta}}_s$, the rate of convergence is quadratic.*

Proof Denote the ith iteration as $\hat{\boldsymbol{\eta}}_{\mathrm{s}}^{i}$, and then we have

$$\begin{aligned} &\hat{\boldsymbol{\eta}}_{\mathrm{s}}^{i} + \boldsymbol{\zeta}_{\mathrm{s}} - \tilde{\boldsymbol{\eta}}_{\mathrm{s}} \\ &= \hat{\boldsymbol{\eta}}_{\mathrm{s}}^{i} - \tilde{\boldsymbol{\eta}}_{\mathrm{s}} - [\mathbf{J}_{\mathrm{s}}^{\mathrm{T}}(\hat{\boldsymbol{\eta}}_{\mathrm{s}}^{i})\mathbf{J}_{\mathrm{s}}(\hat{\boldsymbol{\eta}}_{\mathrm{s}}^{i})]^{-1}\nabla f_{\mathrm{s}}(\hat{\boldsymbol{\eta}}_{\mathrm{s}}^{i}) \\ &= [\mathbf{J}_{\mathrm{s}}^{\mathrm{T}}(\hat{\boldsymbol{\eta}}_{\mathrm{s}}^{i})\mathbf{J}_{\mathrm{s}}(\hat{\boldsymbol{\eta}}_{\mathrm{s}}^{i})]^{-1}\{[\mathbf{J}_{\mathrm{s}}^{\mathrm{T}}(\hat{\boldsymbol{\eta}}_{\mathrm{s}}^{i})\mathbf{J}_{\mathrm{s}}(\hat{\boldsymbol{\eta}}_{\mathrm{s}}^{i})](\hat{\boldsymbol{\eta}}_{\mathrm{s}}^{i} - \tilde{\boldsymbol{\eta}}_{\mathrm{s}}) + \nabla f_{\mathrm{s}}(\tilde{\boldsymbol{\eta}}_{\mathrm{s}}) - \nabla f_{\mathrm{s}}(\hat{\boldsymbol{\eta}}_{\mathrm{s}}^{i})\}, \end{aligned} \tag{2.31}$$

where $f_{\mathrm{s}}(\hat{\boldsymbol{\eta}}_{\mathrm{s}}^{i}) = (\boldsymbol{\varphi}_{\mathrm{s}} - \mathbf{h}_{\mathrm{s}}(\boldsymbol{\eta}_{\mathrm{s}}, \boldsymbol{\xi}))^{\mathrm{T}}(\boldsymbol{\varphi}_{\mathrm{s}} - \mathbf{h}_{\mathrm{s}}(\boldsymbol{\eta}_{\mathrm{s}}, \boldsymbol{\xi})\Big|_{\boldsymbol{\eta}_{\mathrm{s}}=\hat{\boldsymbol{\eta}}_{\mathrm{s}}^{i}}$.

Based on Taylor's Theorem, we have

$$\begin{aligned} &\nabla f_{\mathrm{s}}(\tilde{\boldsymbol{\eta}}_{\mathrm{s}}) - \nabla f_{\mathrm{s}}(\hat{\boldsymbol{\eta}}_{\mathrm{s}}^{i}) \\ &= \int_0^1 [\mathbf{J}_{\mathrm{s}}^{\mathrm{T}}(\hat{\boldsymbol{\eta}}_{\mathrm{s}}^{i}+\lambda(\tilde{\boldsymbol{\eta}}_{\mathrm{s}}-\hat{\boldsymbol{\eta}}_{\mathrm{s}}^{i}))\mathbf{J}_{\mathrm{s}}(\hat{\boldsymbol{\eta}}_{\mathrm{s}}^{i}+\lambda(\tilde{\boldsymbol{\eta}}_{\mathrm{s}}-\hat{\boldsymbol{\eta}}_{\mathrm{s}}^{i}))](\tilde{\boldsymbol{\eta}}_{\mathrm{s}} - \hat{\boldsymbol{\eta}}_{\mathrm{s}}^{i})d\lambda. \end{aligned} \tag{2.32}$$

Notice that $\mathbf{J}_{\mathrm{s}}$ is Lipschitz continuous, then we obtain

$$\mathbf{J}_{\mathrm{s}}^{\mathrm{T}}(\hat{\boldsymbol{\eta}}_{\mathrm{s}}^{i})\mathbf{J}_{\mathrm{s}}(\hat{\boldsymbol{\eta}}_{\mathrm{s}}^{i}) - \mathbf{J}_{\mathrm{s}}^{\mathrm{T}}(\hat{\boldsymbol{\eta}}_{\mathrm{s}}^{i}+\lambda(\tilde{\boldsymbol{\eta}}_{\mathrm{s}}-\hat{\boldsymbol{\eta}}_{\mathrm{s}}^{i}))\mathbf{J}_{\mathrm{s}}(\hat{\boldsymbol{\eta}}_{\mathrm{s}}^{i}+\lambda(\tilde{\boldsymbol{\eta}}_{\mathrm{s}}-\hat{\boldsymbol{\eta}}_{\mathrm{s}}^{i})) \leq L\lambda(\hat{\boldsymbol{\eta}}_{\mathrm{s}}^{i} - \tilde{\boldsymbol{\eta}}_{\mathrm{s}}), \tag{2.33}$$

where L is a positive constant.

Based on (2.32) and (2.33), one has

$$\begin{aligned}
&\left\| [\mathbf{J}_s^{\mathrm{T}}(\hat{\boldsymbol{\eta}}_s^i)\mathbf{J}_s(\hat{\boldsymbol{\eta}}_s^i)](\hat{\boldsymbol{\eta}}_s^i - \tilde{\boldsymbol{\eta}}_s) + \nabla f_s(\tilde{\boldsymbol{\eta}}_s) - \nabla f_s(\hat{\boldsymbol{\eta}}_s^i) \right\| \\
&= \left\| [\mathbf{J}_s^{\mathrm{T}}(\hat{\boldsymbol{\eta}}_s^i)\mathbf{J}_s(\hat{\boldsymbol{\eta}}_s^i)](\hat{\boldsymbol{\eta}}_s^i - \tilde{\boldsymbol{\eta}}_s) + \int_0^1 [\mathbf{J}_s^{\mathrm{T}}(\hat{\boldsymbol{\eta}}_s^i + \lambda(\tilde{\boldsymbol{\eta}}_s - \hat{\boldsymbol{\eta}}_s^i))\mathbf{J}_s(\hat{\boldsymbol{\eta}}_s^i \right. \\
&\quad \left. + \lambda(\tilde{\boldsymbol{\eta}}_s - \hat{\boldsymbol{\eta}}_s^i))](\tilde{\boldsymbol{\eta}}_s - \hat{\boldsymbol{\eta}}_s^i) d\lambda \right\| \\
&= \left\| \int_0^1 \{\mathbf{J}_s^{\mathrm{T}}(\hat{\boldsymbol{\eta}}_s^i)\mathbf{J}_s(\hat{\boldsymbol{\eta}}_s^i) - [\mathbf{J}_s^{\mathrm{T}}(\hat{\boldsymbol{\eta}}_s^i + \lambda(\tilde{\boldsymbol{\eta}}_s - \hat{\boldsymbol{\eta}}_s^i))\mathbf{J}_s(\hat{\boldsymbol{\eta}}_s^i + \lambda(\tilde{\boldsymbol{\eta}}_s - \hat{\boldsymbol{\eta}}_s^i))]\}(\hat{\boldsymbol{\eta}}_s^i - \tilde{\boldsymbol{\eta}}_s) d\lambda \right\| \\
&\leq \left\| \int_0^1 \{\mathbf{J}_s^{\mathrm{T}}(\hat{\boldsymbol{\eta}}_s^i)\mathbf{J}_s(\hat{\boldsymbol{\eta}}_s^i) - [\mathbf{J}_s^{\mathrm{T}}(\hat{\boldsymbol{\eta}}_s^i + \lambda(\tilde{\boldsymbol{\eta}}_s - \hat{\boldsymbol{\eta}}_s^i))\mathbf{J}_s(\hat{\boldsymbol{\eta}}_s^i + \lambda(\tilde{\boldsymbol{\eta}}_s - \hat{\boldsymbol{\eta}}_s^i))]\} \right\| \left\| (\hat{\boldsymbol{\eta}}_s^i - \tilde{\boldsymbol{\eta}}_s) \right\| d\lambda \\
&\leq \frac{1}{2} L \left\| (\hat{\boldsymbol{\eta}}_s^i - \tilde{\boldsymbol{\eta}}_s) \right\|^2 .
\end{aligned} \tag{2.34}$$

Substituting (2.34) into (2.31), we obtain

$$\begin{aligned}
&\left\| \hat{\boldsymbol{\eta}}_s^i + \boldsymbol{\zeta}_s - \tilde{\boldsymbol{\eta}}_s \right\| \\
&\leq \left\| [\mathbf{J}_s^{\mathrm{T}}(\hat{\boldsymbol{\eta}}_s^i)\mathbf{J}_s(\hat{\boldsymbol{\eta}}_s^i)]^{-1} \right\| \left\| [\mathbf{J}_s^{\mathrm{T}}(\hat{\boldsymbol{\eta}}_s^i)\mathbf{J}_s(\hat{\boldsymbol{\eta}}_s^i)](\hat{\boldsymbol{\eta}}_s^i - \tilde{\boldsymbol{\eta}}_s) + \nabla f_s(\tilde{\boldsymbol{\eta}}_s) - \nabla f_s(\hat{\boldsymbol{\eta}}_s^i) \right\| \\
&= \frac{1}{2} L \left\| [\mathbf{J}_s^{\mathrm{T}}(\hat{\boldsymbol{\eta}}_s^i)\mathbf{J}_s(\hat{\boldsymbol{\eta}}_s^i)]^{-1} \right\| \left\| (\hat{\boldsymbol{\eta}}_s^i - \tilde{\boldsymbol{\eta}}_s) \right\|^2 \\
&= \frac{1}{2} \tilde{L} \left\| (\hat{\boldsymbol{\eta}}_s^i - \tilde{\boldsymbol{\eta}}_s) \right\|^2 ,
\end{aligned} \tag{2.35}$$

where $\tilde{L} = L \left\| [\mathbf{J}_s^{\mathrm{T}}(\hat{\boldsymbol{\eta}}_s^i)\mathbf{J}_s(\hat{\boldsymbol{\eta}}_s^i)]^{-1} \right\|$, i.e., the rate of convergence for the iterative least squares estimator (2.23) is quadratic. □

Similarly, the convergence rate of (2.30) can also be provided.

Theorem 2.2 *Consider the iterative least squares estimator (2.30) with Assumption 2.1, if the starting iterative vector $\hat{\boldsymbol{\eta}}_p^0$ sufficiently closes to $\tilde{\boldsymbol{\eta}}_p$, the rate of convergence is quadratic.*

Proof The proof process is similar with the one of Theorem 2.1, and hence it is omitted due to page limitation. □

2.4.3 *Cramér-Rao Lower Bound*

In this section, we drive the Cramér-Rao lower bound (CRLB) for the localization algorithm. Given an unknown vector $\boldsymbol{\eta}$, the log-likelihood function is denoted as $\ln \Lambda(\boldsymbol{\eta})$, and the ground truth is $\tilde{\boldsymbol{\eta}}$. Based on the above denotations, the Fisher Information Matrix (FIM) can be defined as

$$\begin{aligned}
\mathbf{L}(\boldsymbol{\eta}) &= \mathbf{E}\left\{ [\nabla_{\boldsymbol{\eta}} \ln \Lambda(\boldsymbol{\eta})] [\nabla_{\boldsymbol{\eta}} \ln \Lambda(\boldsymbol{\eta})]^{\mathrm{T}} \right\} |_{\boldsymbol{\eta}=\tilde{\boldsymbol{\eta}}} \\
&= -\mathbf{E}\left\{ \nabla_{\boldsymbol{\eta}} \nabla_{\boldsymbol{\eta}}^{\mathrm{T}} \ln \Lambda(\boldsymbol{\eta}) \right\} |_{\boldsymbol{\eta}=\tilde{\boldsymbol{\eta}}} .
\end{aligned} \tag{2.36}$$

Correspondingly, the CRLB can be defined as $\mathrm{CRLB}(\boldsymbol{\eta}) = \mathbf{L}^{-1}(\boldsymbol{\eta})$. For facilitation of the description, the real-time positions of the active sensor node, passive sensor node and AUV n in vertical and horizontal axes are denoted as $\boldsymbol{\eta}'_{\mathrm{s}} = [x'_{\mathrm{s}}, y'_{\mathrm{s}}]^{\mathrm{T}}$, $\boldsymbol{\eta}'_{\mathrm{p}} = [x'_{\mathrm{p}}, y'_{\mathrm{p}}]^{\mathrm{T}}$ and $\boldsymbol{\xi}_n = [\mathcal{X}_n, \mathcal{Y}_n]^{\mathrm{T}}$, respectively. We analyze the estimation performance in vertical and horizontal axes.

(1) The CRLB for Active Sensor Node

For active sensor node, $\boldsymbol{\xi}_n = [\mathcal{X}_n, \mathcal{Y}_n]^{\mathrm{T}}$ is the collected measurement. In addition, $\boldsymbol{\eta}_{\mathrm{s}}$ and $\boldsymbol{\eta}'_{\mathrm{s}}$ satisfy the relationship in (2.11), wherein $\boldsymbol{\eta}_{\mathrm{s}}$ is required to be calculated. The log likelihood function is constructed as

$$\begin{aligned}&\ln \Lambda_{\mathrm{s}}(\boldsymbol{\eta}_{\mathrm{s}})\\ &= \frac{1}{4\delta_{\mathrm{d}}^2}\sum_{n=1}^{3}(\mathrm{c}\Delta\mathcal{T}_{n,\mathrm{s}} - \|\boldsymbol{\xi}_n - \boldsymbol{\eta}_{\mathrm{s}}\| - \|\boldsymbol{\xi}_n - \boldsymbol{\eta}'_{\mathrm{s}}\|)^2 + \frac{1}{6\delta_{\mathrm{d}}^2}\sum_{n=2}^{3}\sum_{k=1}^{n-1}[\mathrm{c}\\ &\times \Delta\mathcal{T}_{k,n} - (\|\boldsymbol{\xi}_k - \boldsymbol{\eta}_{\mathrm{s}}\| + \|\boldsymbol{\xi}_k - \boldsymbol{\xi}_n\| - \|\boldsymbol{\xi}_n - \boldsymbol{\eta}_{\mathrm{s}}\|)]^2.\end{aligned} \tag{2.37}$$

Based on (2.36), the FIM for $\ln \Lambda_{\mathrm{s}}(\boldsymbol{\eta}_{\mathrm{s}})$ is defined as $\mathbf{L}_{\mathrm{s}}(\boldsymbol{\eta}) \in \mathcal{R}^{2\times 2}$. With straightforward derivation, we have

$$\begin{aligned}[\mathbf{L}_{\mathrm{s}}]_{1,1} &= \frac{1}{2\delta_{\mathrm{d}}^2}\sum_{n=1}^{3}\left[\frac{(\mathcal{X}_n - x_{\mathrm{s}})}{\|\boldsymbol{\xi}_n - \boldsymbol{\eta}_{\mathrm{s}}\|} + \frac{(\mathcal{X}_n - x'_{\mathrm{s}})}{\|\boldsymbol{\xi}_n - \boldsymbol{\eta}'_{\mathrm{s}}\|}\right]^2\\ &+ \frac{1}{3\delta_{\mathrm{d}}^2}\sum_{n=2}^{3}\sum_{k=1}^{n-1}\left[\frac{(\mathcal{X}_k - x_{\mathrm{s}})}{\|\boldsymbol{\xi}_k - \boldsymbol{\eta}_{\mathrm{s}}\|} - \frac{(\mathcal{X}_n - x_{\mathrm{s}})}{\|\boldsymbol{\xi}_n - \boldsymbol{\eta}_{\mathrm{s}}\|}\right]^2,\\ [\mathbf{L}_{\mathrm{s}}]_{2,2} &= \frac{1}{2\delta_{\mathrm{d}}^2}\sum_{n=1}^{3}\left[\frac{(\mathcal{Y}_n - y_{\mathrm{s}})}{\|\boldsymbol{\xi}_n - \boldsymbol{\eta}_{\mathrm{s}}\|} + \frac{(\mathcal{Y}_n - y'_{\mathrm{s}})}{\|\boldsymbol{\xi}_n - \boldsymbol{\eta}'_{\mathrm{s}}\|}\right]^2\\ &+ \frac{1}{3\delta_{\mathrm{d}}^2}\sum_{n=2}^{3}\sum_{k=1}^{n-1}\left[\frac{(\mathcal{Y}_k - y_{\mathrm{s}})}{\|\boldsymbol{\xi}_k - \boldsymbol{\eta}_{\mathrm{s}}\|} - \frac{(\mathcal{Y}_n - y_{\mathrm{s}})}{\|\boldsymbol{\xi}_n - \boldsymbol{\eta}_{\mathrm{s}}\|}\right]^2,\\ [\mathbf{L}_{\mathrm{s}}]_{1,2} = [\mathbf{L}_{\mathrm{s}}]_{2,1} &= \frac{1}{2\delta_{\mathrm{d}}^2}\sum_{n=1}^{3}\left[\frac{(\mathcal{X}_n - x_{\mathrm{s}})}{\|\boldsymbol{\xi}_n - \boldsymbol{\eta}_{\mathrm{s}}\|} + \frac{(\mathcal{X}_n - x'_{\mathrm{s}})}{\|\boldsymbol{\xi}_n - \boldsymbol{\eta}'_{\mathrm{s}}\|}\right]\\ &\times\left[\frac{(\mathcal{Y}_n - y_{\mathrm{s}})}{\|\boldsymbol{\xi}_n - \boldsymbol{\eta}_{\mathrm{s}}\|} + \frac{(\mathcal{Y}_n - y'_{\mathrm{s}})}{\|\boldsymbol{\xi}_n - \boldsymbol{\eta}'_{\mathrm{s}}\|}\right] + \frac{1}{3\delta_{\mathrm{d}}^2}\sum_{n=2}^{3}\sum_{k=1}^{n-1}\left[\frac{(\mathcal{X}_k - x_{\mathrm{s}})}{\|\boldsymbol{\xi}_k - \boldsymbol{\eta}_{\mathrm{s}}\|}\right.\\ &\left.- \frac{(\mathcal{X}_n - x_{\mathrm{s}})}{\|\boldsymbol{\xi}_n - \boldsymbol{\eta}_{\mathrm{s}}\|}\right]\left[\frac{(\mathcal{Y}_k - y_{\mathrm{s}})}{\|\boldsymbol{\xi}_k - \boldsymbol{\eta}_{\mathrm{s}}\|} - \frac{(\mathcal{Y}_n - y_{\mathrm{s}})}{\|\boldsymbol{\xi}_n - \boldsymbol{\eta}_{\mathrm{s}}\|}\right].\end{aligned} \tag{2.38}$$

The CRLB for x_{s} and y_{s} are given by $[\mathbf{L}_{\mathrm{s}}^{-1}(\boldsymbol{\eta})]_{1,1}$ and $[\mathbf{L}_{\mathrm{s}}^{-1}(\boldsymbol{\eta})]_{2,2}$, respectively. The corresponding CRLB for $\boldsymbol{\eta}_{\mathrm{s}}$, denoted by CRLB_s, is given by

$$\mathrm{CRLB}_{\mathrm{s}} = [\mathbf{L}_{\mathrm{s}}^{-1}(\boldsymbol{\eta})]_{1,1} + [\mathbf{L}_{\mathrm{s}}^{-1}(\boldsymbol{\eta})]_{2,2} = \mathrm{tr}\left\{\mathbf{L}_{\mathrm{s}}^{-1}(\boldsymbol{\eta})\right\}_{\boldsymbol{\eta}=\tilde{\boldsymbol{\eta}}_{\mathrm{s}}}. \tag{2.39}$$

Hence, the solution from (2.15) has localization error

$$\mathbf{E}\left\{\left\|\hat{\boldsymbol{\eta}}_{\mathrm{s}}-\tilde{\boldsymbol{\eta}}_{\mathrm{s}}\right\|^{2}\right\} \geq \operatorname{tr}\left\{\mathbf{L}_{\mathrm{s}}^{-1}(\boldsymbol{\eta})\right\}\Big|_{\boldsymbol{\eta}=\tilde{\boldsymbol{\eta}}_{\mathrm{s}}}. \tag{2.40}$$

(2) The CRLB for Passive Sensor Node

For passive sensor node, $\boldsymbol{\eta}_{\mathrm{p}}$ and $\boldsymbol{\eta}_{\mathrm{p}}'$ satisfy the relationship in (2.12), wherein $\boldsymbol{\eta}_{\mathrm{p}}$ needs to be calculated. Similar to (2.37), the log likelihood function is

$$\begin{aligned}
&\ln \Lambda_{\mathrm{p}}(\boldsymbol{\eta}_{\mathrm{p}}) \\
&= \frac{1}{6\delta_{\mathrm{d}}^{2}} \sum_{n=2}^{3} \sum_{k=1}^{n-1} [\mathrm{c}\Delta \mathcal{T}_{k,n,\mathrm{p}} - (\left\|\boldsymbol{\xi}_{k}-\boldsymbol{\xi}_{n}\right\| + \left\|\boldsymbol{\xi}_{n}-\boldsymbol{\eta}_{\mathrm{p}}'\right\| \\
&- \left\|\boldsymbol{\xi}_{k}-\boldsymbol{\eta}_{\mathrm{p}}'\right\|)]^{2} + \frac{1}{6\delta_{\mathrm{d}}^{2}} \sum_{n=1}^{3} [\mathrm{c}\Delta \mathcal{T}_{\mathrm{s},n,\mathrm{p}} - (\left\|\boldsymbol{\xi}_{n}-\boldsymbol{\eta}_{\mathrm{s}}\right\| \\
&+ \left\|\boldsymbol{\xi}_{n}-\boldsymbol{\eta}_{\mathrm{p}}'\right\| - \left\|\boldsymbol{\eta}_{\mathrm{s}}-\boldsymbol{\eta}_{\mathrm{p}}\right\|)]^{2}.
\end{aligned} \tag{2.41}$$

Hence, the FIM for $\ln \Lambda_{\mathrm{p}}(\boldsymbol{\eta}_{\mathrm{p}})$ can be defined as $\mathbf{L}_{\mathrm{p}}(\boldsymbol{\eta}) \in \mathcal{R}^{2\times 2}$. With straightforward derivation, one obtains

$$\begin{aligned}
\left[\mathbf{L}_{\mathrm{p}}\right]_{1,1} &= \frac{1}{3\delta_{\mathrm{d}}^{2}} \sum_{n=2}^{3} \sum_{k=1}^{n-1} \left[\frac{\left(\mathcal{X}_{n}-x_{\mathrm{p}}'\right)}{\left\|\boldsymbol{\xi}_{n}-\boldsymbol{\eta}_{\mathrm{p}}'\right\|} - \frac{\left(\mathcal{X}_{k}-x_{\mathrm{p}}'\right)}{\left\|\boldsymbol{\xi}_{k}-\boldsymbol{\eta}_{\mathrm{p}}'\right\|}\right]^{2} \\
&+ \frac{1}{3\delta_{\mathrm{d}}^{2}} \sum_{n=1}^{3} \left[\frac{\left(\mathcal{X}_{n}-x_{\mathrm{p}}'\right)}{\left\|\boldsymbol{\xi}_{n}-\boldsymbol{\eta}_{\mathrm{p}}'\right\|} - \frac{\left(x_{\mathrm{s}}-x_{\mathrm{p}}\right)}{\left\|\boldsymbol{\eta}_{\mathrm{s}}-\boldsymbol{\eta}_{\mathrm{p}}\right\|}\right]^{2}, \\
\left[\mathbf{L}_{\mathrm{p}}\right]_{2,2} &= \frac{1}{3\delta_{\mathrm{d}}^{2}} \sum_{n=2}^{3} \sum_{k=1}^{n-1} \left[\frac{\left(\mathcal{Y}_{n}-y_{\mathrm{p}}'\right)}{\left\|\boldsymbol{\xi}_{n}-\boldsymbol{\eta}_{\mathrm{p}}'\right\|} - \frac{\left(\mathcal{Y}_{k}-y_{\mathrm{p}}'\right)}{\left\|\boldsymbol{\xi}_{k}-\boldsymbol{\eta}_{\mathrm{p}}'\right\|}\right]^{2} \\
&+ \frac{1}{3\delta_{\mathrm{d}}^{2}} \sum_{n=1}^{3} \left[\frac{\left(\mathcal{Y}_{n}-y_{\mathrm{p}}'\right)}{\left\|\boldsymbol{\xi}_{n}-\boldsymbol{\eta}_{\mathrm{p}}'\right\|} - \frac{\left(y_{\mathrm{s}}-y_{\mathrm{p}}\right)}{\left\|\boldsymbol{\eta}_{\mathrm{s}}-\boldsymbol{\eta}_{\mathrm{p}}\right\|}\right]^{2}, \\
\left[\mathbf{L}_{\mathrm{p}}\right]_{1,2} &= \left[\mathbf{L}_{\mathrm{p}}\right]_{2,1} \\
&= \frac{1}{3\delta_{\mathrm{d}}^{2}} \sum_{n=2}^{3} \sum_{k=1}^{n-1} \left[\frac{\left(\mathcal{X}_{n}-x_{\mathrm{p}}'\right)}{\left\|\boldsymbol{\xi}_{n}-\boldsymbol{\eta}_{\mathrm{p}}'\right\|} - \frac{\left(\mathcal{X}_{k}-x_{\mathrm{p}}'\right)}{\left\|\boldsymbol{\xi}_{k}-\boldsymbol{\eta}_{\mathrm{p}}'\right\|}\right] \\
&\left[\frac{\left(\mathcal{Y}_{n}-y_{\mathrm{p}}'\right)}{\left\|\boldsymbol{\xi}_{n}-\boldsymbol{\eta}_{\mathrm{p}}'\right\|} - \frac{\left(\mathcal{Y}_{k}-y_{\mathrm{p}}'\right)}{\left\|\boldsymbol{\xi}_{k}-\boldsymbol{\eta}_{\mathrm{p}}'\right\|}\right] + \frac{1}{3\delta_{\mathrm{d}}^{2}} \sum_{n=1}^{3} \left[\frac{\left(\mathcal{X}_{n}-x_{\mathrm{p}}'\right)}{\left\|\boldsymbol{\xi}_{n}-\boldsymbol{\eta}_{\mathrm{p}}'\right\|}\right. \\
&\left. - \frac{\left(x_{\mathrm{s}}-x_{\mathrm{p}}\right)}{\left\|\boldsymbol{\eta}_{\mathrm{s}}-\boldsymbol{\eta}_{\mathrm{p}}\right\|}\right] \left[\frac{\left(\mathcal{Y}_{n}-y_{\mathrm{p}}'\right)}{\left\|\boldsymbol{\xi}_{n}-\boldsymbol{\eta}_{\mathrm{p}}'\right\|} - \frac{\left(y_{\mathrm{s}}-y_{\mathrm{p}}\right)}{\left\|\boldsymbol{\eta}_{\mathrm{s}}-\boldsymbol{\eta}_{\mathrm{p}}\right\|}\right].
\end{aligned} \tag{2.42}$$

The CRLB for x_p and y_p are given by $[\mathbf{L}_\mathrm{p}^{-1}(\boldsymbol{\eta})]_{1,1}$ and $[\mathbf{L}_\mathrm{p}^{-1}(\boldsymbol{\eta})]_{2,2}$, respectively. The CRLB for $\boldsymbol{\eta}_\mathrm{p}$, denoted by CRLB_p, is given as

$$\mathrm{CRLB}_\mathrm{p} = [\mathbf{L}_\mathrm{p}^{-1}(\boldsymbol{\eta})]_{1,1} + [\mathbf{L}_\mathrm{p}^{-1}(\boldsymbol{\eta})]_{2,2} = \mathrm{tr}\left\{\mathbf{L}_\mathrm{p}^{-1}(\boldsymbol{\eta})\right\}_{\eta=\tilde{\eta}_\mathrm{p}}. \tag{2.43}$$

Therefore, the solution from (2.18) has localization error

$$\mathbf{E}\left\{\left\|\hat{\boldsymbol{\eta}}_\mathbf{p} - \tilde{\boldsymbol{\eta}}_\mathrm{p}\right\|^2\right\} \geq \mathrm{tr}\left\{\mathbf{L}_\mathrm{p}^{-1}(\boldsymbol{\eta})\right\}\Big|_{\eta=\tilde{\eta}_\mathrm{p}}. \tag{2.44}$$

2.5 Simulation Results

2.5.1 Simulation Settings

The simulations are implemented in MATLAB 2016b. Several assumptions with regard to the capabilities of the modems and the communications channel are made. It is assumed that AUVs can efficiently communicate with active and passive sensor nodes, while the data packets during the communication process can be successfully received and decoded. Without loss of generality, we only consider the timing estimation noises at the receiver nodes. In the simulation, three AUVs, one active sensor node and eight passive sensor nodes are deployed in an area of $1000\,\mathrm{m}\times1000\,\mathrm{m}\times1000\,\mathrm{m}$.

At the beginning, AUVs float at the water surface and acquire their accurate locations through GPS. When the localization procedure begins, AUVs submerge and act as anchor nodes to provide localization information for sensor nodes. The following underwater kinematic model (Beerens et al. 1994; Liu et al. 2016) is adopted to provide initial velocities $\boldsymbol{\mu}_n(t)$ for AUVs, i.e.,

$$\begin{cases} \mu_{n,\mathrm{x}}(t) = \mathrm{k}_1\vartheta v \sin(\mathrm{k}_2 x)\cos(\mathrm{k}_3 y) + \mathrm{k}_1\vartheta\cos(2\mathrm{k}_1 t) + \mathrm{k}_4 \\ \mu_{n,\mathrm{y}}(t) = -\vartheta v \sin(\mathrm{k}_3 x)\cos(\mathrm{k}_2 y) + \mathrm{k}_5, \end{cases} \tag{2.45}$$

where $\mu_{n,\mathrm{x}}$ and $\mu_{n,\mathrm{y}}$ are the velocities in X axis and Y axis, respectively. k_1, k_2, k_3, ϑ and v are random variables which are related to the underwater environment factors such as tides and bathymetry. k_4 and k_5 are positive parameters. In special, it is assumed that k_1, k_2, k_3, ϑ and v are subject to the normal distributions, while k_4 and k_5 are constants. Of note, the detailed values of these parameters are shown in Table 2.1.

Table 2.1 Parameters used in the simulation

Parameter	Value	Parameter	Value
k_1	$\mathcal{N}(0.001\pi, 0.0001\pi)$	k_2	$\mathcal{N}(0.01\pi, 0.001\pi)$
k_3	$\mathcal{N}(0.02\pi, 0.002\pi)$	k_4	0.015
k_5	0.01	ϑ	$\mathcal{N}(1, 0.1)$
v	$\mathcal{N}(0.1, 0.01)$	c	1500 m/s

2.5.2 *Results and Analysis*

(1) Localization Trajectory and Error of an Active Sensor Node
As mentioned above, sensor nodes can move passively due to the effect of current. In this chapter, the movements of active and passive sensor nodes are closely related to their nearby AUVs according to the spatial correlations existed in underwater environments (Bagtzoglou and Andrei 2007; Novikov and Bagtzoglou 2006). Under this case, the real trajectory of an active sensor node is shown in Fig. 2.4a, where 18 points are required to be localized. By solving the asynchronous localization problems in Sect. 2.3.3, the localized trajectory is also given in Fig. 2.4a. To show more clearly, a localization error function $\text{DEV}_\text{s} = \sqrt{(x_\text{s} - \hat{x}_\text{s})^2 + (y_\text{s} - \hat{y}_\text{s})^2 + (z_\text{s} - \hat{z}_\text{s})^2}$ is defined for the 18 points, where $(\hat{x}_\text{s}, \hat{y}_\text{s}, \hat{z}_\text{s})$ is the calculated location and $(x_\text{s}, y_\text{s}, z_\text{s})$ is the real location. Notice that z_s can be acquired by pressure sensor, and the main focus in this chapter is the localization algorithm. Hence, it is assumed that pressure sensor can accurately measure the depth information, i.e., $z_\text{s} = \hat{z}_\text{s}$. Correspondingly, the localization errors for the 18 points are shown in Fig. 2.4b. Clearly, the localization task for an active sensor node can be achieved.

(2) Localization Trajectory and Error of a Passive Sensor Node
With the localization method in this chapter, the real and localized trajectories of a passive sensor node are shown in Fig. 2.4c, where 18 points are required to be localized. The localization error is shown in Fig. 2.4d. Obviously, the localization task for a passive sensor node can be achieved.

(3) Localization for the USNs at an Arbitrary Time Instant
In our simulation, the USNs consist of one active sensor node and 8 passive sensor nodes. At an arbitrary time instant, we apply the asynchronous localization approach to localize all sensor nodes. Thereby, the real and localized positions of these nodes are shown in Fig. 2.5a, where the reversed triangle (i.e., ∇) denotes the real position of active sensor node, the regular triangle (i.e., Δ) is for the real position of passive sensor nodes, the plus sign (i.e., $+$) shows the localized position of active sensor node, and the cross sign (i.e., $\times$) shows the localized position of passive sensor node. Correspondingly, the localization errors are provided in Fig. 2.5b, where the ID number '0' denotes the active sensor node, and the other ones are for passive sensor nodes. It is seen that the localization task for the USNs can be achieved.

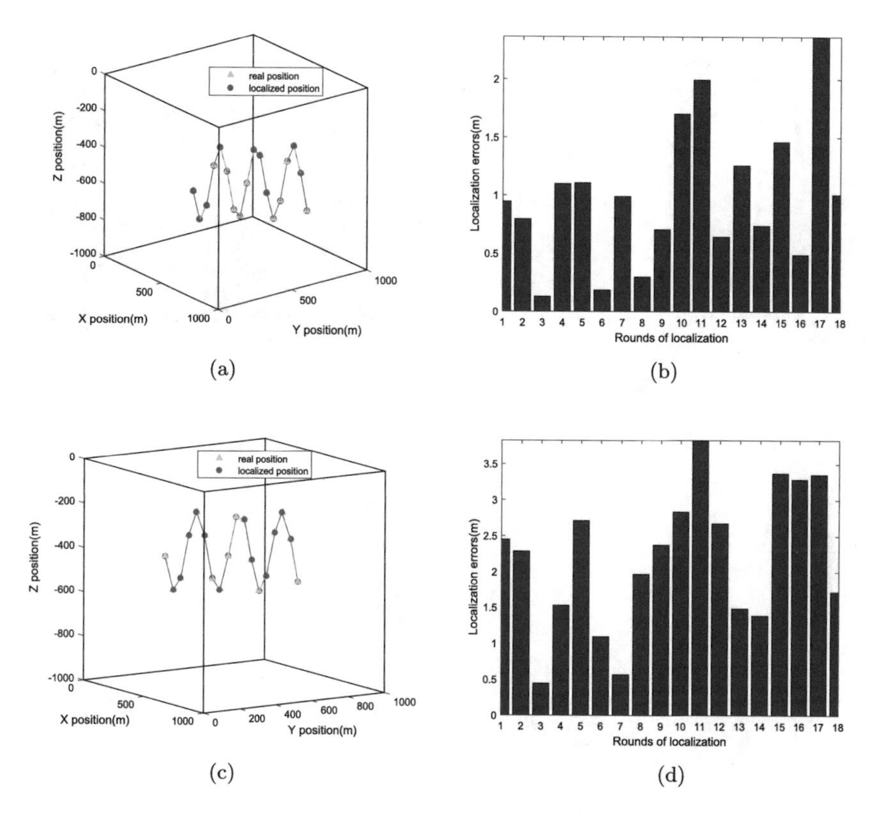

Fig. 2.4 Localization trajectory and error of active sensor node and passive sensor node. (**a**) Trajectories of an active sensor node. (**b**) Localization error for active sensor node. (**c**) Trajectories of a passive sensor node. (**d**) Localization error for passive sensor node

With the iterative least squares estimators in Sect. 2.4.1, Fig. 2.5c shows the convergence process of the localization for active sensor node. Similarly, Fig. 2.5d describes the convergence process of the localization for passive sensor nodes. From Fig. 2.5c, d, we know the average number of iterations is 6, which can satisfy the localization requirement.

(4) Error Analysis of Reference AUVs

In the localization process, AUVs act as anchor nodes to provide auxiliary localization information, and the reference messages provided by AUV affect the accuracy of the localization results. Specially, the reference messages include time stamps, AUV's position and velocity. The error of time stamps is associated with the measurement error and underwater communication noise. The errors of position and velocity influence the accuracy of mobility prediction in Sect. 2.3.2. To verify this conclusion, noises are added to the velocity and position information provided by AUVs. Figure 2.6a shows the localization errors of an active sensor nodes when the velocity information provided by AUV is not accurate. When the position

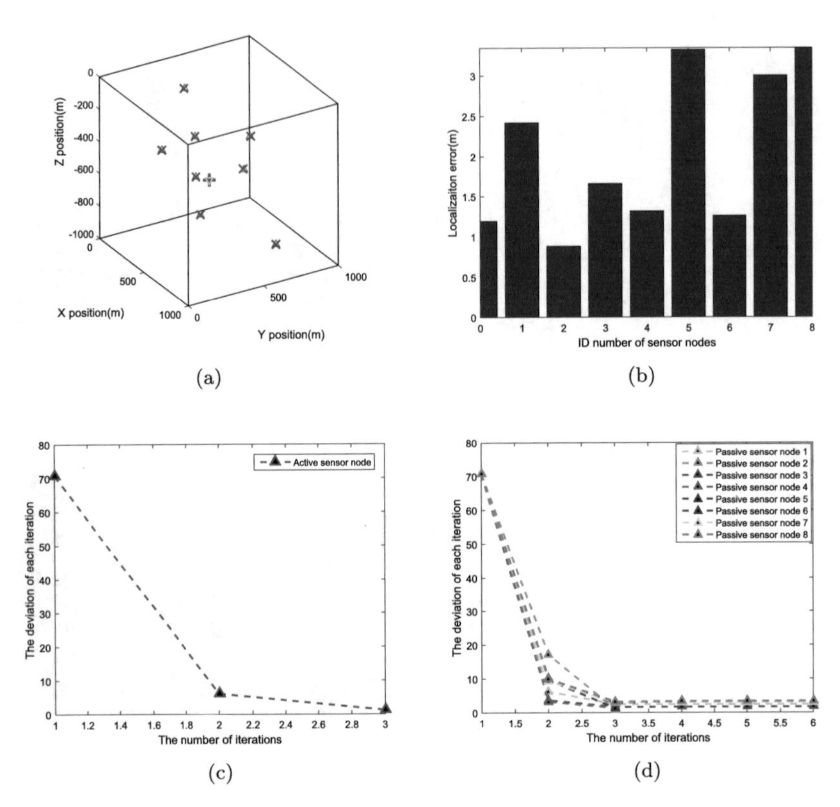

Fig. 2.5 Localization for the USNs at an arbitrary time instant. (**a**) Localization of active and passive sensors. (**b**) Localization errors. (**c**) Iteration process for active sensor node. (**d**) Iteration process for passive sensor nodes

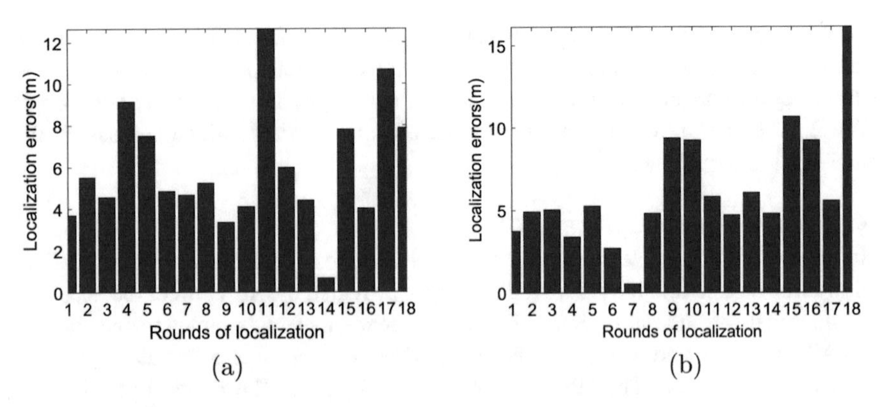

Fig. 2.6 Localization errors when the information provided by AUVs is not accurate. (**a**) Velocity is not accurate. (**b**) Position is not accurate

information provided by AUV is not accurate, the localization errors of an active sensor node are shown in Fig. 2.6b. Clearly, the accuracy of localization is low. The similar validation can also be obtained by the passive sensor nodes. In order to reduce these negative impacts, AUVs (or anchor nodes) are required to surface and re-localize themselves with the assistance of GPS.

(5) Comparison with Synchronous Localization Algorithm

In Zhou et al. (2011), a scalable localization algorithm with mobility prediction (SLMP) was proposed to localize the sensor nodes, where the clock is assumed to be synchronized. The SLMP-based localization approach can achieve localization task when the clock is synchronous, however it cannot obtain the accurate location information when the clock is asynchronous. To verify this conclusion, we apply the SLMP-based localization approach to localize an active sensor node under synchronous assumption. The localization errors are shown in Fig. 2.7a. Subsequently, under the same assumption and noises, we apply the localization approach in this chapter to localize the active sensor node, and the localization errors are also given in Fig. 2.7a. Clearly, the above two methods can both achieve the localization task with synchronous clock. When the clock is asynchronous, the clock offset is set to be randomly initialized from 1 to 200 ms. Hence, the localized trajectories of the USNs by using the SLMP-based localization approach are shown in Fig. 2.7b. To show more clearly, the localization errors are given in Fig. 2.7c. Clearly, the asynchronous clock seriously affects the localization accuracy and the localization task cannot be achieved eventually. Alternatively, the asynchronous algorithm in this chapter can effectively eliminate the impact of the clock asynchronization, i.e., the localization task can be achieved by using the method in this chapter (see Fig. 2.4a, c).

(6) Comparison with Asynchronous Localization Algorithm

In Carroll et al. (2014), an on-demand asynchronous localization approach (ODAL) was proposed for USNs, where exhaustive search method is adopted to find the optimal solution. However, the nodes in Carroll et al. (2014) are assumed to be static, while the duration of localization in exhaustive search is long because a large amount of computation is required. Under static assumption, we apply the ODAL-based localization approach to localize an active sensor node, and the localization errors are shown in Fig. 2.8a. Under the same assumption and noises, we apply the localization approach in this chapter to localize the active sensor node, and the localization errors are also given in Fig. 2.8a. Clearly, the localization task can be achieved with the above two methods. The durations of the two approaches are shown in Fig. 2.8b. Comparing with the exhaustive search-based localization method in Carroll et al. (2014), the localization time with iterative least squares in this chapter can be reduced. On the other hand, mobility in underwater environment must be considered, as the sensor nodes often have passive mobility caused by water current or tides. In such a situation, we apply the ODAL algorithm to a mobile USNs. Figure 2.8c shows the localization trajectory of an active sensor node, and the localization errors are given in Fig. 2.8d. Clearly, the localization accuracy of ODAL declines and the localization task cannot be achieved.

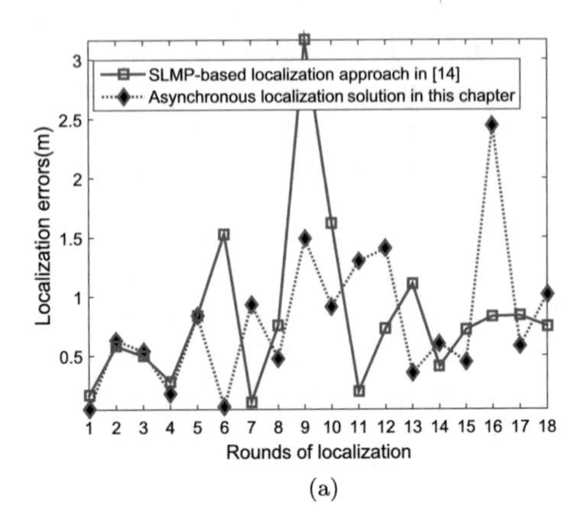

(a)

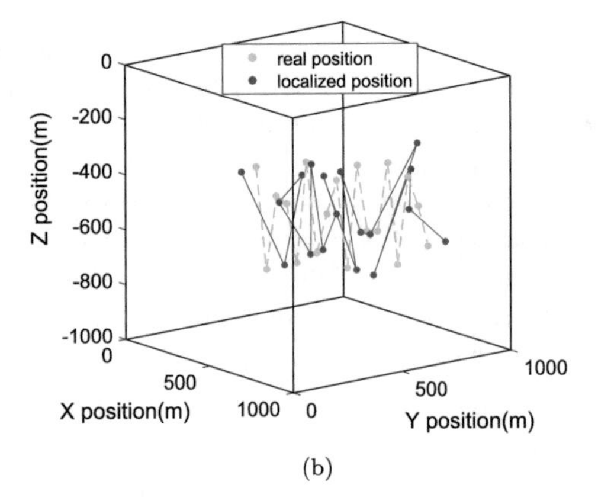

(b)

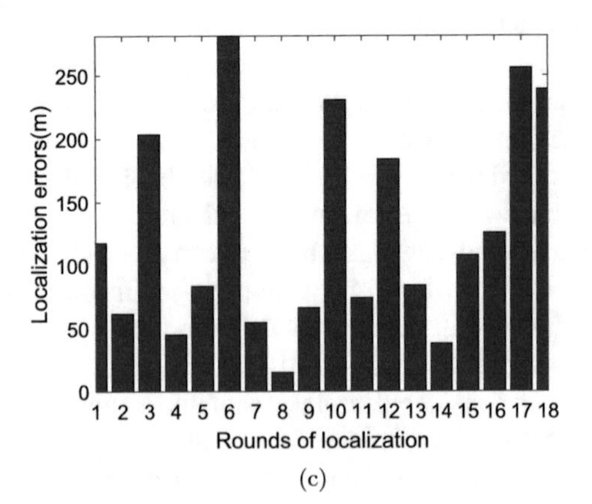

(c)

Fig. 2.7 Comparison with the mobility-based localization algorithm in Zhou et al. (2011). (**a**) Errors under synchronous clock. (**b**) Localization trajectories under asynchronous clock. (**c**) Errors under asynchronous clock

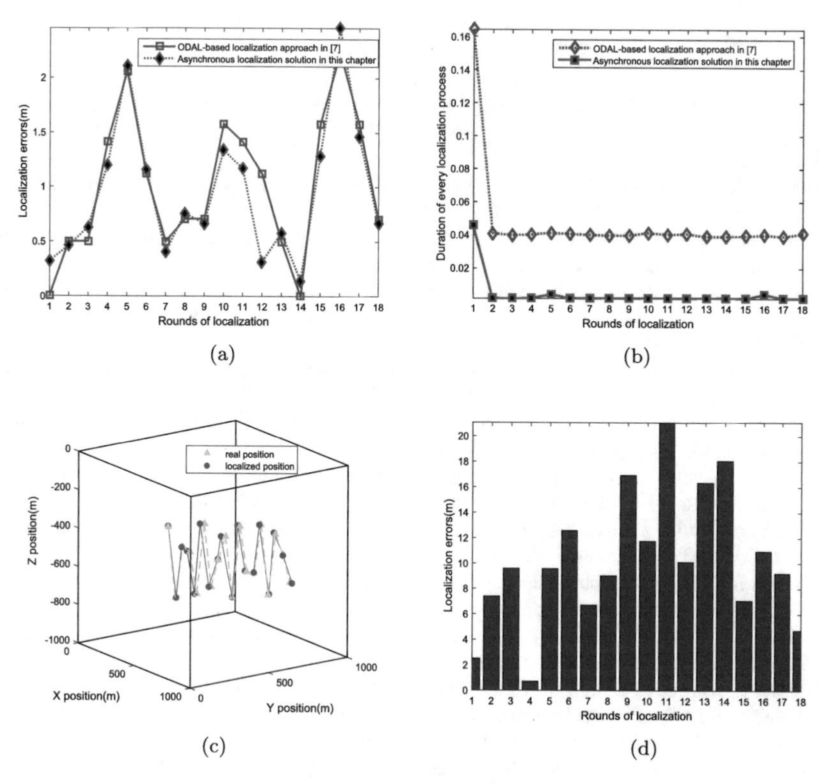

Fig. 2.8 Comparison with the asynchronous localization algorithm in Carroll et al. (2014). (**a**) Localization errors under static USNs. (**b**) Durations under static USNs. (**c**) Localized trajectories in mobile USNs. (**d**) Localization errors in mobile USNs

2.6 Conclusion

This chapter investigates the localization problem for USNs under clock asynchronization and node mobility constraints. Specifically, a hybrid architecture that includes AUVs, active and passive sensor nodes is designed. We construct the relationship between propagation delay and position, and then a mobility prediction strategy is proposed to estimate the future positions of AUV and sensor nodes. With the timestamp measurements, an asynchronous localization algorithm with mobility prediction is provided to estimate the locations of active and passive sensor nodes, where iterative least squares estimators are carried out to solve the optimization problems. Moreover, the convergence analysis and CRLB for the proposed localization algorithm are also given. Simulation results show that the localization accuracy in this chapter can be improved by comparing with the synchronous localization algorithms. Comparing with the exhaustive search-based localization method, the localization time in this chapter is reduced.

References

Bagtzoglou A, Andrei N (2007) Chaotic behaviour and pollution dispersion characteristics in engineered tidal embayments: a numerial investigation. Jawra J Am Water Resour Assoc 43(1):207–219

Beerens S, Ridderinkhof H, Zimmerman J (1994) An analytical study of chaotic stirring in tidal areas. Chaos Solitons Fractals 4(6):1011–1029

Carroll P, Mahmood K, Zhou S, Zhou H, Xu X, Cui J (2014) On-demand asynchronous localization for underwater sensor networks. IEEE Trans Signal Process 62(13):3337–3348

Caruso A, Paparella F, Vieira L, Erol M (2008) The meandering current mobility model and its impact on underwater mobile sensor networks. In: Proc. IEEE conf. comput. commun. (INFOCOM), Phoenix, pp 771–779

Cheng X, Shu H, Liang Q, Du D (2008) Silent positioning in underwater acoustic sensor networks. IEEE Trans Veh Technol 57(3):1756–1766

Isbitiren G, Akan O (2011) Three-dimensional underwater target tracking with acoustic sensor network. IEEE Trans Veh Technol 60(8):3897–3906

Liu J, Wang Z, Peng Z, Cui J, Fiondella L (2014) Suave: swarm underwater autonomous vehicle localization. In: Proc. IEEE conf. comput. commun. (INFOCOM), Toronto, pp 64–72

Liu J, Wang Z, Cui J, Zhou S, Yang B (2016) A joint time synchronization and localization design for mobile underwater sensor networks. IEEE Trans Mob Comput 15(3):530–543

Luo J, Fan L (2017) A two-phase time synchronization-free localization algorithm for underwater sensor networks. Sensors 17(4):726–750

Luo H, Wu K, Gong Y, Ni L (2016) Localization for drifting restricted floating ocean sensor networks. IEEE Trans Veh Technol 65(12):9968–9981

Mortazavi E, Javidan R, Dehghani M, Kavoosi V (2017) A robust method for underwater wireless sensor joint localization and synchronization. Ocean Eng 137(2):276–286

Novikov A, Bagtzoglou A (2006) Hydrodynamic model of the lower Hudson river estuarine system and its application for water quality management. Water Resour Manag 20(2):257–276

Rowden C (1992) Speech processing. McGraw-Hill, London

Wang W, Xie G (2015) Online high-precision probabilistic localization of robotic fish using visual and inertial cues. IEEE Trans Ind Electron 62(2):1113–1124

Wang S, Chen L, Gu D, Hu H (2014) An optimization based moving horizon estimation with application to localization of autonomous underwater vehicles. Robot Auton Syst 62(10):1581–1596

Xia Y, Wang Y, Ma X, Chen C, Guan X (2015) Joint time synchronization and localization for underwater acoustic sensor networks. In: Pro. int. conf. underwater netw. & syst., Washington DC, pp 22–24

Yan J, Xu Z, Wan Y, Chen C, Luo X (2017) Consensus estimation-based target localization in underwater acoustic sensor networks. Int J Robust Nonlinear Control 27(9):1607–1627

Zhang Y, Liang J, Jiang S, Chen W (2016) A localization method for underwater wireless sensor networks based on mobility prediction and particle swarm optimization algorithms. Sensors 16(2):212–229

Zhou Z, Cui J, Zhou S (2010) Efficient localization for large-scale underwater sensor networks. Ad Hoc Netw 8(3):267–279

Zhou Z, Peng Z, Cui J, Shi Z, Bagtzoglou A (2011) Scalable localization with mobility prediction for underwater sensor networks. IEEE Trans Mobile Comput 10(3):335–348

Chapter 3
Async-Localization of USNs with Consensus-Based Unscented Kalman Filtering

Abstract With consideration of the asynchronous clock, the stratification effect and the strong-noise measurement, an asynchronous localization issue for underwater targets is studied. Specifically, the relationship between the propagation delay and the position can be constructed to eliminate the impacts of asynchronous clocks. Afterwards, a localization optimization problem is built to minimize the sum of all measurement errors, and then a consensus-based unscented Kalman filtering (UKF) localization algorithm is developed to solve this localization optimization problem. In addition, the Cramér-Rao lower bounds and convergence conditions are also analyzed. Finally, simulation results show that the proposed localization algorithm can reduce the localization time as compared with the exhaustive search method. Meanwhile, the proposed localization algorithm can improve localization accuracy by comparing with other works.

Keywords Localization · Stratification effect · Asynchronous clock · Underwater Sensor Networks (USNs)

3.1 Introduction

Due to the challenges of asynchronous clock, stratification effect and strong-noise measurement, the localization schemes developed in wireless sensor networks are not suitable for USNs. In order to achieve effective localization in USNs, some time-based localization approaches have been presented (Bayat et al. 2016; Tomczak 2011; Ramezani et al. 2013; Yan et al. 2017a; Zhou et al. 2011; Moreno-Salinas et al. 2016; Zhang et al. 2017b). However, these studies rely on clock synchronization assumption. Considering the existence of the asynchronous clock, some researchers attempt to employ RSS to measure the distances. Particularly, the RSS adopt path loss to measure the relative distances, where clock synchronization assumption is not required. Inspired by this, an attenuation model of underwater acoustic channel was developed in Chang et al. (2018), through which an RSS-based localization algorithm was designed. In Zhang et al. (2016), the stratification compensation was integrated into the design of RSS-based localization algorithm. Nevertheless, the

J. Yan et al., *Localization in Underwater Sensor Networks*, Wireless Networks,
https://doi.org/10.1007/978-981-16-4831-1_3

RSS signals in water are usually inaccurate (Erol-Kantarci et al. 2011), which in turn limit the applications of these algorithms. To obtain the accurate localization information, some joint localization and synchronization algorithms have been proposed. For example, a joint solution to localization and synchronization was provided in Liu et al. (2016), where the stratification effect is compensated with Fermat's principle. In Mortazavi et al. (2017), the sound velocity profile was employed to compensate the stratification effect. A follow-up work (Zhang et al. 2017a) was presented to provide a unified framework. In these studies, the first-order linearization is required to calculate the Jacobian matrix, through which least squares estimators are designed. However, the first-order linearization can introduce large model errors (Li et al. 2008). How to relax the linearization requirement and design a high accuracy localization approach are largely unexplored.

On the other hand, most of the existing works rely on a single measurement. As it is pointed in Shen et al. (2014), the fusion of all available measurements can produce improved estimator robustness and accuracy. In view of this, the *consensus*-based strategies have been widely employed to estimation fusion of wireless sensor networks (Charalambous et al. 2015; Ren and Beard 2005). For instance, Olfati-Saber and Shamma (2006) employed the consensus-based Kalman filtering to improve the tracking accuracy of sensor networks. Several other consensus-based solution were developed in Yan et al. (2017b), Kolomvatsos et al. (2017), and Li et al. (2016). Nevertheless, these consensus fusion-based solutions are not developed in the context of underwater target localization. With regarding to target localization, a consensus-based approach was presented in Chai et al. (2013) to overcome the possible measurement failures of sensors. However, the weak acoustic communication property was not considered in Chai et al. (2013). How to design a localization approach that jointly considers the asynchronous clock, the stratification effect and the strong-noise measurements is not well investigated.

In this chapter, the problem of underwater target localization considering asynchronous clock, the stratification effect and the strong-noise measurements is studied. It is worth mentioning that, the localization solution in Chap. 2 requires to calculate the Jacobian matrix, however it is not an easy task. In order to relax the linearization requirement and improve the localization accuracy, a consensus-based UKF localization algorithm is developed in this chapter. Main contributions of this chapter are shown as:

1. **Asynchronous localization strategy with the consideration of the stratification effect.** In order to eliminate the influences of asynchronous clock, the relationship between the propagation delay and the position is constructed. Meanwhile, the ray tracing approach is applied to model the stratification effect. Compared with the works (Carroll et al. 2014; Yan et al. 2019a,b), the proposed method in this chapter can compensate the effect of asynchronous clock, the stratification effect and the strong-noise measurements.
2. **Consensus-based UKF localization algorithm.** In order to estimate the position of an underwater target, a consensus-based UKF localization algorithm is established. Instead of linearization technique required by Liu et al. (2016),

Mortazavi et al. (2017), and Zhang et al. (2017a), the unscented transform in this chapter can reduce the linearization errors during the localization procedure. Besides, the influence of malicious measurements can also be reduced by the consensus fusion.

3.2 Network Architecture and Overview of the Localization Procedure

3.2.1 *Network Architecture*

The network architecture considered in this chapter is mainly comprised of three different types of nodes, as shown in Fig. 3.1.

- **Surface Buoys**. Surface buoys are equipped with GPS to obtain their positions and global time references. They act as the role of "satellite" nodes, whose objective is to provide self-localization for sensor nodes.
- **Sensor Nodes**. Sensor nodes make direct communication with surface buoys, while their clocks are synchronized and positions are accurately pre-known. The role of sensor nodes is to locate the target. It is assumed that sensor nodes are connected with each other via acoustic communication.
- **Target**. Static network is considered in this chapter, i.e., the position of target is fixed. Of note, the clocks between sensor nodes and target are asynchronous. Of note, the target is installed with a depth sensor, thereby the positions on X axis and Y axis are required to be calculated.

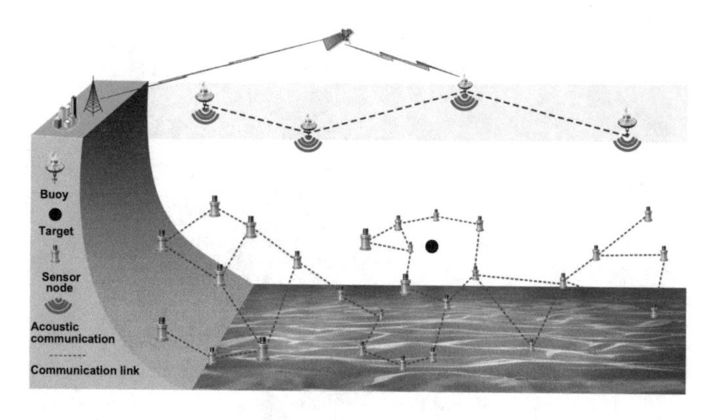

Fig. 3.1 Network architecture of USNs

3.2.2 Overview of the Localization Procedure

The position vector of target is denoted by (x, y, z), where x, y and z denote the positions on X, Y and Z axes, respectively. In order to accurately locate the target, the localization process can be divided into the following two stages: (1) timestamp collection; (2) consensus-based localization. In Stage 1, target sends out an initiator message to its neighboring sensor nodes. Without loss of generality, the neighboring sensor nodes are labelled as $1, \ldots, N$, where $N \geq 3$. Through a two-way transmission process, the timestamps are collected by target. It is worth mentioning that, three non-collinear sensor nodes can locate the target. In view of this, the timestamps from any three sensor nodes form a group, through which $N(N-1)(N-2)/6$ groups can be defined. With the timestamps in different groups, a consensus-based localization algorithm is presented in Stage 2 to locate the target. An example of the localization procedure with $N = 4$ is shown in Fig. 3.2. Correspondingly, the localization process can be presented as follows.

1. At time $T_{\mathrm{T,T}}$, target sends out an initiator message to the sensor nodes. The initiator message includes the sending orders for the neighboring sensor nodes, indexed by 1, ..., 4. In the following, target goes into the listening mode, waiting for the replies from its neighboring sensor nodes.
2. At time $t_{\mathrm{T},n}$, sensor node n receives the initiator message, where $n \in [1, 2, 3, 4]$. After receiving the initiator message from target, sensor node n switches into the waiting mode and receives message from sensor node 1 to $n-1$. The arrival time is denoted by $t_{\bar{k},n}$, where $\bar{k} \in [1, ..., n-1]$. Subsequently, sensor node n sends out its localization message at time $t_{n,n}$. In special, the localization message includes the position of sensor node n, $t_{\mathrm{T},n}$, $\{t_{\bar{k},n}\}_{\forall \bar{k}}$ and the transmission time $t_{n,n}$.
3. At time $T_{N,\mathrm{T}}$, target receives the reply from sensor node N, and then a complete round of localization transmission is finished. The timestamp measurements are obtained as

$$T_{\mathrm{T,T}}, \left\{t_{\mathrm{T},n}, t_{n,n}, T_{n,\mathrm{T}}\right\}_{n=1}^{4}, \{t_{\bar{k},n}\}_{n=2,\bar{k}=1}^{4,n-1}, \tag{3.1}$$

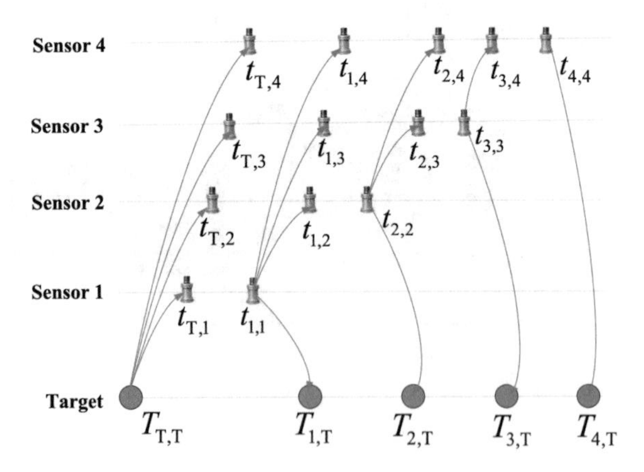

Fig. 3.2 Example of the localization procedure

where the timestamps in group 1 is $\{T_{\mathrm{T,T}}, t_{\mathrm{T,1}}, t_{\mathrm{T,2}}, t_{\mathrm{T,3}}, t_{1,1}, t_{2,2}, t_{3,3}, T_{1,\mathrm{T}}, T_{2,\mathrm{T}}, T_{3,\mathrm{T}}, t_{1,2}, t_{1,3}, t_{2,3}\}$, the ones in group 2 is $\{T_{\mathrm{T,T}}, t_{\mathrm{T,1}}, t_{\mathrm{T,2}}, t_{\mathrm{T,4}}, t_{1,1}, t_{2,2}, t_{4,4}, T_{1,\mathrm{T}}, T_{2,\mathrm{T}}, T_{4,\mathrm{T}}, t_{1,2}, t_{1,4}, t_{2,4}\}$, the ones in group 3 is $\{T_{\mathrm{T,T}}, t_{\mathrm{T,1}}, t_{\mathrm{T,3}}, t_{\mathrm{T,4}}, t_{1,1}, t_{3,3}, t_{4,4}, T_{1,\mathrm{T}}, T_{3,\mathrm{T}}, T_{4,\mathrm{T}}, t_{1,3}, t_{1,4}, t_{3,4}\}$, and the ones in group 4 is $\{T_{\mathrm{T,T}}, t_{\mathrm{T,2}}, t_{\mathrm{T,3}}, t_{\mathrm{T,4}}, t_{2,2}, t_{3,3}, t_{4,4}, T_{2,\mathrm{T}}, T_{3,\mathrm{T}}, T_{4,\mathrm{T}}, t_{2,3}, t_{2,4}, t_{3,4}\}$.

4. With the measurements in (3.1), a consensus-based UKF localization approach is given to locate the target, as provided in Sect. 3.3.

Similar to Liu et al. (2016), Mortazavi et al. (2017), and Zhang et al. (2017a), the clock skew and offset are both considered in this chapter. Hence, the following asynchronous clock model is given, i.e.,

$$T = \alpha t + \beta, \tag{3.2}$$

where T denotes the measured time on the target, t denotes the real time, α denotes the clock skew, and β denotes the clock offset.

Localization Objective With consideration of the stratification effect and strong-noise measurement in underwater environment, we attempt to design an asynchronous localization strategy. In order to improve the localization accuracy, we aim to develop a consensus-based UKF localization algorithm, wherein the timestamp measurements in different groups can be fused to reduce the influence of malicious measurements.

3.3 Consensus-Based UKF Localization Approach

3.3.1 Relationship Between Delay and Position

The sound rays in water do not propagate along straight lines, as illustrated in Fig. 3.3. For that reason, we assume that the sound speed profile is depth dependent, which can be formulated as Ramezani et al. (2013)

$$C(\bar{z}) = \bar{a}\bar{z} + \bar{b}, \tag{3.3}$$

where $\bar{a}$ represents a constant depending on the environment, $\bar{b}$ represents the sound speed at surface, and $\bar{z}$ represents the depth.

In view of (3.3), the one-way propagation delay between sensor node n and target is denoted as $\tau_{\mathrm{T},n}$, while the relative distance between sensor node n and target is represented as $d_{\mathrm{T},n}$. Similarly, the one-way propagation delay between sensor node $\bar{k}$ and sensor node n is represented as $\tau_{\bar{k},n}$, while the relative distance between sensor

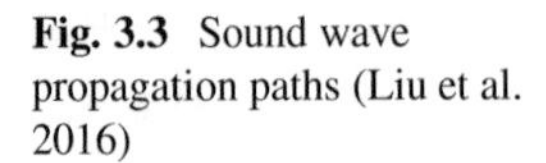

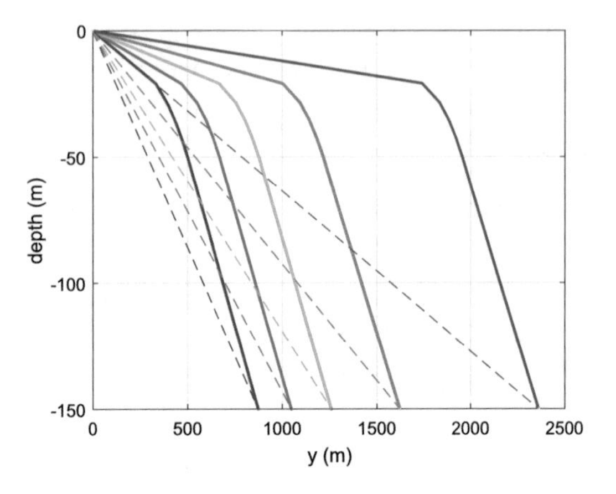

Fig. 3.3 Sound wave propagation paths (Liu et al. 2016)

node $\bar{k}$ and sensor node n is represented as $d_{\bar{k},n}$. With the ray tracing approach (Ramezani et al. 2013), $\tau_{\mathrm{T},n}$ and $\tau_{\bar{k},n}$ can be denoted as

$$\begin{aligned}\tau_{\mathrm{T},n} &= -\frac{1}{\bar{a}}(\ln\frac{1+\sin\theta_{\mathrm{T}}}{\cos\theta_{\mathrm{T}}} - \ln\frac{1+\sin\theta_n}{\cos\theta_n}),\ n \in [1, ..., N] \\ \tau_{\bar{k},n} &= -\frac{1}{\bar{a}}(\ln\frac{1+\sin\bar{\theta}_{\bar{k}}}{\cos\bar{\theta}_{\bar{k}}} - \ln\frac{1+\sin\bar{\theta}_n}{\cos\bar{\theta}_n}),\ \bar{k} \in [1, ..., n-1]\end{aligned} \tag{3.4}$$

where the arcs for $d_{\mathrm{T},n}$ and $d_{\bar{k},n}$ are denoted by $\breve{d}_{\mathrm{T},n}$ and $\breve{d}_{\bar{k},n}$, respectively. For arc $\breve{d}_{\mathrm{T},n}$, $\theta_{\mathrm{T}} = \beta_0 - \alpha_0$ is the ray angle at the location of target, and $\theta_n = \beta_0 + \alpha_0$ is the ray angle at the location of sensor node n. Particularly, $\beta_0 = \arctan\left(\frac{z-z_n}{\sqrt{(x-x_n)^2+(y-y_n)^2}}\right)$ denotes the angle of the straight line between target and sensor node n, w.r.t. the horizontal axis. $\alpha_0 = \arctan\left(\frac{0.5\bar{a}\sqrt{(x-x_n)^2+(y-y_n)^2}}{\bar{b}+0.5\bar{a}(z+z_n)}\right)$ denotes the angle, at which the ray trajectory deviates from this straight line. For arc $\breve{d}_{\bar{k},n}$, $\bar{\theta}_{\bar{k}} = \bar{\beta}_0 - \bar{\alpha}_0$ is the ray angle at the location of sensor node $\bar{k}$, and $\bar{\theta}_n = \bar{\beta}_0 + \bar{\alpha}_0$ is the ray angle at the location of sensor node n. Meanwhile, $\bar{\beta}_0 = \arctan\left(\frac{z_{\bar{k}}-z_n}{\sqrt{(x_{\bar{k}}-x_n)^2+(y_{\bar{k}}-y_n)^2}}\right)$ denotes the angle of the straight line between sensor nodes $\bar{k}$ and n, w.r.t. the horizontal axis. $\bar{\alpha}_0 = \arctan\left(\frac{0.5\bar{a}\sqrt{(x_{\bar{k}}-x_n)^2+(y_{\bar{k}}-y_n)^2}}{\bar{b}+0.5\bar{a}(z_{\bar{k}}+z_n)}\right)$ denotes the angle, at which the ray trajectory deviates from this straight line. In addition, the position vectors of sensor node n and sensor node $\bar{k}$ are represented by (x_n, y_n, z_n) and $(x_{\bar{k}}, y_{\bar{k}}, z_{\bar{k}})$, respectively.

3.3.2 Asynchronous Localization Optimization Problem

Different from the synchronization assumption in Bayat et al. (2016), Liu et al. (2014), and Zhou et al. (2011), the clock in this chapter is asynchronous. In order to remove the effect of asynchronous clock, we define the following time differences, i.e.,

$$\begin{aligned} \Delta\mathcal{T}_{n,\bar{k}} &= t_{\mathrm{T},n} - t_{\mathrm{T},\bar{k}},\ \bar{k} \in [1, ..., n-1], \forall n \\ \Delta\mathcal{T}_{\bar{k},n} &= (t_{\bar{k},n} - t_{\mathrm{T},n}) - (t_{\bar{k},\bar{k}} - t_{\mathrm{T},\bar{k}}). \end{aligned} \tag{3.5}$$

We assume that each node has the same measurement quality, whose measurement noise is zero-mean Gaussian with a variance σ^2_{mea}. Specially, Fig. 3.4 is presented to show the transmission process of sensor node n, sensor node $\bar{k}$ and target. As a result, we have $t_{\mathrm{T},n} - t_{\mathrm{T},\bar{k}} = \tau_{\mathrm{T},n} - \tau_{\mathrm{T},\bar{k}} + \omega_{n,\bar{k}}$ and $t_{\bar{k},n} - t_{\mathrm{T},n} = \tau_{\bar{k},n} + (t_{\bar{k},\bar{k}} - t_{\mathrm{T},\bar{k}}) + \tau_{\mathrm{T},\bar{k}} - \tau_{\mathrm{T},n} + \omega_{\bar{k},n}$. By moving $(t_{\bar{k},\bar{k}} - t_{\mathrm{T},\bar{k}})$ to the left side, we have $(t_{\bar{k},n} - t_{\mathrm{T},n}) - (t_{\bar{k},\bar{k}} - t_{\mathrm{T},\bar{k}}) = \tau_{\bar{k},n} + \tau_{\mathrm{T},\bar{k}} - \tau_{\mathrm{T},n} + \omega_{\bar{k},n}$. In the following, it is defined that $\Delta T_{n,\bar{k}} = t_{\mathrm{T},n} - t_{\mathrm{T},\bar{k}}$ and $\Delta T_{\bar{k},n} = (t_{\bar{k},n} - t_{\mathrm{T},n}) - (t_{\bar{k},\bar{k}} - t_{\mathrm{T},\bar{k}})$. Thereby, the relationship of time differences and propagation delays is calculated as

$$\begin{aligned} \Delta\mathcal{T}_{n,\bar{k}} &= \tau_{\mathrm{T},n} - \tau_{\mathrm{T},\bar{k}} + \varpi_{n,\bar{k}}, \\ \Delta\mathcal{T}_{\bar{k},n} &= \tau_{\mathrm{T},\bar{k}} + \tau_{\bar{k},n} - \tau_{\mathrm{T},n} + \varpi_{\bar{k},n}, \end{aligned} \tag{3.6}$$

where $\varpi_{n,\bar{k}}$ and $\varpi_{\bar{k},n}$ represent the measurement noises. Meanwhile, the measurement noises $\varpi_{n,\bar{k}}$ and $\varpi_{\bar{k},n}$ satisfy the following distributions: $\varpi_{n,\bar{k}} \sim \mathcal{N}(0, 2\sigma^2_{\mathrm{mea}})$ and $\varpi_{\bar{k},n} \sim \mathcal{N}(0, 3\sigma^2_{\mathrm{mea}})$.

Ignoring the multi-path effect and the self-localization error of sensor nodes, this chapter focuses on the influences of asynchronous clock, stratification effect and strong-noise. Hence, the following asynchronous localization optimization problem is formulated on account of (3.5) and (3.6).

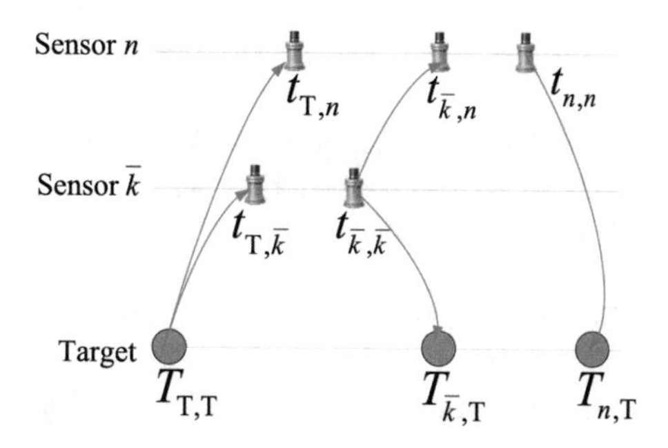

Fig. 3.4 Transmission of sensor nodes n, $\bar{k}$ and target

Problem 3.1 With consideration of stratification effect and asynchronous clock constraints, the localization optimization problem for target can be formulated as minimizing the sum of all measurement errors, i.e.,

$$(\tilde{x}, \tilde{y}) = \underset{(x,y)}{\operatorname{argmin}} \sum_{n=1}^{N} \sum_{\bar{k}=1}^{n-1} \left\{ \frac{1}{4\sigma_{\text{mea}}^2} (\Delta \mathcal{T}_{n,\bar{k}} - \tau_{\text{T},n} + \tau_{\text{T},\bar{k}})^2 \right.$$
$$\left. + \frac{1}{6\sigma_{\text{mea}}^2} [\Delta \mathcal{T}_{\bar{k},n} - (\tau_{\text{T},\bar{k}} + \tau_{\bar{k},n} - \tau_{\text{T},n})]^2 \right\}. \tag{3.7}$$

3.3.3 Consensus-Based UKF Localization Algorithm

For Problem 3.1, three non-collinear sensor nodes can locate the target. Particularly, the timestamps from any three sensor nodes form a group, and then $N(N-1)(N-2)/6$ groups can be defined. For simplicity and without loss of generality, the sensor nodes in group j are labelled as 1_j, 2_j and 3_j. Correspondingly, the time measurements in group j are described as $\mathbf{D}^j = [\Delta\mathcal{T}_{2_j,1_j}, \Delta\mathcal{T}_{3_j,1_j}, \Delta\mathcal{T}_{3_j,2_j}, \Delta\mathcal{T}_{1_j,2_j}, \Delta\mathcal{T}_{1_j,3_j}, \Delta\mathcal{T}_{2_j,3_j}]^{\text{T}}$. Based on this, the time measurement $\mathbf{D}^j(k)$ at time step k is rearranged as

$$\mathbf{D}^j(k) = \mathbf{H}^j(\mathbf{X}(k)) + \boldsymbol{\nu}^j(k) \tag{3.8}$$

with

$$\mathbf{H}^j(\mathbf{X}(k)) = \begin{bmatrix} \tau_{\text{T},2_j}(k) - \tau_{\text{T},1_j}(k) \\ \tau_{\text{T},3_j}(k) - \tau_{\text{T},1_j}(k) \\ \tau_{\text{T},3_j}(k) - \tau_{\text{T},2_j}(k) \\ \tau_{\text{T},1_j}(k) + \tau_{1_j,2_j}(k) - \tau_{\text{T},2_j}(k) \\ \tau_{\text{T},1_j}(k) + \tau_{1_j,3_j}(k) - \tau_{\text{T},3_j}(k) \\ \tau_{\text{T},2_j}(k) + \tau_{2_j,3_j}(k) - \tau_{\text{T},3_j}(k) \end{bmatrix} \tag{3.9}$$

where $\mathbf{X}(k) = [x(k), y(k), z(k)]^{\text{T}}$ is the position of target at time step k. $\boldsymbol{\nu}^j(k)$ is the measurement noise at time step k, which mean value and variance are $\mathbf{r}^j(k)$ and $\mathbf{R}^j(k)$, respectively.

When $\mathbf{D}^j(k)$ is not accurate, the localization errors can be increased. In the following, we design a consensus-based UKF localization algorithm to bias the weights of each estimator in the fusion process. Our objective is to bias the weights of multi-group measurements in the fusion process in accordance with signal stability. Due to the nonlinear dynamics of the measurement model, conventional Kalman Filtering is not applicable, UKF can provide us an effective way for the state estimation of nonlinear systems by using sampling techniques (Julier and Uhlmann 1997), in which a set of sigma points is chosen with initial conditions. However, the initial conditions can be selected arbitrarily, i.e., the initial information for $\mathbf{X}$ is not required to be known previously.

The time measurement is denoted as $\mathbf{D} = [\mathbf{D}^1, \ldots, \mathbf{D}^M]$, where $M = N(N-1)(N-2)/6$ is the number of groups. Based on the time measurement $\mathbf{D}^j(k)$, the estimated value of $\mathbf{X}(k)$ is denoted by $\hat{\mathbf{X}}^j(k)$, and $\mathbf{P}^j(k)$ is its error covariance matrix. Through fusion process (see Algorithm 1), the fusion-based estimation can be denoted by $\hat{\mathbf{X}}(k)$, and $\mathbf{P}(k)$ is its error covariance matrix. Particularly, when $M = 1$, there is no fusion process, and this case is similar to the one in Liu et al. (2016), Zhang et al. (2017a), Mortazavi et al. (2017), and Yan et al. (2019b). Alternatively, we are more interested in designing a fusion-based localization algorithm, and hence the case of $M = 1$ is omitted. Now, we give the consensus-based UKF localization algorithm.

Step 1: Initialization
At $k = 0$, we have the following initial conditions

$$\hat{\mathbf{X}}(0) = \mathbf{E}\left[\mathbf{X}(0)\right], \ \mathbf{P}(0) = \mathbf{E}\left[(\mathbf{X}(0) - \hat{\mathbf{X}}(0))(\mathbf{X}(0) - \hat{\mathbf{X}}(0))^{\mathrm{T}}\right]. \tag{3.10}$$

Step 2: Selection of Sigma Points
At the start of time step k, a set of $2\bar{n} + 1$ sigma points are selected based on the state estimation $\hat{\mathbf{X}}(k-1)$ and the error covariance matrix $\mathbf{P}(k-1)$,

$$\begin{aligned}
\chi_s(k-1) &= \hat{\mathbf{X}}(k-1), \ s = 0, \\
\chi_s(k-1) &= \hat{\mathbf{X}}(k-1) + (\sqrt{(\bar{n}+\lambda)\mathbf{P}(k-1)})_s, \ s = 1, ..., \bar{n}, \\
\chi_s(k-1) &= \hat{\mathbf{X}}(k-1) - (\sqrt{(\bar{n}+\lambda)\mathbf{P}(k-1)})_{s-\bar{n}}, \ s = \bar{n}+1, ..., 2\bar{n},
\end{aligned} \tag{3.11}$$

where $\bar{n} = 3$ is the dimension of the state. λ is a composite scaling parameter, whose aim is to reduce the predication error. The $(\cdot)_s$ denotes the sth column of a matrix. In addition, λ can be any number (positive or negative) providing that $(\bar{n}+\lambda)\mathbf{P}(k-1)$ is a semi-positive definite matrix.

Step 3: Iteration Update
Introducing sigma points into the state of target, the prediction state $\chi_s(k|k-1)$ can be expressed as

$$\chi_s(k|k-1) = \chi_s(k-1), s = 0, ..., 2\bar{n}. \tag{3.12}$$

Using a weighted sample mean, the priori state estimation can be calculated as

$$\hat{\mathbf{X}}(k|k-1) = \sum\nolimits_{s=0}^{2\bar{n}} W_s \chi_s(k|k-1), \tag{3.13}$$

Algorithm 1: Consensus-based UKF localization

Input: Initiation $\hat{\mathbf{X}}(0)$ and $\mathbf{P}(0)$.
for $k = 1 : k_*$ **do**
1. Generate the Sigma point $\boldsymbol{\chi}_s(k-1)$ with (3.11).
2. Calculate $\boldsymbol{\chi}_s(k|k-1)$, $\hat{\mathbf{X}}(k|k-1)$ and $\mathbf{P}(k|k-1)$ with (3.12)–(3.14), respectively.
3. With (3.15), the prediction $\boldsymbol{\chi}_s(k|k-1)$ is updated.
4. Mapping $\boldsymbol{\chi}_s(k|k-1)$ into measurement (3.8), we have $\mathbf{D}_s^j(k)$ and $\hat{\mathbf{D}}^j(k)$ through (3.16) and (3.17).
5. Update $\mathbf{P}^j_{\hat{\mathbf{D}}^j(k)\hat{\mathbf{D}}^j(k)}$ and $\mathbf{P}^j_{\hat{\mathbf{X}}(k)\hat{\mathbf{D}}^j(k)}$ with (3.18).
6. According to (3.21), the state estimation $\hat{\mathbf{X}}(k)$ and the covariance matrix $\mathbf{P}(k)$ are calculated.

if $|\hat{\mathbf{X}}(k) - \hat{\mathbf{X}}(k-1)|$ is less than a small value ε **then**
break
end if
end for
Output: The position estimation $\hat{\mathbf{X}}$ for target.

and the covariance matrix is calculated as

$$\mathbf{P}(k|k-1) = \sum\nolimits_{s=0}^{2\bar{n}} W_s \left[\boldsymbol{\chi}_s(k|k-1) - \hat{\mathbf{X}}(k|k-1)\right] \times \left[\boldsymbol{\chi}_s(k|k-1) - \hat{\mathbf{X}}(k|k-1)\right]^{\mathrm{T}}, \tag{3.14}$$

where $W_s = \lambda/(\bar{n}+\lambda)$ if $s=0$, otherwise, $W_s = 1/(2(\bar{n}+\lambda))$.

With (3.11), (3.13) and (3.14), the prediction state in (3.12) is updated as

$$\begin{aligned}
&\boldsymbol{\chi}_s(k|k-1) = \hat{\mathbf{X}}(k|k-1),\ s=0\\
&\boldsymbol{\chi}_s(k|k-1) = \hat{\mathbf{X}}(k|k-1) + (\sqrt{(\bar{n}+\lambda)\mathbf{P}(k|k-1)})_s,\ s=1,\ldots,\bar{n}\\
&\boldsymbol{\chi}_s(k|k-1) = \hat{\mathbf{X}}(k|k-1) - (\sqrt{(\bar{n}+\lambda)\mathbf{P}(k|k-1)})_{s-\bar{n}},\ s=\bar{n}+1,\ldots,2\bar{n}.
\end{aligned} \tag{3.15}$$

Step 4: Measurement Update

For group j, mapping $\boldsymbol{\chi}_s(k|k-1)$ into measurement (3.8), one has

$$\mathbf{D}_s^j(k) = \mathbf{H}^j\left[\boldsymbol{\chi}_s(k|k-1)\right] + \mathbf{r}^j(k),\ s=0,\ldots,2\bar{n}. \tag{3.16}$$

Priori measurement estimation is approximated by weighted mean value

$$\hat{\mathbf{D}}^j(k) = \sum\nolimits_{s=0}^{2\bar{n}} W_s \mathbf{D}_s^j(k) + \mathbf{r}^j(k). \tag{3.17}$$

Thus, the predicted measurement covariance matrix $\mathbf{P}^j_{\hat{\mathbf{D}}^j(k)\hat{\mathbf{D}}^j(k)}$ and the state-measurement cross-covariance matrix $\mathbf{P}^j_{\hat{\mathbf{X}}(k)\hat{\mathbf{D}}^j(k)}$ can be expressed as

$$\begin{aligned}\mathbf{P}^j_{\hat{\mathbf{D}}^j(k)\hat{\mathbf{D}}^j(k)} &= \sum_{s=0}^{2\bar{n}} W_s\left[\mathbf{D}^j_s(k)-\hat{\mathbf{D}}^j(k)\right]\left[\mathbf{D}^j_s(k)-\hat{\mathbf{D}}^j(k)\right]^{\mathrm{T}}+\mathbf{R}^j(k),\\ \mathbf{P}^j_{\hat{\mathbf{X}}(k)\hat{\mathbf{D}}^j(k)} &= \sum_{s=0}^{2\bar{n}} W_s\left[\boldsymbol{\chi}_s(k|k-1)-\hat{\mathbf{X}}(k|k-1)\right]\left[\mathbf{D}^j_s(k)-\hat{\mathbf{D}}^j(k)\right]^{\mathrm{T}}.\end{aligned} \tag{3.18}$$

It is noted that, the knowledge of the measurement noise $\boldsymbol{\nu}^j(k)$ cannot be exactly pre-known. Then, the Sage-Husa estimator (Shi et al. 2009) is applied to estimate the statistical parameters, i.e.

$$\begin{aligned}\mathbf{r}^j(k) &= (1-\mu(k))\mathbf{r}^j(k-1)+\mu(k)(\mathbf{D}^j(k)-\sum_{s=0}^{2\bar{n}} W_s\mathbf{H}^j\left[\boldsymbol{\chi}_s(k|k-1)\right])\\ \mathbf{R}^j(k) &= (1-\mu(k))\mathbf{R}^j(k-1)+\mu(k)[(\mathbf{D}^j(k)-\hat{\mathbf{D}}^j(k))(\mathbf{D}^j(k)-\hat{\mathbf{D}}^j(k))^{\mathrm{T}}\\ &\quad -\sum_{s=0}^{2\bar{n}} W_s(\mathbf{D}^j_s(k)-\hat{\mathbf{D}}^j(k))(\mathbf{D}^j_s(k)-\hat{\mathbf{D}}^j(k))^{\mathrm{T}}],\end{aligned} \tag{3.19}$$

where $\mu(k)\in[0,1]$ is the estimation weight.

Step 5: Kalman Filtering Update Based Consensus Fusion
The Kalman gain $\mathbf{K}^j(k)$, the state estimation $\hat{\mathbf{X}}^j(k)$ and its covariance matrix $\mathbf{P}^j(k)$ are updated as

$$\begin{aligned}\mathbf{K}^j(k) &= \mathbf{P}^j_{\hat{\mathbf{X}}(k)\hat{\mathbf{D}}^j(k)}\left[\mathbf{P}^j_{\hat{\mathbf{D}}^j(k)\hat{\mathbf{D}}^j(k)}\right]^{-1},\\ \hat{\mathbf{X}}^j(k) &= \hat{\mathbf{X}}(k|k-1)+\mathbf{K}^j(k)\left[\mathbf{D}^j(k)-\hat{\mathbf{D}}^j(k)\right],\\ \mathbf{P}^j(k) &= \mathbf{P}(k|k-1)-\mathbf{K}^j(k)\mathbf{P}^j_{\hat{\mathbf{D}}^j(k)\hat{\mathbf{D}}^j(k)}\left[\mathbf{K}^j(k)\right]^{\mathrm{T}}.\end{aligned} \tag{3.20}$$

The estimated state and state covariance matrix in the other groups can be acquired with the similar way. Accordingly, we adopt the idea of consensus to fuse the results of $\hat{\mathbf{X}}^j(k)$ and $\mathbf{P}^j(k)$, where $j\in[1,\ldots,M]$. The fusion-based estimated state $\hat{\mathbf{X}}(k)$ and the variance matrix $\mathbf{P}(k)$ are

$$\begin{aligned}\hat{\mathbf{X}}(k) &= \sum_{j=1}^{M}\alpha^j(k)\hat{\mathbf{X}}^j(k),\\ \mathbf{P}(k) &= \sum_{j=1}^{M}\alpha^j(k)\mathbf{P}^j(k),\end{aligned} \tag{3.21}$$

where $\alpha^j(k)\geqslant 0$ represent the weights for group j. It is worth mentioning that, the weights satisfy the condition of $\sum_{j=1}^{M}\alpha^j(k)=1$.

Difference from the stochastic weights in Li et al. (2016), the weights in this chapter are deterministic. For clear description, we take the case of two groups as an example. The virtual measurements for groups j and m are defined as

$$\begin{aligned} q^j(k) &= [\boldsymbol{\gamma}^j(k)]^{\mathrm{T}}[\mathbf{P}^j_{\hat{\mathbf{D}}^j(k)\hat{\mathbf{D}}^j(k)}]^{-1}[\boldsymbol{\gamma}^j(k)], \\ q^m(k) &= [\boldsymbol{\gamma}^m(k)]^{\mathrm{T}}[\mathbf{P}^m_{\hat{\mathbf{D}}^m(k)\hat{\mathbf{D}}^m(k)}]^{-1}[\boldsymbol{\gamma}^m(k)], \end{aligned} \tag{3.22}$$

where $\boldsymbol{\gamma}^j(k) = \mathbf{D}^j(k) - \hat{\mathbf{D}}^j(k)$ and $\boldsymbol{\gamma}^m(k) = \mathbf{D}^m(k) - \hat{\mathbf{D}}^m(k)$. Of note, $j, m \in \{1, ..., M\}$ and $j \neq m$.

For $q^j(k)$ and $q^m(k)$, the threshold measurements are denoted by $q^j_{\max}$ and $q^m_{\max}$, respectively. When $q^j(k) > q^j_{\max}$ and $q^m(k) \leq q^m_{\max}$, only the time measurement in group m is feasible, and the weights can be designed as $\alpha^j(k) = 0$ and $\alpha^m(k) = 1$. Similarly, when $q^j(k) \leq q^j_{\max}$ and $q^m(k) > q^m_{\max}$, only the time measurement obtained by group j is reliable, the weights can be designed as $\alpha^j(k) = 1$ and $\alpha^m(k) = 0$. If $q^j(k) \leq q^j_{\max}$ and $q^m(k) \leq q^m_{\max}$, the time measurements in group j and group m are both useful. Thus, the weights are designed as $\alpha^j(k) = \frac{q^j_{\max} - q^j(k)}{q^m_{\max} - q^m(k) + q^j_{\max} - q^j(k)}$ and $\alpha^m(k) = \frac{q^m_{\max} - q^m(k)}{q^m_{\max} - q^m(k) + q^j_{\max} - q^j(k)}$, respectively. When $q^j(k) > q^j_{\max}$ and $q^m(k) > q^m_{\max}$, the measurements are questionable, then the estimation in (3.21) is rewritten as $\hat{\mathbf{X}}(k) = \hat{\mathbf{X}}(k|k-1)$. It is noted that, the last case is out of the scope of this chapter, so it can be omitted here.

With the definitions in (3.22), one knows the quadratic form q^j (or q^m) is theoretically a χ^2 distribution with 6 degrees of freedom (Li et al. 2008). Referring to the standard χ^2 table, the values of $q^j_{\max}$ and $q^m_{\max}$ can be selected. For example, considering a 95% confidence level, we can set $q^j_{\max} = q^m_{\max} = 12.592$.

3.4 Performance Analysis

3.4.1 Convergence Conditions

In Julier and Uhlmann (1997), it is pointed out that the prerequisite condition of using UKF is state observability. In the following, Corollary 3.1 is given to analyze the observability of measurement (3.8).

Corollary 3.1 *It is assumed that the locations of sensor nodes are not co-linear, whose measurement is denoted by (3.8). Then, the system state* (x, y, z) *is observable with the availability of measurement (3.8).*

Proof With consideration of stratification effect, the time information in (3.4) can be described as

$$\begin{aligned}\tau_{T,i}(k) &= -\frac{1}{\bar{a}}(\ln\frac{1+\sin\theta_T}{\cos\theta_T} - \ln\frac{1+\sin\theta_i}{\cos\theta_i}) \\ &= -\frac{1}{\bar{a}}(\ln\frac{1+\sin(\arctan(\xi_i)-\arctan(\gamma_i))}{\cos(\arctan(\xi_i)-\arctan(\gamma_i))} \\ &\quad - \ln\frac{1+\sin(\arctan(\xi_i)+\arctan(\gamma_i))}{\cos(\arctan(\xi_i)+\arctan(\gamma_i))}),\end{aligned} \tag{3.23}$$

where $\gamma_i = \frac{0.5\bar{a}\sqrt{(x-x_i)^2+(y-y_i)^2}}{\bar{b}+0.5\bar{a}(z+z_i)}$ and $\xi_i = \frac{z-z_i}{\sqrt{(x-x_i)^2+(y-y_i)^2}}$ for $i \in [1_j, 2_j, 3_j]$.

Taking straightforward derivation, one has

$$\begin{aligned}&\frac{\partial\tau_{\mathrm{T},i}(k)}{\partial x} \\ &= -\frac{1}{\bar{a}}\{(1+\sin(\arctan(\xi_i)-\arctan(\gamma_i)))(\frac{1}{1+\xi_i^2}\frac{\partial\xi_i}{\partial x} - \frac{1}{1+\gamma_i^2}\frac{\partial\gamma_i}{\partial x}) \\ &\quad\times[(1+\sin(\arctan(\xi_i)-\arctan(\gamma_i)))\cos(\arctan(\xi_i)-\arctan(\gamma_i))]^{-1} \\ &\quad-(1+\sin(\arctan(\xi_i)+\arctan(\gamma_i)))(\frac{1}{1+\xi_i^2}\frac{\partial\xi_i}{\partial x} + \frac{1}{1+\gamma_i^2}\frac{\partial\gamma_i}{\partial x}) \\ &\quad\times[(1+\sin(\arctan(\xi_i)+\arctan(\gamma_i)))\cos(\arctan(\xi_i)+\arctan(\gamma_i))]^{-1}\},\end{aligned} \tag{3.24}$$

$$\begin{aligned}&\frac{\partial\tau_{\mathrm{T},i}(k)}{\partial y} \\ &= -\frac{1}{\bar{a}}\{(1+\sin(\arctan(\xi_i)-\arctan(\gamma_i)))(\frac{1}{1+\xi_i^2}\frac{\partial\xi_i}{\partial y} - \frac{1}{1+\gamma_i^2}\frac{\partial\gamma_i}{\partial y}) \\ &\quad\times[(1+\sin(\arctan(\xi_i)-\arctan(\gamma_i)))\cos(\arctan(\xi_i)-\arctan(\gamma_i))]^{-1} \\ &\quad-(1+\sin(\arctan(\xi_i)+\arctan(\gamma_i)))(\frac{1}{1+\xi_i^2}\frac{\partial\xi_i}{\partial y} + \frac{1}{1+\gamma_i^2}\frac{\partial\gamma_i}{\partial y}) \\ &\quad\times[(1+\sin(\arctan(\xi_i)+\arctan(\gamma_i)))\cos(\arctan(\xi_i)+\arctan(\gamma_i))]^{-1}\},\end{aligned}$$

$$\begin{aligned}&\frac{\partial\tau_{\mathrm{T},i}(k)}{\partial z} \\ &= -\frac{1}{\bar{a}}\{(1+\sin(\arctan(\xi_i)-\arctan(\gamma_i)))(\frac{1}{1+\xi_i^2}\frac{\partial\xi_i}{\partial z} - \frac{1}{1+\gamma_i^2}\frac{\partial\gamma_i}{\partial z}) \\ &\quad\times[(1+\sin(\arctan(\xi_i)-\arctan(\gamma_i)))\cos(\arctan(\xi_i)-\arctan(\gamma_i))]^{-1} \\ &\quad-(1+\sin(\arctan(\xi_i)+\arctan(\gamma_i)))(\frac{1}{1+\xi_i^2}\frac{\partial\xi_i}{\partial z} + \frac{1}{1+\gamma_i^2}\frac{\partial\gamma_i}{\partial z}) \\ &\quad\times[(1+\sin(\arctan(\xi_i)+\arctan(\gamma_i)))\cos(\arctan(\xi_i)+\arctan(\gamma_i))]^{-1}\},\end{aligned} \tag{3.25}$$

with

$$
\begin{aligned}
\frac{\partial \gamma_i}{\partial x} &= \frac{0.5\bar{a}(x-x_i)}{(\bar{b}+0.5\bar{a}(z+z_i))\sqrt{(x-x_i)^2+(y-y_i)^2}},\\
\frac{\partial \gamma_i}{\partial y} &= \frac{0.5\bar{a}(y-y_i)}{(\bar{b}+0.5\bar{a}(z+z_i))\sqrt{(x-x_i)^2+(y-y_i)^2}},\\
\frac{\partial \gamma_i}{\partial z} &= \frac{-0.25\bar{a}^2\sqrt{(x-x_i)^2+(y-y_i)^2}}{(\bar{b}+0.5\bar{a}(z+z_i))^2},\\
\frac{\partial \xi_i}{\partial x} &= -\frac{(x-x_i)(z-z_i)}{[(x-x_i)^2+(y-y_i)^2]^{1.5}},\\
\frac{\partial \xi_i}{\partial y} &= -\frac{(y-y_i)(z-z_i)}{[(x-x_i)^2+(y-y_i)^2]^{1.5}},\\
\frac{\partial \xi_i}{\partial z} &= \frac{1}{\sqrt{(x-x_i)^2+(y-y_i)^2}}.
\end{aligned}
\tag{3.26}
$$

According to Hermann and Krener (1977), a system is observable if $\text{rank}[\nabla\mathbf{D}^j(k);\ \nabla L_f\mathbf{D}^j(k);\ldots;\ \nabla L_f^{\bar{n}-1}\mathbf{D}^j(k)] = \bar{n}$ is satisfied. Of note, f is a vector field, $\nabla\mathbf{D}^j(k)$ denotes the gradient of $\mathbf{D}^j(k)$ with respect to the state, $L_f\mathbf{D}^j(k) = \frac{\partial \mathbf{D}^j(k)}{\partial \mathbf{X}} f$ denotes the Lie derivative of $\mathbf{D}^j(k)$ to f, and $L_f^{\bar{n}-1}\mathbf{D}^j(k) = \frac{\partial(L_f^{\bar{n}-2}\mathbf{D}^j(k))}{\partial \mathbf{X}} f$.

In measurement (3.8), one knows $\bar{n} = 3$. By combining with the above derivations, it is obtained that $\text{rank}[\nabla\mathbf{D}^j(k);\nabla L_f\mathbf{D}^j(k);\ldots;\nabla L_f^2\mathbf{D}^j(k)]= 3$. Therefore, the state (x, y, z) is observable. □

With Corollary 3.1, we investigate the convergence conditions for the consensus-based UKF localization algorithm. Considering the measurement output (3.8), we use the technique provided in Lefebvre et al. (2002) to derive a pseudo measurement matrix $\breve{\mathbf{H}}^j(k)$. Based on this, $\breve{\mathbf{H}}^j(k)$ can be calculated as

$$
\breve{\mathbf{H}}^j(k) \triangleq \left[\mathbf{P}^j_{\hat{\mathbf{X}}(k)\hat{\mathbf{D}}^j(k)}\right]^{\mathrm{T}} \left[\mathbf{P}^j(k|k-1)\right]^{-1}. \tag{3.27}
$$

As a result, we have $\mathbf{D}^j(k) = \breve{\mathbf{H}}^j(k)\mathbf{X}(k) + \boldsymbol{\nu}^j(k)$ and $\mathbf{P}^j_{\hat{\mathbf{X}}(k)\hat{\mathbf{D}}^j(k)} = \mathbf{P}^j(k|k-1)(\breve{\mathbf{H}}^j(k))^{\mathrm{T}}$. In view of $\mathbf{P}^j_{\hat{\mathbf{D}}^j(k)\hat{\mathbf{D}}^j(k)} = \breve{\mathbf{H}}^j(k)\mathbf{P}^j(k|k-1)(\breve{\mathbf{H}}^j(k))^{\mathrm{T}} + \mathbf{R}^j(k)$ and (3.27), Eq. (3.20) is rearranged as

$$
\begin{aligned}
&\mathbf{K}^j(k) = \mathbf{P}^j(k|k-1)(\breve{\mathbf{H}}^j(k))^{\mathrm{T}}\left[\breve{\mathbf{H}}^j(k)\mathbf{P}^j(k|k-1)(\breve{\mathbf{H}}^j(k))^{\mathrm{T}} + \mathbf{R}^j(k)\right]^{-1},\\
&\mathbf{P}^j(k) = (\mathbf{I} - \mathbf{K}^j(k)\breve{\mathbf{H}}^j(k))\mathbf{P}^j(k|k-1).
\end{aligned}
\tag{3.28}
$$

Substituting (3.28) into (3.21), the state estimate $\hat{\mathbf{X}}(k)$ and covariance matrix $\mathbf{P}(k)$ are updated as

$$\begin{aligned}\hat{\mathbf{X}}(k) &= \hat{\mathbf{X}}(k|k-1) + \sum_{j=1}^{M} \alpha^j(k)\mathbf{K}^j(k)(\breve{\mathbf{H}}^j(k)\tilde{\mathbf{X}}(k|k-1) + \boldsymbol{\nu}^j(k)),\\ \mathbf{P}(k) &= \left(\mathbf{I} - \sum_{j=1}^{M} \alpha^j(k)\mathbf{K}^j(k)\breve{\mathbf{H}}^j(k)\right)\mathbf{P}(k|k-1).\end{aligned} \tag{3.29}$$

Define the prediction error as $\tilde{\mathbf{X}}(k+1|k) = \mathbf{X}(k+1) - \hat{\mathbf{X}}(k+1|k)$, and the estimation error as $\tilde{\mathbf{X}}(k) = \mathbf{X}(k) - \hat{\mathbf{X}}(k)$. Then, Theorem 3.1 is given to illustrate the effectiveness of update rule (3.21).

Theorem 3.1 *Consider the observable state* (x, y, z) *with measurement (3.8). If there exist real numbers* $h_{\min}$ *and* $h_{\max}$, *positive real constants* $p_{\min}$, $r_{\min}$, $p_{\max}$ *and* $r_{\max}$, *such that the following conditions hold, i.e., (a):* $h_{\min}^2\mathbf{I} \leqslant \breve{\mathbf{H}}^j(k)(\breve{\mathbf{H}}^j(k))^T \leqslant h_{\max}^2\mathbf{I}$ *; (b):* $p_{\min}\mathbf{I} \leqslant \mathbf{P}(k) \leqslant p_{\max}\mathbf{I}$, $r_{\min}\mathbf{I} \leqslant \mathbf{R}^j(k) \leqslant r_{\max}\mathbf{I}$, *then the estimation error* $\hat{\mathbf{X}}(k)$ *is bounded.*

Proof Define the following Lyapunov function

$$\mathcal{V}(\tilde{\mathbf{X}}(k+1|k)) = \tilde{\mathbf{X}}^{\mathrm{T}}(k+1|k)\mathbf{P}^{-1}(k+1|k)\tilde{\mathbf{X}}(k+1|k). \tag{3.30}$$

With (3.30) and the condition (b), we have $\mathcal{V}(\tilde{\mathbf{X}}(k+1|k)) > 0$ and $\frac{\|\tilde{\mathbf{X}}(k+1|k)\|^2}{p_{\max}} \leqslant \mathcal{V}(\tilde{\mathbf{X}}(k+1|k)) \leqslant \frac{\|\tilde{\mathbf{X}}(k+1|k)\|^2}{p_{\min}}$. Then, $\tilde{\mathbf{X}}(k+1|k)$ can be expressed as

$$\begin{aligned}&\tilde{\mathbf{X}}(k+1|k)\\ &= \mathbf{X}(k+1) - \hat{\mathbf{X}}(k+1|k)\\ &= \left[\mathbf{I} - \sum_{j=1}^{M} \alpha^j(k)\mathbf{K}^j(k)\breve{\mathbf{H}}^j(k)\right]\tilde{\mathbf{X}}(k|k-1)\\ &\quad + \left[-\sum_{j=1}^{M} \alpha^j(k)\mathbf{K}^j(k)\boldsymbol{\nu}^j(k)\right].\end{aligned} \tag{3.31}$$

Defining $\boldsymbol{\Gamma} = \mathbf{I} - \sum_{j=1}^{M} \alpha^j(k)\mathbf{K}^j(k)\breve{\mathbf{H}}^j(k)$ and substituting (3.31) into (3.30), one has

$$\mathbf{E}\left\{\mathcal{V}(\tilde{\mathbf{X}}(k+1|k))|\tilde{\mathbf{X}}(k|k-1)\right\} = \Phi_{\mathrm{X}}(k+1) + \Phi_{\mathrm{Y}}(k+1), \tag{3.32}$$

where $\Phi_{\mathrm{X}}(k+1) = \mathbf{E}\{(\boldsymbol{\Gamma}\tilde{\mathbf{X}}(k|k-1))^{\mathrm{T}}\mathbf{P}^{-1}(k+1|k)(\boldsymbol{\Gamma}\tilde{\mathbf{X}}(k|k-1))|\tilde{\mathbf{X}}(k|k-1)\}$ and $\Phi_{\mathrm{Y}}(k+1) = \mathbf{E}\{[-\sum_{j=1}^{M} \alpha^j(k)\mathbf{K}^j(k)\boldsymbol{\nu}^j(k)]^{\mathrm{T}}\mathbf{P}^{-1}(k+1|k)[-\sum_{j=1}^{M} \alpha^j(k)$ $\mathbf{K}^j(k)\boldsymbol{\nu}^j(k)]|\tilde{\mathbf{X}}(k|k-1)\}$.

Notice that $\mathbf{P}(k+1|k) = \int[\mathbf{X}(k) - \hat{\mathbf{X}}(k+1|k)][\mathbf{X}(k) - \hat{\mathbf{X}}(k+1|k)]^{\mathrm{T}}\mathcal{N}_{\mathbf{X}(k)}d\mathbf{X}(k)$ and $\mathbf{P}(k) = \int[\mathbf{X}(k) - \hat{\mathbf{X}}(k+1|k)][\mathbf{X}(k) - \hat{\mathbf{X}}(k+1|k)]^{\mathrm{T}}\mathcal{N}_{\mathbf{X}(k)}d\mathbf{X}(k)$, where $\mathcal{N}_{\mathbf{X}(k)}$ is the probability density for $\mathbf{X}(k)$. Thus, we obtain $\mathbf{P}(k+1|k) = \mathbf{P}(k)$. For $\Phi_{\mathrm{X}}(k+1)$, we have

$$\begin{aligned}
&\Phi_{\mathrm{X}}(k+1) \\
&= \mathbf{E}\left\{(\boldsymbol{\Gamma}\tilde{\mathbf{X}}(k|k-1))^{\mathrm{T}}\mathbf{P}^{-1}(k)(\boldsymbol{\Gamma}\tilde{\mathbf{X}}(k|k-1))|\tilde{\mathbf{X}}(k|k-1)\right\} \\
&= \mathbf{E}\{\tilde{\mathbf{X}}^{\mathrm{T}}(k|k-1)[\mathbf{I}-\sum_{j=1}^{M}\alpha^j(k)\mathbf{K}^j(k)\breve{\mathbf{H}}^j(k)]\mathbf{P}^{-1}(k|k-1) \\
&\quad\times \tilde{\mathbf{X}}(k|k-1)|\tilde{\mathbf{X}}(k|k-1)\}.
\end{aligned} \tag{3.33}$$

With (3.29), we obtain $\Phi_{\mathrm{X}}(k+1) \leq \mathbf{E}\left\{\mathcal{V}(\tilde{\mathbf{X}}(k|k-1)\right\}$. Denote $\mathbf{R}(k) = \sum_{j=1}^{M}\mathbf{R}^j(k)$, then one has

$$\begin{aligned}
\Phi_{\mathrm{Y}}(k+1) &= \mathbf{E}\{\left[-\sum_{j=1}^{M}\alpha^j(k)\mathbf{K}^j(k)\boldsymbol{\nu}^j(k)\right]^{\mathrm{T}}\mathbf{P}^{-1}(k) \\
&\quad\times\left[-\sum_{j=1}^{M}\alpha^j(k)\mathbf{K}^j(k)\boldsymbol{\nu}^j(k)\right]\left|\tilde{\mathbf{X}}(k|k-1)\right.\} \\
&\leq p_{\min}^{-1}(k)\mathbf{R}(k)E\left\{\sum_{j=1}^{M}[\mathbf{K}^j(k)]^{\mathrm{T}}\mathbf{K}^j(k)|\tilde{\mathbf{X}}(k|k-1)\right\}.
\end{aligned} \tag{3.34}$$

It is defined that $\mathcal{L}_{\max} = \frac{p_{\max}h_{\max}}{p_{\min}h_{\min}^2+r_{\min}}$. If the conditions (a) and (b) are satisfied, $\mathbf{K}^j(k)$ in (3.28) has the following boundedness, i.e., $\mathbf{K}^j(k) = \frac{\mathbf{P}^j(k|k-1)(\breve{\mathbf{H}}^j(k))^{\mathrm{T}}}{\breve{\mathbf{H}}^j(k)\mathbf{P}^j(k|k-1)\left[\breve{\mathbf{H}}^j(k)\right]^{\mathrm{T}}+\mathbf{R}^j(k)} \leq \frac{p_{\max}h_{\max}}{p_{\min}h_{\min}^2+r_{\min}}\mathbf{I} = \mathcal{L}_{\max}\mathbf{I}$.

Thereby, $\Phi_{\mathrm{Y}}(k+1)$ can be rewritten as $\Phi_{\mathrm{Y}}(k+1) \leq p_{\min}^{-1}\mathcal{L}_{\max}^2\mathbf{R}(k)$. Meanwhile, the boundedness of prediction error $\tilde{\mathbf{X}}(k+1|k)$ can be obtained as

$$\mathbf{E}\{\mathcal{V}(\tilde{\mathbf{X}}(k+1|k))|\tilde{\mathbf{X}}(k|k-1)\} \leq \mathbf{E}\{\mathcal{V}(\tilde{\mathbf{X}}(k|k-1))\} + p_{\min}^{-1}\mathcal{L}_{\max}^2\mathbf{R}(k). \tag{3.35}$$

With these results, one has $\mathbf{E}\{\mathcal{V}(\tilde{\mathbf{X}}(k+1|k))|\tilde{\mathbf{X}}(k|k-1)\} - \mathbf{E}\{\mathcal{V}(\tilde{\mathbf{X}}(k|k-1))\}$ is bounded, i.e., $\tilde{\mathbf{X}}(k+1|k)$ is bounded. From (3.31), one has $\tilde{\mathbf{X}}(k+1|k) = \tilde{\mathbf{X}}(k)$, thus $\tilde{\mathbf{X}}(k)$ is also bounded. □

3.4.2 *Cramér-Rao Lower Bound*

Let $\mathbf{X}_0(k) = \{\mathbf{X}(0), \mathbf{X}(1), ..., \mathbf{X}(k)\}$ be the state vector series. $\mathbf{D}_0(k) = \{\mathbf{D}^j(1), \mathbf{D}^j(2), ..., \mathbf{D}^j(k)\}$ is the observation value for group j, and $p(\mathbf{D}_0(k), \mathbf{X}_0(k))$ is the joint probability density of the pair $(\mathbf{D}_0(k), \mathbf{X}_0(k))$. Then, the estimation error covariance $\breve{\mathbf{P}}(k)$ has the following form, i.e.,

$$\breve{\mathbf{P}}(k) = \mathbf{E}\{[\mathbf{X}(k) - \hat{\mathbf{X}}^j(k)][\mathbf{X}(k) - \hat{\mathbf{X}}^j(k)]^{\mathrm{T}}\} \geqslant \mathbf{J}^{-1}(k), \tag{3.36}$$

where $\mathbf{J}(k) = -\mathbf{E}\left[\frac{\partial^2 \ln p(\mathbf{D}_0(k), \mathbf{X}_0(k))}{\partial \mathbf{X}^2(k)}\right]$ denotes the Fisher information matrix.

With the knowledge of CRLB, the sequence $\{\mathbf{J}(k)\}$ of posterior information matrices for estimating state vectors $\{\mathbf{X}(k)\}$ obeys the recursion

$$\mathbf{J}(k) = I_{2,2}(k) - I_{2,1}(k)[\mathbf{J}(k-1) + I_{1,1}(k)]^{-1} I_{1,2}(k), \tag{3.37}$$

where

$$\begin{aligned} I_{1,1}(k) &= -\mathbf{E}\left\{\frac{\partial^2 \ln p(\mathbf{X}(k)|\mathbf{X}(k-1))}{\partial \mathbf{X}^2(k-1)}\right\}, \\ I_{1,2}(k) &= -\mathbf{E}\left\{\frac{\partial^2 \ln p(\mathbf{X}(k)|\mathbf{X}(k-1))}{\partial \mathbf{X}(k)\partial \mathbf{X}(k-1)}\right\}, \\ I_{2,1}(k) &= -\mathbf{E}\left\{\frac{\partial^2 \ln p(\mathbf{X}(k)|\mathbf{X}(k-1))}{\partial \mathbf{X}(k-1)\partial \mathbf{X}(k)}\right\}, \\ I_{2,2}(k) &= -\mathbf{E}\left\{\frac{\partial^2 \ln p(\mathbf{X}(k)|\mathbf{X}(k-1))}{\partial \mathbf{X}^2(k)} + \frac{\partial^2 \ln p(\mathbf{D}^j(k)|\mathbf{X}(k))}{\partial \mathbf{X}^2(k)}\right\}. \end{aligned} \tag{3.38}$$

and the conditional density function is expressed as

$$\begin{aligned} p(\mathbf{D}^j(k)|\mathbf{X}(k)) = \frac{1}{\sqrt{2\pi|\mathbf{R}^j(k)|}} \exp\{-1/2[\mathbf{D}^j(k) - \breve{\mathbf{H}}^j(k)\mathbf{X}(k) - \mathbf{r}^j(k)]^{\mathrm{T}} \\ \times (\mathbf{R}^j(k))^{-1}[\mathbf{D}^j(k) - \breve{\mathbf{H}}^j(k)\mathbf{X}(k) - \mathbf{r}^j(k)]\}. \end{aligned} \tag{3.39}$$

As the target is fixed, then $p(\mathbf{X}(k)|\mathbf{X}(k-1)) = 1$. With straightforward derivation, one has

$$\begin{aligned} &I_{1,1}(k) = 0,\ I_{1,2}(k) = 0,\ I_{2,1}(k) = 0, \\ &I_{2,2}(k) = (\breve{\mathbf{H}}^j(k))^{\mathrm{T}}(\mathbf{R}^j(k))^{-1}\breve{\mathbf{H}}^j(k). \end{aligned} \tag{3.40}$$

Thus, one has $\mathbf{J}(k) = (\breve{\mathbf{H}}^j(k))^{\mathrm{T}}(\mathbf{R}^j(k))^{-1}\breve{\mathbf{H}}^j(k)$. The Fisher information matrix $\mathbf{J}(k)$ can be rearranged as

$$\mathbf{J}(k) = \mathbf{J}(k-1) + (\breve{\mathbf{H}}^j(k))^{\mathrm{T}}(\mathbf{R}^j(k))^{-1}\breve{\mathbf{H}}^j(k), \tag{3.41}$$

where $\mathbf{J}(0)$ can be calculated by the inverse of initial covariance matrix $\mathbf{P}(0)$, i.e., $\mathbf{J}(0) = \mathbf{P}^{-1}(0)$.

Therefore, the CRLB can be obtained by $\breve{\mathbf{P}}(k) \geqslant \mathbf{J}^{-1}(k)$.

3.4.3 Error of Acoustic Wave Speed

In this chapter, the acoustic wave speed C is depth dependent. In the following, we analyze the sound speed error and find out the error bound for the distance measurement. With respect to (3.3), the speed error ΔC in C can be given as $\Delta C \leq \frac{\partial C}{\partial \bar{z}} \Delta \bar{z} = \bar{a} \Delta \bar{z}$, where $\Delta \bar{z}$ is the error of depth.

We take the time measurement in group j as an example, and the first element of $\mathbf{D}^j$ in (3.8) is transformed into distance measurement, i.e., $d_1 = c_{2_j,1_j} \Delta \mathcal{T}_{2_j,1_j} = c_{2_j,1_j} \tau_{\mathrm{T},2_j} - c_{2_j,1_j} \tau_{\mathrm{T},1_j} + c_{2_j,1_j} \varpi_{2_j,1_j}$, where $c_{2_j,1_j}$ is computed by arithmetic mean speed, i.e., $c_{2_j,1_j} = \frac{1}{\bar{z}_{2_j} - \bar{z}_{1_j}} \int_{\bar{z}_{1_j}}^{\bar{z}_{2_j}} C(\bar{z}) d\bar{z}$. Accordingly, the error bound is

$$
\begin{aligned}
\Delta d_1 &\leq \frac{\partial d_1}{\partial c_{2_j,1_j}} \Delta c_{2_j,1_j} + \frac{\partial d_1}{\partial \tau_{T,2_j}} \Delta \tau_{\mathrm{T},2_j} + \frac{\partial d_1}{\partial \tau_{T,1_j}} \Delta \tau_{\mathrm{T},1_j} \\
&= (\tau_{\mathrm{T},2_j} - \tau_{\mathrm{T},1_j} + \varpi_{2_j,1_j}) \Delta c_{2_j,1_j} + c_{2_j,1_j} (\Delta \tau_{\mathrm{T},2_j} - \Delta \tau_{\mathrm{T},1_j}),
\end{aligned}
\tag{3.42}
$$

where $\Delta c_{2_j,1_j}$ is the error of acoustic wave speed $c_{2_j,1_j}$, $\Delta \tau_{\mathrm{T},2_j}$ is the error of the one-way propagation delay between sensor node 2_j and target, $\Delta \tau_{\mathrm{T},1_j}$ is the error of the one-way propagation delay between sensor node 1_j and target. Because the change of $\Delta \tau_{\mathrm{T},2_j} - \Delta \tau_{\mathrm{T},1_j}$ is much slower than $\Delta c_{2_j,1_j}$, Δd_1 is mainly related to $\Delta c_{2_j,1_j}$, i.e., $\Delta c_{2_j,1_j}$ changes with respect to d_1.

3.4.4 Computational Complexity Analysis

Finally, the computational complexity analysis is presented. According to Arasaratnam and Haykin (2008), an effective way to analyze the computational complexity is to count the flops (floating point operations). Inspired by this, the flops for consensus-based UKF algorithm and exhaustive search method are calculated here.

(1) Flops for Consensus-Based UKF Algorithm

As the measurement in each group has the same update process, we only give the computational complexity for group j. In *Step* 2, sigma points are selected with (3.11), thus $\frac{1}{3}\bar{n}^3 + 3\bar{n}^2$ flops are required. In *Step* 3, the prediction state $\boldsymbol{\chi}_s(k|k-1)$ in (3.12) requires $4\bar{n}^3 - \bar{n}$ flops. Meanwhile, the priori state estimation $\hat{\mathbf{X}}(k|k-1)$ and the covariance matrix $\mathbf{P}(k|k-1)$ require $2\bar{n}^2 + 2\bar{n}$ flops and $4\bar{n}^3 + 5\bar{n}^2 + 2\bar{n}$ flops, respectively. In (3.15), the update of prediction state takes $\frac{1}{3}\bar{n}^3 + 3\bar{n}^2$ flops. In *Step* 4, the measurement prediction $\mathbf{D}^j(k)$ in (3.16) takes $4\bar{n}^2 l - l$ flops, where l is the dimension of measurement. In (3.17), the priori measurement estimation takes $2\bar{n}l + 2l$ flops. In (3.18), the predicted measurement covariance matrix $P^j_{\hat{\mathbf{D}}^j(k)\hat{\mathbf{D}}^j(k)}$ and the state-measurement cross-covariance matrix $\mathbf{P}^j_{\hat{\mathbf{X}}(k)\hat{\mathbf{D}}^j(k)}$ take $4\bar{n}l^2 + 2\bar{n}l + 3l^2 + 2l$ flops and $4\bar{n}^2 l + 2\bar{n}^2 + 2\bar{n}l + 2\bar{n}$ flops, respectively. With (3.20), the Kalman gain $\mathbf{K}^j(k)$ requires $l^3 + (2\bar{n}l^2 - \bar{n}l)$ flops, the estimated state $\hat{\mathbf{X}}^j(k)$ requires $2\bar{n}l + l$

flops, and state covariance matrix $\mathbf{P}^j(k)$ requires $2\bar{n}l^2 - \bar{n}l + 2\bar{n}^2l$ flops. With these results, the computational complexity F_{ukf} for group j is

$$F_{\text{ukf}} = \frac{26}{3}\bar{n}^3 + 15\bar{n}^2 + 10\bar{n}^2l + 5\bar{n} + 8\bar{n}l^2 + 6\bar{n}l + l^3 + 3l^2 + 4l. \tag{3.43}$$

(2) Flops for Exhaustive Search Method
As mentioned above, the exhaustive search method was applied by Carroll et al. (2014) to locate the node. In this section, the computational complexity of exhaustive search method is given. For X axis, the search range and precision are designed as ϑ_1 and ρ_1, respectively. Similarly, the search range and precision for Y axis are designed as ϑ_2 and ρ_2, respectively. With these designs, the grids in X axis and Y axis can be calculated as $\breve{n}_1 = \left\lceil \frac{\vartheta_1}{\rho_1} \right\rceil$ and $\breve{n}_2 = \left\lceil \frac{\vartheta_2}{\rho_2} \right\rceil$, where $\lceil \cdot \rceil$ denotes the Ceiling function. With the knowledge of exhaustive search method, the complexity of exhaustive search method F_{esm} can be given by

$$F_{\text{esm}} = \breve{n}_1 \times \breve{n}_2 = \left\lceil \frac{\vartheta_1}{\rho_1} \right\rceil \left\lceil \frac{\vartheta_2}{\rho_2} \right\rceil. \tag{3.44}$$

In this chapter, the dimension of state $\bar{n}$ and the dimension of measurement l are invariant. Hence, it is obtained that the flops for consensus-based UKF algorithm are fixed, because F_{ukf} in (3.43) is invariant. By referring to (3.44), we know the computational complexity of exhaustive search method is variable with respect to the search range and precision, i.e., ϑ_1, ρ_1, ϑ_2 and ρ_2. Thus, we believe that the $F_{\text{ukf}} << F_{\text{esm}}$ if a high accurate localization is required. Due to this property, the consensus-based UKF algorithm is selected in this chapter to locate the underwater target.

3.5 Simulation Results

3.5.1 Simulation Settings

Without loss of generality, it is assumed that four sensor nodes are deployed to locate the target. Specially, the topology relationship of underwater sensor nodes is described by Fig. 3.5. Some parameters used for the simulation are given as: $\bar{a} = 0.017$, $\bar{b} = 1473$ m/s, $\mathbf{P}(0) = 0.1\mathbf{I}$, and $q_{\text{max}}^1 = q_{\text{max}}^2 = 12.592$. In addition, 60 points are required to be localized, i.e., target is sequently located at 60 different points. Particularly, the actual position for the first target point is set as $[501, 201, -700.2339]^{\text{T}}$, while the positions of sensor nodes are given as $[600.5, 101, -650.5]^{\text{T}}$, $[300.5, 301.5, -601]^{\text{T}}$, $[651, 152, -652]^{\text{T}}$, and $[600.5, 401.5, -500.5]^{\text{T}}$.

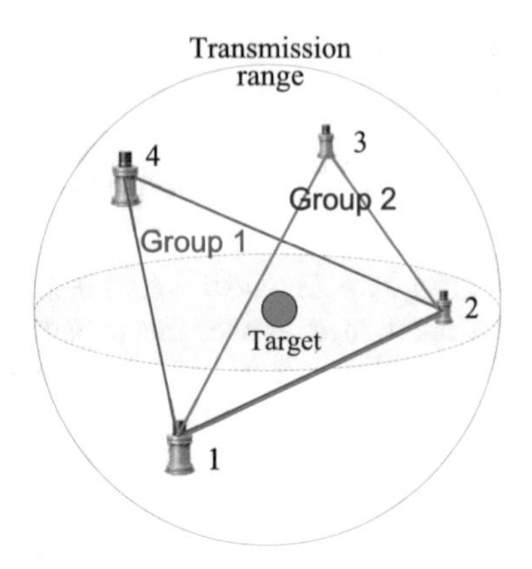

Fig. 3.5 The topology relationship of sensor nodes

3.5.2 Results and Analysis

(1) Comparison Between UKF and Exhaustive Search Method

We first investigate the results in one group, i.e., the consensus-based fusion is not considered. It is noted that, the time noise $\varpi_{n,\bar{k}}$ satisfies the following distribution: $\varpi_{n,\bar{k}} \sim \mathcal{N}(0, 2\sigma_{\text{mea}}^2)$. Thus, the distance noise between node n and node $\bar{k}$ can approximately obey the distribution $\mathcal{N}(0, 2\text{c}^2\sigma_{\text{mea}}^2)$, while the distance standard deviation can approximately reach to $\sqrt{2}\text{c}\sigma_{\text{mea}}$, where $\text{c} \approx 1500\,\text{m/s}$. As a result, the distance standard deviation between node n and node $\bar{k}$ can approximately reach to 6.363 m when σ_{mea} is set as 0.003. Similarly, the distance standard deviation is 2.121 m when σ_{mea} is set as 0.001. Clearly, the value of 6.363 m is more unreliable than the value of 2.121 m. Based on this, the scenario of $\sigma_{\text{mea}} = 0.003$ is considered as strong noise case, while the scenario of $\sigma_{\text{mea}} = 0.001$ is weak noise case.

When the values of noise measurements in (3.8) are relative strong, e.g., $\sigma_{\text{mea}} = 0.003$, we use UKF algorithm to locate the target. In Carroll et al. (2014), an on-demand asynchronous localization approach was proposed, where exhaustive search method was given to find the optimal position. For comparison, we apply the ODAL-based localization approach to locate the target under the same parameters. In Case 1, the actual and estimated positions of target are shown in Fig. 3.6a. As the target is equipped with a depth sensor, the localization errors for the two approaches can be defined as $\text{error}_1 = \sqrt{(x - \hat{x}_1)^2 + (y - \hat{y}_1)^2}$ and $\text{error}_2 = \sqrt{(x - \hat{x}_2)^2 + (y - \hat{y}_2)^2}$, respectively. Of note, (x, y) is the actual position of target, $(\hat{x}_1, \hat{y}_1)$ is estimation via UKF algorithm, and $(\hat{x}_2, \hat{y}_2)$ is estimation through exhaustive search method. Correspondingly, the localization error are shown in Fig. 3.6b. Clearly, the UKF algorithm in this chapter can improve the localization accuracy by comparing with the exhaustive search method (Carroll et al. 2014).

Meanwhile, we consider the weak noise scenario in Case 2, e.g., $\sigma_{\text{mea}} = 0.001$. In Case 2, the localized positions and localization error are shown in Fig. 3.6c, d, respectively. Similar to Case 1, the performance of UKF algorithm is better than the exhaustive search method.

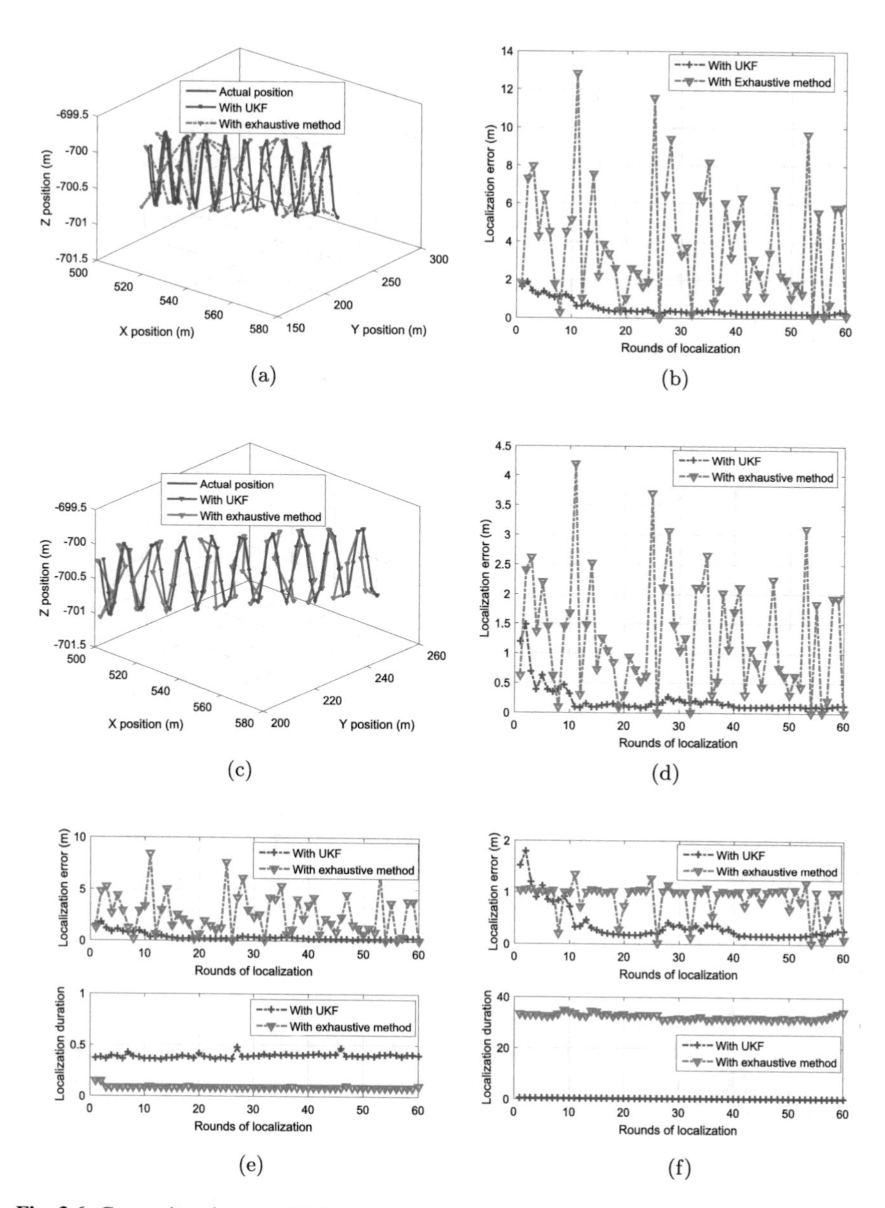

Fig. 3.6 Comparison between UKF and exhaustive search method (Carroll et al. 2014). (**a**) Case 1: Actual and localized positions. (**b**) Case 1: Localization error. (**c**) Case 2: Actual and localized positions. (**d**) Case 2: Localization error. (**e**) Localization accuracy is weakened. (**f**) Localization accuracy is emphasized

As mentioned above, the computational complexity of exhaustive search method is highly related with the search range and precision, i.e., ϑ_1, ρ_1, ϑ_2 and ρ_2. In order to acquire a high precision localization, a larger amount of search operations should be employed, which can increase the duration of localization process. To verify this conclusion, we let $\vartheta_1 = \vartheta_2 = 20$ and $\rho_1 = \rho_2 = 0.2$. Then, the localization error and localization duration are shown in Fig. 3.6e. Clearly, the localization duration of UKF is longer than exhaustive search method when localization accuracy is not considered. For the requirement of high accurate localization, we let $\vartheta_1 = \vartheta_2 = 2$ and $\rho_1 = \rho_2 = 0.001$, i.e., the localization accuracy is considered. Then, the localization error and localization duration are shown in Fig. 3.6f. Through comparison, it is obtained that the localization duration of UKF algorithm is shorter than the one of exhaustive search method, while the localization accuracy of UKF algorithm can also be guaranteed.

(2) Effect of Stratification Effect

In Wang et al. (2011), asynchronous localization algorithm was given to locate terrestrial sensor networks. It should be stressed that, the method in Wang et al. (2011) cannot be directly applied to underwater environment, due to the stratification effect. In the following, we make a loose assumption, i.e., the stratification effect does not exist in water. Under this assumption, we apply the method in Wang et al. (2011) to locate the target, and the localization error is shown in Fig. 3.7a. Clearly, the method in Wang et al. (2011) can achieve the localization task, as the errors are within an acceptable range. However, this assumption is not supposed, as the stratification effect cannot be ignored in water. Without this loose assumption, the actual and estimated positions of target are shown in Fig. 3.7b. Correspondingly, the localization error are shown in Fig. 3.7c. Compared with Wang et al. (2011), the localization error in this chapter is more smaller. This comparison study reflects that the consideration of stratification effect in this chapter is meaningful and necessary for USNs.

(3) Effect of the Asynchronous Clock

In Yan et al. (2018), we applied the iterative least squares estimators to achieve underwater localization, and the clock skew was ignored, i.e., $\alpha = 1$. When $\alpha = 1$, we set the clock offset as random value, e.g., it is from 1 to 200 ms. Based on this, the method in Yan et al. (2018) is applied to locate the target. The estimated positions and localization errors are also shown in Fig. 3.7d, e, respectively. Clearly, the above two methods can both achieve the localization task when the clock skew is ignored. Next, we consider a general case, i.e., $\alpha \neq 1$. Inspired by this, the clock skew is set to be 1.05. Thereby, the localization errors by using the iterative least squares localization approach is shown in Fig. 3.7f. It is clear that the clock skew seriously affects the localization accuracy and the localization task cannot be achieved eventually. Alternatively, the localization scheme in this chapter can eliminate the effect of asynchronous clocks (i.e. clock offset and clock skew), and the localization task can be achieved. By the above comparisons, we know the

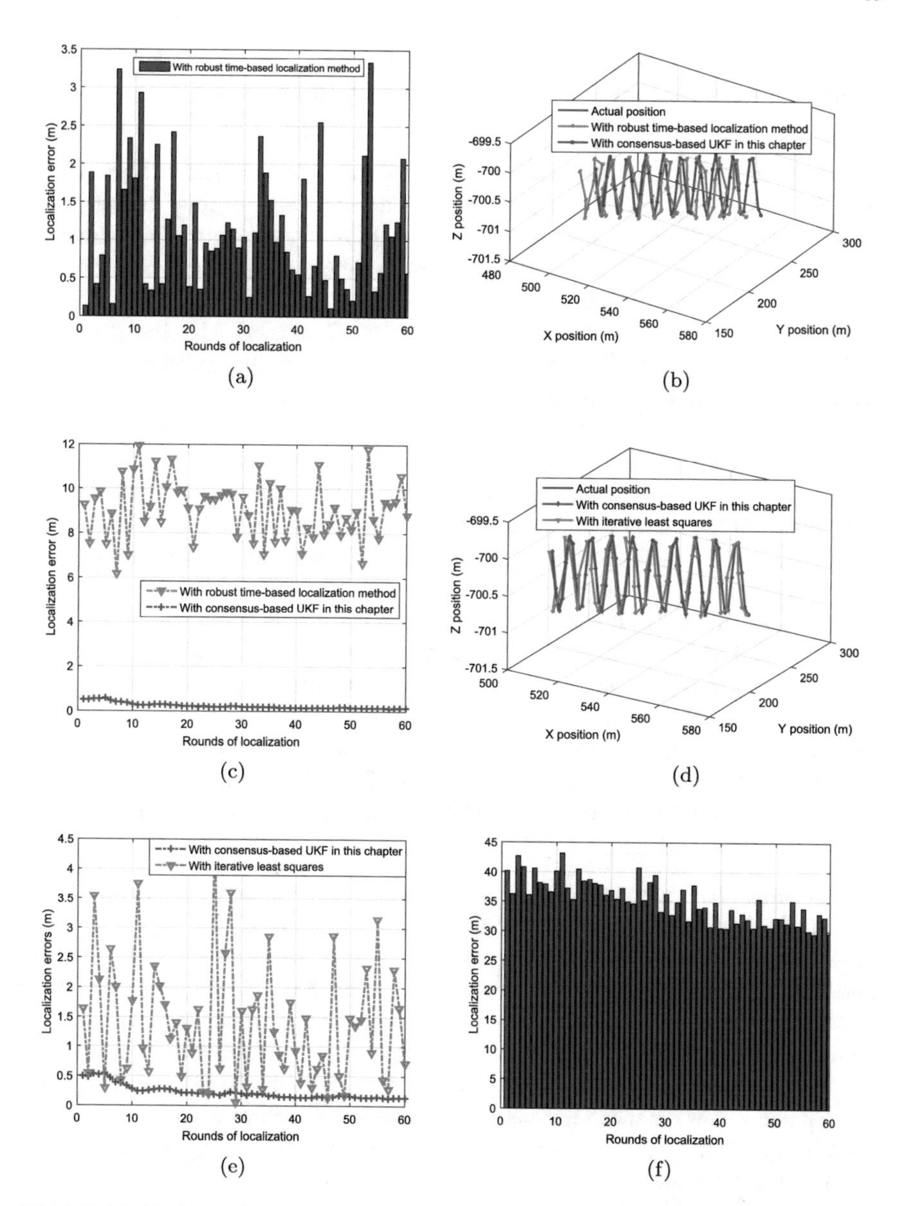

Fig. 3.7 (**a**)–(**c**) Comparison with robust time-based localization algorithm in Wang et al. (2011); (**d**)–(**f**) Comparison with the localization algorithm in Yan et al. (2018). (**a**) Errors without stratification effect. (**b**) Positions with stratification effect. (**c**) Errors with stratification effect. (**d**) The actual and localized positions. (**e**) Errors when clock skew is ignored. (**f**) Errors when clock skew is not ignored

consideration of clock skew is necessary, and the consensus-based UKF algorithm in this chapter can effectively eliminate the impact of the clock skew.

(4) Simulation for the Consensus-Based UKF Algorithm

This chapter adopts the consensus-based UKF algorithm to fuse two groups of measurements. The time measurement can be obtained by group 1 and group 2. To test the advantage of deterministic weights in our algorithm, malicious measurements are added at the fifth and the twentieth rounds of localization for group 1. The actual and localized positions of target are shown in Fig. 3.8a. In addition, Fig. 3.8b shows the localization errors. Compared with the stochastic weights in Li et al. (2016), the deterministic weights in this chapter can reduce the impact of malicious measurements, as it maximizes the contributions of stable measurements while minimizing the contributions of those that are less stable. Of interest, at the moments of malicious measurements, the weights are $\alpha^1(5) = 0$, $\alpha^2(5) = 1$, $\alpha^1(20) = 0.0051$ and $\alpha^2(20) = 0.9949$. Clearly, the influence of malicious measurements is reduced, i.e., $\alpha^1(5) = 0$ and $\alpha^1(20) = 0.0051$. This comparison study shows that the fusion of measurements is important, while the deterministic weights are meaningful.

In order to illustrate the lower bound of the consensus-based UKF algorithm, we define the RMSE, i.e., root mean square error. The RMSE are calculated by averaging the square error between $x(k)$ and $\hat{x}(k)$. The error performance of the three methods is compared with the theoretically derived CRLB, which is calculated by using (3.36). Particularly, the initial information matrix for computation of the CRLB is set to $\mathbf{J}(0) = \mathbf{P}^{-1}(0)$. From Fig. 3.8c, the localization accuracy by consensus-based UKF algorithm with deterministic fusion in this chapter is closest to the theoretical CRLB.

(5) Simulation for the Error of Wave Speed and the Error of Sensor Position

As mentioned in Sect. 3.4.3, the distance error is highly related to the error of acoustic wave speed. When $\Delta\tau_{\mathrm{T},2_j} - \Delta\tau_{\mathrm{T},1_j}$ is ignored, the relationship between Δd_1 and d_1 is shown in Fig. 3.8d. As shown in Fig. 3.8d, Δd_1 increases as d_1 is increased. With different $\Delta c_{2_j,1_j}$, the bigger of the $\Delta c_{2_j,1_j}$, the bigger of the Δd_1 with the same distance measurement d_1. When $\Delta\tau_{\mathrm{T},2_j} - \Delta\tau_{\mathrm{T},1_j}$ is not ignored, the results are shown in Fig. 3.8e. Comparing Fig. 3.8d with Fig. 3.8e, we know the error of acoustic wave speed is much sensitive to the estimation accuracy. However, the effect of $\Delta\tau_{\mathrm{T},2_j} - \Delta\tau_{\mathrm{T},1_j}$ can not be ignored, due to the requirement of high accurate localization.

It is noted that, the localization messages provided by sensor nodes can affect the accuracy of target localization. To verify this judgement, measurement noises are added to the position information of sensor nodes. For instance, the positions of sensor nodes in group j are updated as: $(x_{1_j}, y_{1_j}, z_{1_j}) \longleftarrow (x_{1_j} + 20\varpi_{1_j}, y_{1_j} + 20\varpi_{1_j}, z_{1_j} + 20\varpi_{1_j})$, $(x_{2_j}, y_{2_j}, z_{2_j}) \longleftarrow (x_{2_j} + 10\varpi_{2_j}, y_{2_j} + 10\varpi_{2_j}, z_{2_j} + 10\varpi_{2_j})$, and $(x_{3_j}, y_{3_j}, z_{3_j}) \longleftarrow (x_{3_j} + 30\varpi_{3_j}, y_{3_j} + 30\varpi_{3_j}, z_{3_j} + 30\varpi_{3_j})$, where ϖ_{1_j}, ϖ_{2_j} and ϖ_{3_j} are zero-mean Gaussian noises with a variance 1. Correspondingly, the localization errors are shown in Fig. 3.8f. It is clear that, the localization accuracy is reduced by comparing the ones in Fig. 3.8a–c. To reduce

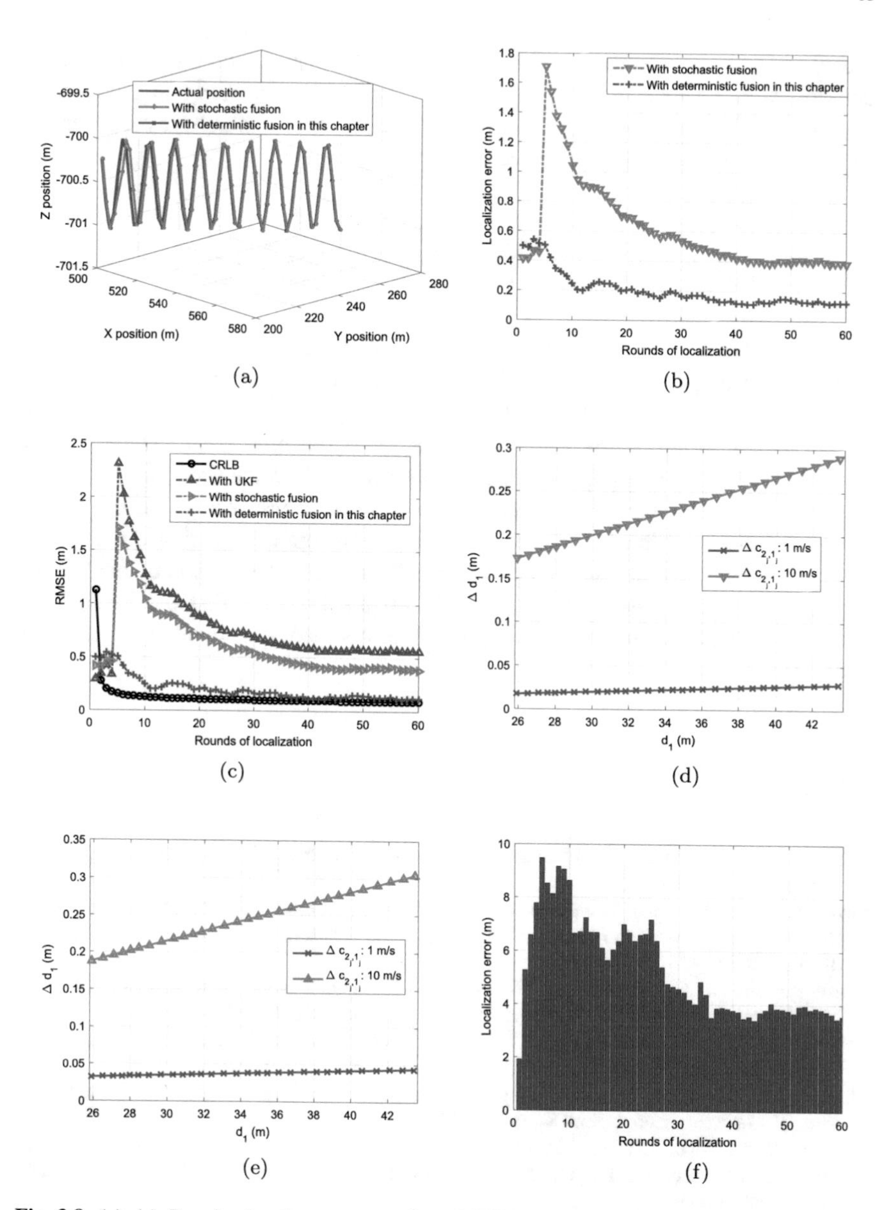

Fig. 3.8 (**a**)–(**c**) Results for the consensus-based UKF algorithm; (**d**)–(**f**) Δd_1 versus d_1 for different $\Delta c_{2_j,1_j}$. (**a**) The actual and localized positions. (**b**) Localization error. (**c**) RMSE of location estimation vs. CRLB. (**d**) $\Delta\tau_{T,2_j} - \Delta\tau_{T,1_j}$ is ignored. (**e**) $\Delta\tau_{T,2_j} - \Delta\tau_{T,1_j}$ is not ignored. (**f**) Errors when the self-localization of sensor nodes is not accurate

these negative impacts, the positions of sensor nodes are required to be accurately acquired with the assistance of surface buoys. It should be stressed that, the research topic of this chapter is the localization of underwater target. Thereby, we assume that the locations of sensor nodes are accurately pre-known.

3.6 Conclusion

In this chapter, we considered the constraints of asynchronous clock, stratification effect, and strong-noise characteristics in the underwater environment. Specifically, a network architecture that included surface buoys, sensor nodes, and a target was designed. To eliminate the effect of asynchronous clocks, we established the relationship between propagation delay and position. With the timestamp measurements, a consensus-based UKF localization approach was proposed to estimate the location of the target, where deterministic fusion was carried out to solve the optimization problems. Meanwhile, the CRLB for the proposed localization algorithm was also given. Finally, simulation results show that the localization accuracy in this chapter can be improved as compared with other localization methods.

References

Arasaratnam I, Haykin S (2008) Square-root quadrature Kalman filtering. IEEE Trans Signal Process 56(6):2589–2593

Bayat M, Crasta N, Aguiar A, Pascoal A (2016) Range-based underwater vehicle localization in the presence of unknown ocean currents: theory and experiments. IEEE Trans Control Syst Technol 24(1):122–139

Carroll P, Mahmood K, Zhou S, Zhou H, Xu X, Cui J (2014) On-demand asynchronous localization for underwater sensor networks. IEEE Trans Signal Process 62(13):3337–3348

Chai G, Lin Z, Fu M (2013) Consensus-based cooperative source localization of multi-agent systems. In: Proc. 32nd CCC, Xi'an, pp 6809–6814

Chang S, Li Y, He Y, Wang H (2018) Target Localization in underwater acoustic sensor networks using RSS measurements. Appl Sci 8(2):1–14

Charalambous T, Yuan Y, Yang T, Pan W, Hadjicostis C, Johansson M (2015) Distributed finite-time average consensus in digraphs in the presence of time delays. IEEE Trans Control Netw Syst 2(4):370–381

Erol-Kantarci M, Mouftah H, Oktug S (2011) A survey of architectures and localization techniques for underwater underwater acoustic sensor networks. IEEE Commun Surv Tut 13(3):487–502

Hermann R, Krener A (1977) Nonlinear controllability and observability. IEEE Trans Autom Control 22(5):728–740

Julier S, Uhlmann J (1997) New extension of the Kalman filter to nonlinear systems. In: Proc. int. society for optics and photonics, Orlando, pp 182–193

Kolomvatsos K, Anagnostopoulos C, Hadjiefthymiades S (2017) Distributed localized contextual event reasoning under uncertainty. IEEE Internet Things J 4(1):183–191

Lefebvre T, Bruyninckx H, Schuller JD (2002) Comment on 'A new method for the nonlinear transformation of means and covariances in filters and estimators' [and authors' reply]. IEEE Trans Autom Control 45(3):477–482
Li M, Hall D, Llinas J (2008) Handbook of multisensor data fusion: theory and practice. Artech House Radar Lib 39(5):180–184
Li W, Wei G, Han F, Liu Y (2016) Weighted average consensus-based unscented Kalman filtering. IEEE Trans Cybern 46(2):558–567
Liu J, Wang Z, Peng Z, Cui J, Fiondella L (2014) Suave: swarm underwater autonomous vehicle localization. In: Proc. IEEE INFOCOM, Toronto, pp 64–72
Liu J, Wang Z, Cui J, Zhou S, Yang B (2016) A joint time synchronization and localization design for mobile underwater sensor networks. IEEE Trans Mob Comput 15(3):530–543
Moreno-Salinas D, Pascoal A, Aranda J (2016) Optimal sensor placement for acoustic underwater target positioning with range-only measurements. IEEE J Oceanic Eng 41(3):620–643
Mortazavi E, Javidan R, Dehghani MJ, Kavoosi V (2017) A robust method for underwater wireless sensor joint localization and synchronization. Ocean Eng 137(2):276–286
Olfati-Saber R, Shamma J (2006) Consensus filters for sensor networks and distributed sensor fusion. In: Proc. 44th IEEE CDC, Seville, pp 6698–6703
Ramezani H, Jamali-Rad H, Leus G (2013) Target localization and tracking for an isogradient sound speed profile. IEEE Trans Signal Process 61(6):1434–1446
Ren W, Beard R (2005) Consensus seeking in multiagent systems under dynamically changing interaction topologies. IEEE Trans Autom Control 50(5):655–661
Shen S, Mulgaonkar Y, Michael N, Kumar V (2014) Multi-sensor fusion for robust autonomous flight in indoor and outdoor environments with a rotorcraft MAV. In: Proc. IEEE int. conf. rob. autom., Hong Kong, pp 4974–4981
Shi Y, Han C, Liang Y (2009) Adaptive UKF for target tracking with unknown process noise statistics. In: Proc. int. conf. inf. fusion, Seattle, WA, pp 1815–1820
Tomczak A (2011) Modern methods of underwater positioning applied in subsea mining. AGH J Min Geoeng 35:381–394
Wang Y, Ma X, Leus G (2011) Robust time-based localization for asynchronous networks. IEEE Trans Signal Process 59(9):4397–4410
Yan J, Li X, Luo X, Guan X (2017a) Virtual-lattice based intrusion detection algorithm over actuator-assisted underwater wireless sensor networks. Sensors 17(5):1168–1185
Yan J, Yang X, Luo X, Chen C, Guan X (2017b) Consensus of teleoperating cyber-physical system via centralized and decentralized controllers. IEEE Access 5(1):17271–17287
Yan J, Zhang X, Luo X, Wang Y, Chen C, Guan X (2018) Asynchronous localization with mobility prediction for underwater acoustic sensor networks. IEEE Trans Veh Technol 67(3):2543–2556
Yan J, Ban H, Luo X, Zhao H, Guan X (2019a) Joint localization and tracking design for AUV with asynchronous clocks and state disturbances. IEEE Trans Veh Technol 68(5):4707–4720
Yan J, Zhao H, Wang Y, Luo X, Guan X (2019b) Asynchronous localization for UASNs: an unscented transform based method. IEEE Signal Process Lett 26(4):602–606
Zhang B, Wang H, Xu T, Zheng L, Yang Q (2016) Received signal strength-based underwater acoustic localization considering stratification effect. In: Proc. IEEE conf. Oceans, Shanghai, pp 1–8
Zhang B, Wang H, Zheng L, Wu J, Zhuang Z (2017a) Joint synchronization and localization for underwater sensor networks considering stratification effect. IEEE Access 8(1):26932–26943
Zhang B, Wang Y, Wang H, Guan X, Zhuang Z (2017b) Tracking a duty-cycled autonomous underwater vehicle by underwater wireless sensor networks. IEEE Access 5(1):18016–18032
Zhou Z, Peng Z, Cui J, Shi Z, Bagtzoglou A (2011) Scalable localization with mobility prediction for underwater sensor networks. IEEE Trans Mob Comput 10(3):335–348

Chapter 4
Reinforcement Learning-Based Asynchronous Localization of USNs

Abstract In this chapter, an autonomous underwater vehicle (AUV) aided localization issue is studied under the constraints of asynchronous time clock, stratification effect and node mobility. Particularly, an asynchronous localization protocol is constructed and then the localization problem is built to minimize the sum of all measurement errors. To solve this localization problem, we propose a reinforcement learning (RL) based localization algorithm to locate the positions of AUVs, active and passive sensor nodes. It is noted that, the proposed localization algorithm employs two neural networks to approximate the increment policy and value function, and more importantly, it is much preferable for nonsmooth and nonconvex underwater localization problem due to its insensitivity to the local optimal. Besides that, the performance analyses of proposed algorithm are given. Finally, simulation and experimental results show that the localization performance in this chapter can be significantly improved as compared with the other works.

Keywords Reinforcement Learning (RL) · Localization · Asynchronous clock · Mobility · Underwater Sensor Networks (USNs)

4.1 Introduction

To acquire accurate location information, range-based localization methods are more beneficial than the range-free ones (Sorbelli et al. 2019; Phoemphon et al. 2018; Hong et al. 2017). Normally, the range-based localization methods rely on various mechanisms such as time of flight (TOF) (Ramezani et al. 2013), TOA (Isbitiren and Akan 2011), and TDOA (Shi and Wu 2018) to measure the relative range. With regard to wireless sensor nodes, the relative range measurement is easy to implement, due to the time clocks are synchronized and the transmission speed is fixed. However, the specific properties of underwater environment, such as asynchronous clock, stratification effect, and mobilities of sensor nodes, make it much more challenging to measure the relative range in USNs.

Affected by the above properties, the localization methods proposed for wireless sensor nodes is not suitable for USNs. To achieve effective localization for USNs,

J. Yan et al., *Localization in Underwater Sensor Networks*, Wireless Networks,
https://doi.org/10.1007/978-981-16-4831-1_4

several range-based underwater localization methods have been developed, e.g., Zhou et al. (2011), Carroll et al. (2014). These methods mainly involve the following procedures: (1) *Anchor discovery*; (2) *Range measurement*; (3) *Location estimation*. With the above procedures, an optimization problem in the form of $\mathbf{X}^* = \text{argmin}\ \mathcal{F}(\mathbf{X})$ can be built to locate the source node, where $\mathcal{F}$, $\mathbf{X}$ and $\mathbf{X}^*$ are the objective function, decision variable and optimized decision, respectively. One solution to this problem is to apply the least squares-based estimators. Nevertheless, the least squares-based estimators can easily fall into local minimum, because the objective functions for underwater localization problem are usually nonconvex. To solve this issue, some relaxation approaches (e.g., Soares et al. 2017; Jia et al. 2019) have been proposed to transform the nonconvex problem into convex optimization problem. However, the relaxation approaches do suffer from poor performance when the underwater localization problem is complicated. Motivated by this fact, we observe that an optimization method called *reinforcement learning* (RL) (Kiumarsi et al. 2018; Lewis and Vrabie 2009; Wen et al. 2019; Hong et al. 2019) can provide us with a promising solution. Specifically, RL is a goal-oriented optimization strategy, and the decision maker (or agent) learns a specified policy to optimize a long term reward via the interface with external environment. We believe that the localization accuracy of USNs can benefit a lot from the RL strategy. Until now, it is still an open challenge to employ RL strategy for the accurate localization of USNs.

Most existing underwater localization algorithms often apply a large amount of anchor nodes to cover the monitoring areas, because the coverage radius of a fixed anchor node is quite limited (Diamant and Lampe 2013; Yan et al. 2019; Parras et al. 2019; Houegnigan et al. 2017; Rauchenstein et al. 2018). Generally, a single anchor node with frequency bandwidths of $20 \sim 50$ kHz can only transmit its signal up to 1 km (Akyildiz et al. 2004). Although the coverage radius can be extended by adding transmission power, the extension of coverage radius is at an expense of increased communication energy consumption, which in turn reduces the working life of anchor node. To deal with this problem, the authors in Liu et al. (2014) and Gong et al. (2018) adopted autonomous underwater vehicles (AUVs) as the mobile anchor nodes. Compared with fixed anchor nodes, the mobility of AUVs can obtain a high degree of spatial reuse mechanism, through which the number of recruited anchor nodes can be significantly reduced. However, the localization methods in Liu et al. (2014) and Gong et al. (2018) depended on the assumption of synchronous clock or constant acoustic speed. With the aid of AUV, how to design an RL-based localization method with consideration of asynchronous clock, stratification effect and mobility of nodes is not well investigated.

Note that the least squares-related localization estimators are employed in Chaps. 2 and 3, however, the least squares-related localization estimators can easily fall into local minimum. In view of this, an AUV-aided localization solution is proposed for USNs in this chapter, where an RL-based localization algorithm is developed to acquire the positions of active and passive sensor nodes. Main contributions lie in two aspects:

1. **AUV-aided asynchronous localization protocol.** Considering asynchronous clock, stratification effect and mobility, we develop an AUV-aided asynchronous localization protocol, where the motion and ray compensations are both employed. Compared with the existing works (Liu et al. 2014; Gong et al. 2018; Yan et al. 2018, 2019b), the proposed localization protocol in this chapter can remove the impacts of asynchronous clock, stratification effect and mobility together.
2. **RL-based localization algorithm**. We develop an RL-based localization algorithm for AUVs, active and passive sensor nodes, where an online value iteration procedure is proposed to seek the optimization value. Different from the least squares based localization methods (Liu et al. 2016; Zhang et al. 2017; Mortazavi et al. 2017), the proposed RL-based localization algorithm can relax the linearization requirement, and more importantly, the global optimal solution can be acquired.

4.2 System Description and Problem Formulation

To achieve AUV-aided localization task for USNs, we provide a hybrid network architecture that includes four types of nodes, as shown in Fig. 4.1.

- **Surface Buoys.** Surface buoys are installed with GPS to obtain their location estimations and time references. Especially, the function of surface buoys is to provide self-localization and synchronization services for AUVs.

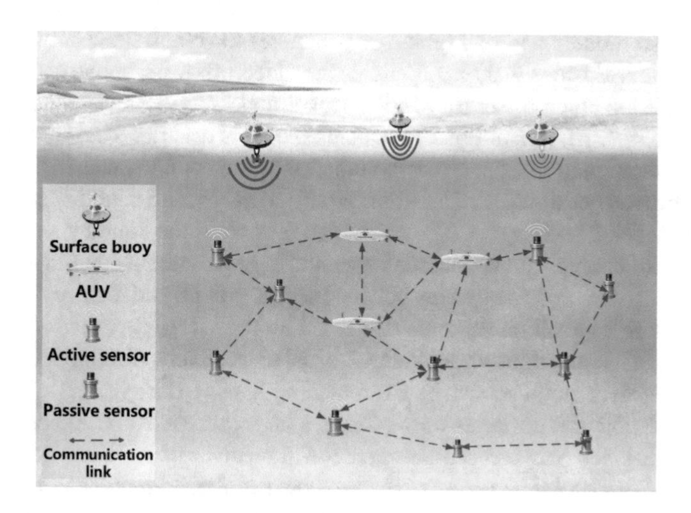

Fig. 4.1 Network architecture for the localization system

- **AUVs.** AUVs serve as the "mobile anchor nodes", whose advantage is to provide a high degree of spatial reuse mechanism for localization task. Particularly, the positions of AUVs can be acquired through the direct interaction with surface buoys. When AUVs receive localization request from active sensor nodes, they move to the monitoring area and provide localization services for active and passive sensor nodes. It is noted that the clocks of AUVs are not required to be synchronized with the real time.
- **Active Sensor Nodes.** By broadcasting timestamps to the networks, active sensor nodes initiate the localization procedure. In addition, the positions of active sensor nodes are required to be estimated, whose time clocks are not well synchronized with the real time.
- **Passive Sensor Nodes.** Different from active sensor nodes, the passive sensor nodes monitor the messages and locate themselves without transmitting any messages to the networks. Similar to active sensor nodes, the time clocks of passive sensor nodes are also asynchronous.

The USNs considered in this chapter is composed of M active sensor nodes, N passive sensor nodes and K AUVs. The index sets of active sensor nodes, passive sensor nodes and AUVs are denoted as $\mathcal{I}_{\mathrm{A}} = \{1, ..., M\}$, $\mathcal{I}_{\mathrm{P}} = \{1, ..., N\}$ and $\mathcal{I}_{\mathrm{V}} = \{1, ..., K\}$, respectively. Similarly, the clock models of AUV, active and passive sensor nodes are modeled as follows, i.e.,

$$\begin{aligned} T_{\mathrm{A},i} &= \alpha_{\mathrm{A},i}t + \beta_{\mathrm{A},i}, \ i \in \mathcal{I}_{\mathrm{A}} \\ T_{\mathrm{P},j} &= \alpha_{\mathrm{P},j}t + \beta_{\mathrm{P},j}, \ j \in \mathcal{I}_{\mathrm{P}} \\ T_{\mathrm{V},l} &= \alpha_{\mathrm{V},l}t + \beta_{\mathrm{V},l}, \ l \in \mathcal{I}_{\mathrm{V}} \end{aligned} \tag{4.1}$$

where $T_{\mathrm{A},i}$, $T_{\mathrm{P},j}$ and $T_{\mathrm{V},l}$ are the local clocks of active sensor node i, passive sensor node j and AUV l, respectively. Specifically, $\alpha_{\mathrm{A},i}$ denotes the clock skew between active sensor node i and real time t, while $\alpha_{\mathrm{P},j}$ denotes the clock skew between passive sensor node j and real time t. $\alpha_{\mathrm{V},l}$ denotes the clock skew between AUV l and real time t. Besides that, $\beta_{\mathrm{A},i}$ denotes the clock offset for active sensor node i, $\beta_{\mathrm{P},j}$ denotes the clock offset for passive sensor node j, and $\beta_{\mathrm{V},l}$ denotes the clock offset for AUV l. An illustration of the clock model is described in Fig. 4.2a.

For the stratification effect, it is assumed that the sound speed is linear to the depth (Ramezani et al. 2013; Liu et al. 2018). Thereby, the sound speed at depth z can be modeled as $C(z) = b + az$, where b denotes the sound speed on water surface, and a denotes the update scalar related to environment condition. Consider the constraint of tracing a sound ray between sender point S and receiver point R. For sender point S, its position vector is expressed by $\mathbf{P}_{\mathrm{S}} = [x_{\mathrm{S}}, y_{\mathrm{S}}, z_{\mathrm{S}}]^{\mathrm{T}}$, where x_{S}, y_{S} and z_{S} denote the positions on X, Y and Z axes, respectively. Similarly, the position vector of receiver point R can be expressed by $\mathbf{P}_{\mathrm{R}} = [x_{\mathrm{R}}, y_{\mathrm{R}}, z_{\mathrm{R}}]^{\mathrm{T}}$, where x_{R}, y_{R} and z_{R} denote the positions on X, Y and Z axes, respectively. Accordingly, the horizontal distance between sender point S and receiver point R can be calculated as $r_{\mathrm{S,R}} = \sqrt{(x_{\mathrm{S}} - x_{\mathrm{R}})^2 + (y_{\mathrm{S}} - y_{\mathrm{R}})^2}$. As shown in Fig. 4.2b, we can further have $\frac{\cos\theta}{C(z)} = \frac{\cos\theta_{\mathrm{S}}}{C(z_{\mathrm{S}})} = \frac{\cos\theta_{\mathrm{R}}}{C(z_{\mathrm{R}})}$ by adopting Snell's law (Ramezani et al. 2013), where

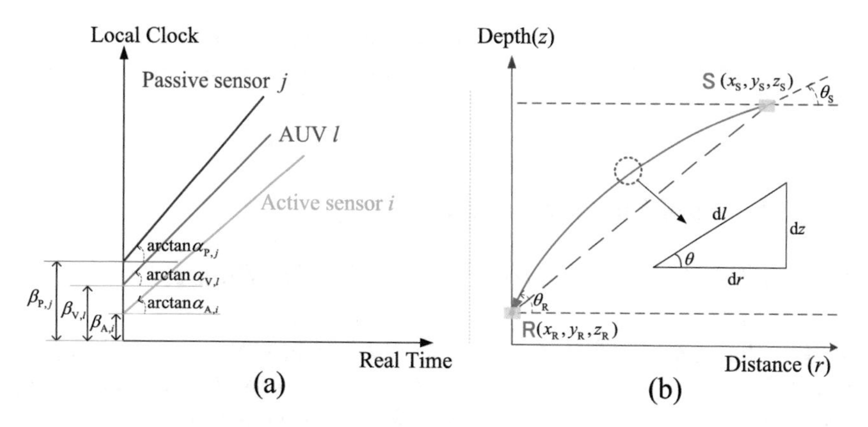

Fig. 4.2 (**a**) Illustration of the asynchronous clock model. (**b**) Description of the sound ray between two nodes

$\theta_S \in [-\frac{\pi}{2}, \frac{\pi}{2}]$ and $\theta_R \in [-\frac{\pi}{2}, \frac{\pi}{2}]$ are the ray angles at sender point S and receiver point R, respectively. Particularly, we have the following equations: $\mathrm{d}r = \frac{\mathrm{d}z}{\tan\theta}$, $\mathrm{d}l = \frac{\mathrm{d}z}{\sin\theta}$ and $\mathrm{d}t = \frac{\mathrm{d}l}{C(z)}$, where l denotes the length of acoustic propagation path between sender point S and receiver point R. Thus, the propagation delay from sender to receiver can be given as

$$\tau_{S,R} = -\frac{1}{a}\left(\ln\frac{1+\sin\theta_S}{\cos\theta_S} - \ln\frac{1+\sin\theta_R}{\cos\theta_R}\right), \tag{4.2}$$

where $\theta_S = \phi_0 - \psi_0$ and $\theta_R = \phi_0 + \psi_0$. To be specific, $\phi_0 = \arctan\frac{z_S - z_R}{r_{S,R}}$ denotes the angle of the straight line between sender point S and receiver point R with regard to the horizontal axis. $\psi_0 = \arctan\frac{0.5ar_{S,R}}{b+0.5a(z_S+z_R)}$ denotes the angle, at which the ray deviates from the straight line.

The following two problems should be solved for the USNs system.

Problem 4.1 (AUV-Aided Asynchronous Localization Protocol Design) From the perspective of acoustic communication, the sensor nodes in USNs have some unique characteristics such as asynchronous clock, stratification effect and mobility. These features make the wireless-suited localization protocols inaccurate or even failed. With regard to this situation, we try to develop an AUV-aided asynchronous localization protocol, where the motion and ray compensations are both employed. This problem is reduced to construct the localization optimization problems for AUVs, active and passive sensor nodes, with the consideration of (4.1) and (4.2).

Problem 4.2 (RL-Based Optimal Estimator Design) From the perspective of signal processing, the underwater localization optimization problems are much more complex than the ones in terrestrial environment, and it is always hard to find the global optimum solution. To solve this issue, an RL-based localization algorithm is developed, where RL-based estimator are sought to find the global optimum solution. This problem is reduced to the estimations of $\mathbf{P}_{A,i}$, $\mathbf{P}_{P,j}$ and $\mathbf{P}_{V,l}$, where $\mathbf{P}_{A,i}$, $\mathbf{P}_{P,j}$ and $\mathbf{P}_{V,l}$ are the positions of active sensor node i, passive sensor node j and AUV l, respectively.

4.3 RL-Based Localization for USNs

4.3.1 AUV-Aided Asynchronous Localization Protocol

To locate the sensor nodes, the locations of AUVs require to be known. Motivated by this, the localization protocol can be divided into two phases: (I) AUV localization; (II) Sensor localization. In Phase I, each AUV sends out an initiator message to surface buoys, and then a localization protocol is designed to build the optimization problem for AUVs. In Phase II, localization optimization problems for active and passive sensor nodes are built.

In the following, the detailed description of Phase I is given. It is worth mentioning that, the depth information of AUV can be accurately acquired by the installed depth test unit (Aras et al. 2012; Jorgensen et al. 2019). Meantime, three non-collinear surface buoys can locate a two-dimensional node. Therefore, it is conducted that $N_{\#} \geq 3$ surface buoys are deployed in the monitoring area. At timestamp $T_{l,l}$, AUV l sends out an initial message to the surface buoys, where $l \in \mathcal{I}_{V}$. Subsequently, AUV l goes into listening mode and waits for the reply messages from surface buoys. At timestamp $t_{l,g}$, surface buoy $g \in \{1, ..., N_{\#}\}$ receives the message from AUV l, and then it switches into the waiting mode to receive messages from the other surface buoys. Specially, the timestamp when surface buoy $\bar{g} \in \{2, ..., N_{\#}\}$ receives message from surface buoy $h \in \{1, ..., \bar{g}-1\}$ is expressed by $t_{l,h,\bar{g}}$. Then, at timestamp $t_{l,g,g}$, surface buoy g sends out its reply message to AUV l, and AUV l receives this message at timestamp $T_{g,l}$. After receiving the reply message from all surface buoys, the transmission procedure between AUV l and surface buoys is completed. Accordingly, the collected timestamps on AUV l can be denoted as $T_{l,l}$, $t_{l,g}$, $t_{l,h,\bar{g}}$,$t_{l,g,g}$ and $T_{g,l}$.

In order to eliminate the influence of clock offset and skew, the following time differences can be defined for AUV $l \in \mathcal{I}_{V}$, i.e.,

$$\begin{aligned}
\Delta\mathcal{T}_{l,\bar{g},h} &= t_{l,\bar{g}} - t_{l,h} \\
&= (T_{l,l} - \beta_{V,l})/\alpha_{V,l} + \tau_{l,\bar{g}} + \omega_{l,\bar{g}} \\
&\quad - ((T_{l,l} - \beta_{V,l})/\alpha_{V,l} + \tau_{l,h} + \omega_{l,h}) \\
&= \tau_{l,\bar{g}} - \tau_{l,h} + \omega_{l,\bar{g},h}, \\
\Delta\mathcal{T}_{l,h,\bar{g}} &= (t_{l,h,\bar{g}} - t_{l,\bar{g}}) - (t_{l,h,h} - t_{l,h}) \\
&= ((t_{l,h,h} + \tau_{l,h,\bar{g}} + \omega_{l,h,\bar{g}}) - ((T_{l,l} - \beta_{V,l})/\alpha_{V,l} + \tau_{l,\bar{g}} \\
&\quad + \omega_{l,\bar{g}})) - (t_{l,h,h} - ((T_{l,l} - \beta_{V,l})/\alpha_{V,l} + \tau_{l,h} + \omega_{l,h})) \\
&= \tau_{l,h} + \tau_{l,h,\bar{g}} - \tau_{l,\bar{g}} + \omega_{l,h,\bar{g}},
\end{aligned} \tag{4.3}$$

where $\tau_{l,\bar{g}}$ represents the propagation delay from AUV l to surface buoy $\bar{g}$, $\tau_{l,h}$ represents the propagation delay from AUV l to surface buoy h, and $\tau_{l,h,\bar{g}}$ represents the propagation delay from surface buoys h to $\bar{g}$. In addition, $\tau_{l,\bar{g}}$, $\tau_{l,h}$ and $\tau_{l,h,\bar{g}}$ can be calculated by (4.2). Besides that, $\omega_{l,\bar{g}}$, $\omega_{l,h}$, and $\omega_{l,h,\bar{g}}$ represent measurement noises, subject to zero-mean Gaussian distribution with variance σ_{mea}^2. Based on this, the reorganized noises $\omega_{l,\bar{g},h} = \omega_{l,\bar{g}} - \omega_{l,h}$ and $\omega_{{l,h,\bar{g}}} = \omega_{l,h,\bar{g}} - \omega_{l,\bar{g}} + \omega_{l,h}$ satisfy the distributions of $\omega_{l,\bar{g},h} \sim \mathcal{N}(0, 2\sigma_{mea}^2)$ and $\omega_{{l,h,\bar{g}}} \sim \mathcal{N}(0, 3\sigma_{mea}^2)$.

Accordingly, the likelihood function for AUV $l \in \mathcal{I}_V$ can be built as

$$p_l = \frac{1}{\Upsilon_l} \exp\left\{ -\frac{1}{4\sigma_{mea}^2} \sum_{\bar{g}=2}^{N_\#} \sum_{h=1}^{\bar{g}-1} (\Delta T_{l,\bar{g},h} - (\tau_{l,\bar{g}} - \tau_{l,h}))^2 \right. \\ \left. - \frac{1}{6\sigma_{mea}^2} \sum_{\bar{g}=2}^{N_\#} \sum_{h=1}^{\bar{g}-1} [\Delta T_{l,h,\bar{g}} - (\tau_{l,h} + \tau_{l,h,\bar{g}} - \tau_{l,\bar{g}})]^2 \right\}, \tag{4.4}$$

where x_l and y_l denote the positions of AUV l on X and Y axes, respectively. Besides that, Υ_l is a constant, which is related with the number of measurements. It is noted that, Υ_l is not required to be known in this chapter.

Referring to Torrieri (1984), one knows the maximum likelihood estimator of (x_l, y_l) can be acquired by maximizing (4.4). Inspired by this, the maximum likelihood estimator minimizes the following quadratic form, i.e.,

$$\mathcal{H}(\hat{x}_l, \hat{y}_l) = \frac{1}{4\sigma_{mea}^2} \sum_{\bar{g}=2}^{N_\#} \sum_{h=1}^{\bar{g}-1} (\Delta T_{l,\bar{g},h} - (\tau_{l,\bar{g}} - \tau_{l,h}))^2 \\ + \frac{1}{6\sigma_{mea}^2} \sum_{\bar{g}=2}^{N_\#} \sum_{h=1}^{\bar{g}-1} [\Delta T_{l,h,\bar{g}} - (\tau_{l,h} + \tau_{l,h,\bar{g}} - \tau_{l,\bar{g}})]^2, \tag{4.5}$$

where $\hat{x}_l$ and $\hat{y}_l$ denote the estimations of x_l and y_l, respectively.

It is worth mentioning that the minimization of (4.5) is a reasonable criterion for determination of an estimator even when the noises cannot be assumed to Gaussian (Torrieri 1984). Based on this, the localization optimization problem for AUV $l \in \mathcal{I}_V$ can be formulated as

$$(x_l^*, y_l^*) = \text{argmin}_{(x_l, y_l)} \mathcal{H}(\hat{x}_l, \hat{y}_l). \tag{4.6}$$

From (4.6), the estimated position of AUV l at timestamp $T_{l,l}$ can be acquired. Due to the mobility property, the position at timestamp $T_{l,l}$ is different from the one at the current timestamp T_l. To handle this issue, a motion compensation strategy is proposed here. Specifically, the horizontal position vector of AUV l at timestamp $T_{l,l}$ is expressed as $\bar{\mathbf{P}}_l = [x_l^*, y_l^*]^{\mathrm{T}}$. At current timestamp T_l, the horizontal position vector of AUV l is denoted as $\bar{\mathbf{P}}_l' = [x_l', y_l']^{\mathrm{T}}$, where x_l' and y_l' represent the current positions of AUV l on X and Y axes, respectively. Without loss of generality, it is assumed that the velocity of AUV l can be acquired by gyro units. Based on this, the relationship between $\bar{\mathbf{P}}_l$ and $\bar{\mathbf{P}}'_l$ is constructed as

$$\bar{\mathbf{P}}'_l = \bar{\mathbf{P}}_l + \delta_l \sum\nolimits_{m_l=0}^{n_l} \mathbf{v}_l(T'_l), \tag{4.7}$$

where $\delta_l \in \mathcal{R}^+$ is the local sampling interval for AUV l. With consideration of the influence of clock skew, δ_l can be denoted as $\delta_l = \delta\alpha_{\mathrm{V},l}$. Based on this, the total updating step n_l can be expressed as $n_l = \lfloor (T_l - T_{l,l})/\delta_l \rfloor$, where $\lfloor \cdot \rfloor$ is the floor function. In addition, $\mathbf{v}_l(T'_l)$ is the horizontal velocity vector of AUV l at timestamp T'_l, where $T'_l = T_{l,l} + m_l\delta_l$.

In Phase II, the localization optimization problems for active and passive sensor nodes can be built. Similar to Carroll et al. (2014) and Yan et al. (2018), we assume that the IDs of AUVs are pre-known to active and passive sensor nodes. First of all, active sensor node $i \in \mathcal{I}_{\mathrm{A}}$ sends out an initiator message to the networks. After receiving it, AUVs record and reply the initiator message, through which a series of timestamps can be collected by active sensor node i. Meantime, the passive sensor node $j \in \mathcal{I}_{\mathrm{P}}$ listens to the networks, and collects timestamps from AUVs. Specifically, an illustration of the timestamp transmission procedure is given in Fig. 4.3. Based on this, an asynchronous localization protocol for active sensor node $i \in \mathcal{I}_{\mathrm{A}}$ and passive sensor node $j \in \mathcal{I}_{\mathrm{P}}$ is given as follows.

1. At time $T^{\mathrm{A}}_{i,0}$, active sensor node i sends its initial message to the networks, which includes the sending order of AUV l, i.e., $l \in \{1, ..., K\}$. Hence, active sensor i goes into listening mode and waits for the replies from AUVs.

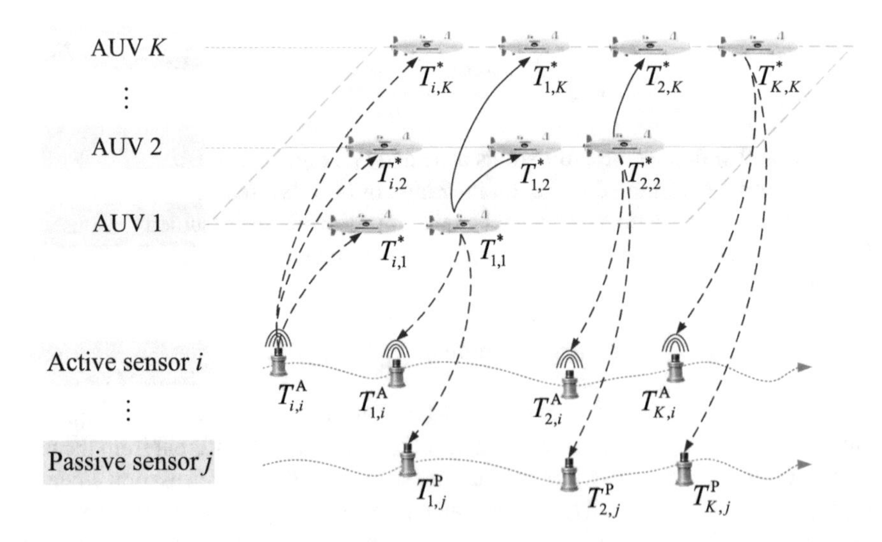

Fig. 4.3 Illustration of the timestamp transmission process

2. At time $T_{i,l}^{\mathrm{V}}$, AUV l receives the message from active sensor node i. Subsequently, it switches into the listening mode whose objective is to receive messages from the other AUVs. At time $T_{m,\bar{l}}^{*}$, AUV $\bar{l}$ receives message from AUVs 1 to $\bar{l}-1$, where $\bar{l} \in \{2, ..., K\}$ and $m \in \{1, ..., \bar{l}-1\}$. Then, AUV l sends out its localization message at time $T_{l,l}^{*}$. Specifically, the localization message includes $T_{i,l}^{\mathrm{V}}$, $T_{m,\bar{l}}^{*}$, $T_{l,l}^{*}$ and the location information of AUV l.
3. At time $T_{l,i}^{\mathrm{A}}$, active sensor node i receives the reply message from AUV l. After acquiring the reply messages from all AUVs, active sensor node i closes its listening mode. Based on this, the collected timestamps on active sensor node i can be expressed as $T_{i,0}^{\mathrm{A}}$, $T_{i,l}^{\mathrm{V}}$, $T_{m,\bar{l}}^{*}$, $T_{l,l}^{*}$ and $T_{l,i}^{\mathrm{A}}$.
4. Different form active sensor node i, passive sensor node j does not transmit any message to the networks. At time $T_{l,j}^{\mathrm{P}}$, passive sensor node j receives localization message from AUV l. When passive sensor node j receives all the AUVs' messages, the transmission process is ended. The collected timestamps by passive sensor node j are denoted as $T_{l,l}^{*}$ and $T_{l,j}^{\mathrm{P}}$.

Through the above procedure, as described in Fig. 4.3, the following time difference can be designed to remove the influence of clock offset and skew for active sensor node $i \in \mathcal{I}_{\mathrm{A}}$, i.e.,

$$
\begin{aligned}
& T_{m,\bar{l}}^{*} - T_{i,\bar{l}}^{\mathrm{V}} \\
&= \alpha_{\mathrm{V},\bar{l}}[\frac{T_{m,m}^{*} - \beta_{\mathrm{V},m}}{\alpha_{\mathrm{V},m}} + \bar{\tau}_{m,\bar{l}} + \bar{\omega}_{m,\bar{l}}] + \beta_{\mathrm{V},\bar{l}} \\
&\quad - \alpha_{\mathrm{V},\bar{l}}[\frac{T_{i,0}^{\mathrm{A}} - \beta_{\mathrm{A},i}}{\alpha_{\mathrm{A},i}} + \tau_{i,\bar{l}}^{*} + \omega_{i,\bar{l}}^{*}] - \beta_{\mathrm{V},\bar{l}} \\
&= \alpha_{\mathrm{V},\bar{l}}(\bar{\tau}_{m,\bar{l}} - \tau_{i,\bar{l}}^{*} + \bar{\omega}_{m,\bar{l}} - \omega_{i,\bar{l}}^{*}) \\
&\quad + \alpha_{\mathrm{V},\bar{l}}[\frac{T_{m,m}^{*} - \beta_{\mathrm{V},m}}{\alpha_{\mathrm{V},m}} - \frac{T_{i,0}^{\mathrm{A}} - \beta_{\mathrm{A},i}}{\alpha_{\mathrm{A},i}}] \\
&= \alpha_{\mathrm{V},\bar{l}}(\bar{\tau}_{m,\bar{l}} - \tau_{i,\bar{l}}^{*} + \tau_{i,m}^{*}) + \varpi_{m,\bar{l}} + \frac{\alpha_{\mathrm{V},\bar{l}}}{\alpha_{\mathrm{V},m}}(T_{m,m}^{*} - T_{i,m}^{\mathrm{V}}).
\end{aligned}
\tag{4.8}
$$

Then, one has

$$
\begin{aligned}
\Delta\Gamma_{m,\bar{l}} &= (T_{m,\bar{l}}^{*} - T_{i,\bar{l}}^{\mathrm{V}}) - \frac{\alpha_{\mathrm{V},\bar{l}}}{\alpha_{\mathrm{V},m}}(T_{m,m}^{*} - T_{i,m}^{\mathrm{V}}) \\
&= \alpha_{\mathrm{V},\bar{l}}(\bar{\tau}_{m,\bar{l}} - \tau_{i,\bar{l}}^{*} + \tau_{i,m}^{*}) + \varpi_{m,\bar{l}},
\end{aligned}
\tag{4.9}
$$

where $\bar{l} \in \{2, ..., K\}$ and $m \in \{1, ..., \bar{l}-1\}$. Especially, $\tau_{i,\bar{l}}^{*}$ denotes the propagation delay from active sensor node i to AUV $\bar{l}$, $\tau_{i,m}^{*}$ denotes the propagation delay from active sensor node i to AUV m, and $\bar{\tau}_{m,\bar{l}}$ denotes the propagation delay from AUV

m to AUV $\bar{l}$. $\tau^*_{i,\bar{l}}$, $\tau^*_{i,m}$ and $\bar{\tau}_{m,\bar{l}}$ can be calculated by (4.2). $\omega^*_{i,\bar{l}}$, $\omega^*_{i,m}$, and $\bar{\omega}_{m,\bar{l}}$ are measurement noises, subject to zero-mean Gaussian distribution with variance σ^2_{mea}. Based on this, $\varpi_{m,\bar{l}} = \bar{\omega}_{m,\bar{l}} - \omega^*_{i,\bar{l}} - \omega^*_{i,m}$ satisfy the distributions of $\varpi_{m,\bar{l}} \sim \mathcal{N}(0, 3\sigma^2_{\text{mea}})$.

Similar to (4.4)–(4.6), the localization optimization problem for the active sensor node $i \in \mathcal{I}_{\text{A}}$ can be formulated as

$$(x^*_{\text{A},i}, y^*_{\text{A},i}) = \operatorname{argmin}_{(x_{\text{A},i}, y_{\text{A},i})} \mathcal{H}_1(\hat{x}_{\text{A},i}, \hat{y}_{\text{A},i}), \tag{4.10}$$

with

$$\mathcal{H}_1(\hat{x}_{\text{A},i}, \hat{y}_{\text{A},i}) = \frac{1}{6\sigma^2_{\text{mea}}} \sum_{\bar{l}=2}^{K} \sum_{m=1}^{\bar{l}-1} [\Delta\Gamma_{m,\bar{l}} - \alpha_{\text{V},\bar{l}}(\bar{\tau}_{m,\bar{l}} - \tau^*_{i,\bar{l}} + \tau^*_{i,m})]^2, \tag{4.11}$$

where $x_{\text{A},i}$ and $y_{\text{A},i}$ are the locations of active sensor node i on X and Y axes, respectively. Besides, $\hat{x}_{\text{A},i}$ and $\hat{y}_{\text{A},i}$ are the estimated positions of active sensor node i on X and Y axes, respectively.

By solving the optimization problem in (4.10), the horizontal position vector of active sensor node i at timestamp $T^{\text{A}}_{i,0}$ can be acquired. Note that the location of active sensor node i at timestamp $T^{\text{A}}_{i,0}$ is different from the one at current timestamp T^{A}_i. Therefore, the motion compensation strategy is employed to handle this issue. Let us denote $\bar{\mathbf{P}}_{\text{A}.i} = [x^*_{\text{A},i}, y^*_{\text{A},i}]^{\text{T}}$ as the horizontal position vector of active sensor node i at timestamp $T^{\text{A}}_{i,0}$. Meantime, the horizontal position vector of active sensor node i at current timestamp T^{A}_i is expressed as $\bar{\mathbf{P}}'_{\text{A},i} = [x'_{\text{A},i}, y'_{\text{A},i}]^{\text{T}}$, where $x'_{\text{A},i}$ and $y'_{\text{A},i}$ denote the locations of active sensor node i on X and Y axes, respectively. Accordingly, the relationship between $\bar{\mathbf{P}}_{\text{A}.i}$ and $\bar{\mathbf{P}}'_{\text{A},i}$ can be formulated as

$$\bar{\mathbf{P}}'_{\text{A},i} = \bar{\mathbf{P}}_{\text{A},i} + \delta_{\text{A},i} \sum_{m_{\text{A},i}=0}^{n_{\text{A},i}} \mathbf{v}_{\text{A},i}(T'_{\text{A},i}), \tag{4.12}$$

where $\delta_{\text{A},i} = \delta\alpha_{\text{A},i}$ represents the local sampling interval for active sensor node i. Besides that, $T'_{\text{A},i} = T^{\text{A}}_{i,0} + m_{\text{A},i}\delta_{\text{A},i}$, and $\mathbf{v}_{\text{A},i}(T'_{\text{A},i})$ denotes the horizontal velocity vector of active sensor node i at timestamp $T'_{\text{A},i}$, which can be measured by a gyro unit. Specifically, the total updating step $n_{\text{A},i}$ can be expressed as $n_{\text{A},i} = \lfloor (T^{\text{A}}_i - T^{\text{A}}_{i,0})/\delta_{\text{A},i} \rfloor$.

Similar to (4.10), the localization optimization problem for passive sensor node $j \in \mathcal{I}_{\text{P}}$ can be formulated as

$$(x^*_{\text{P},j}, y^*_{\text{P},j}, \alpha^*_{\text{P},j}, \beta^*_{\text{P},j}) = \operatorname{argmin}_{(x_{\text{P},j}, y_{\text{P},j}, \alpha_{\text{P},j}, \beta_{\text{P},j})} \mathcal{H}_2(\hat{x}_{\text{P},j}, \hat{y}_{\text{P},j}, \hat{\alpha}_{\text{P},j}, \hat{\beta}_{\text{P},j}), \tag{4.13}$$

with

$$\mathcal{H}_2(\hat{x}_{\mathrm{P},j}, \hat{y}_{\mathrm{P},j}, \hat{\alpha}_{\mathrm{P},j}, \hat{\beta}_{\mathrm{P},j}) \\ = \frac{1}{2\sigma_{\mathrm{mea}}^2} \sum_{l=1}^{K} \{T_{l,j}^{\mathrm{P}} - \hat{\alpha}_{\mathrm{P},j}[\frac{T_{l,l}^{*} - \beta_{\mathrm{V},l}}{\alpha_{\mathrm{V},l}} + \tau_{l,j}^{\#}] - \hat{\beta}_{\mathrm{P},j}\}^2, \tag{4.14}$$

where $\tau_{l,j}^{\#}$ denotes the one-way propagation delay from AUV l to passive sensor node j. It is noted that, $\tau_{l,j}^{\#}$ can be calculated via (4.2). $x_{\mathrm{P},j}$ and $y_{\mathrm{P},j}$ are the positions of passive sensor node j on X and Y axes, respectively. Besides that, $\hat{x}_{\mathrm{p},j}$ and $\hat{y}_{\mathrm{P},j}$ represent the estimations of $x_{\mathrm{P},j}$ and $y_{\mathrm{P},j}$, respectively. Especially, $\hat{\alpha}_{\mathrm{P},j}$ and $\hat{\beta}_{\mathrm{P},j}$ denote the estimations of $\alpha_{\mathrm{P},j}$ and $\beta_{\mathrm{P},j}$.

From (4.13), the estimated position information of passive sensor node j at timestamp $T_{1,j}^{\mathrm{P}}$ can be acquired. Of note, the position at timestamp $T_{1,j}^{\mathrm{P}}$ is different from the one at current timestamp T_j^{P}. Thereby, we denote the horizontal position vector of passive sensor node j at timestamp $T_{1,j}^{\mathrm{P}}$ as $\bar{\mathbf{P}}_{\mathrm{P},j} = [x_{\mathrm{P},j}^{*}, y_{\mathrm{P},j}^{*}]^{\mathrm{T}}$. At current timestamp T_j^{P}, the horizontal position vector of passive sensor node j is expressed as $\bar{\mathbf{P}}_{\mathrm{P},j}' = [x_{\mathrm{P},j}', y_{\mathrm{P},j}']^{\mathrm{T}}$, where $x_{\mathrm{P},j}'$ and $y_{\mathrm{P},j}'$ denote the positions of passive sensor node j on X and Y axes, respectively. Without loss of generality, it is assumed that the velocity of passive sensor node j can be measured via gyro units. Based on this, the relationship between $\bar{\mathbf{P}}_{\mathrm{P},j}$ and $\bar{\mathbf{P}}_{\mathrm{P},j}'$ can be given as

$$\bar{\mathbf{P}}_{\mathrm{P},j}' = \bar{\mathbf{P}}_{\mathrm{P},j} + \delta_{\mathrm{P},j} \sum_{m_{\mathrm{P},j}=0}^{n_{\mathrm{P},j}} \mathbf{v}_{\mathrm{P},j}(T_{\mathrm{P},j}'), \tag{4.15}$$

where $\delta_{\mathrm{P},j} \in \mathcal{R}^{+}$ represents the local sampling interval for passive sensor node j. Influenced by the clock skew, $\delta_{\mathrm{P},j}$ can be expressed as $\delta_{\mathrm{P},j} = \delta\alpha_{\mathrm{P},j}$. Based on this, the total updating step $n_{\mathrm{P},j}$ can be expressed as $n_{\mathrm{P},j} = \lfloor (T_j^{\mathrm{P}} - T_{1,j}^{\mathrm{P}})/\delta_{\mathrm{P},j} \rfloor$, and $T_{\mathrm{P},j}' = T_{1,j}^{\mathrm{P}} + m_{\mathrm{P},j}\delta_{\mathrm{P},j}$. In addition, $\mathbf{v}_{\mathrm{P},j}(T_{\mathrm{P},j}')$ denotes the horizontal velocity vector of passive sensor node j at timestamp $T_{\mathrm{P},j}'$.

4.3.2 *RL-Based Localization Algorithm*

In this section, the approximate dynamic programming (ADP), which is a core classification of RL strategies, is applied to seek the solutions for the optimization problems in (4.6), (4.10) and (4.13). Especially, ADP uses function approximation to calculate the value functions, and hence, the dynamic programming problems can be solved by adaptive critics (Kiumarsi et al. 2018; Lewis and Vrabie 2009). With this end in view, the estimation increment in a state iterative manner can be updated, an illustration is given in Fig. 4.4.

Now, we first present an RL-based localization algorithm for the optimization of (4.6). Let $\hat{\mathbf{P}}_l = [\hat{x}_l, \hat{y}_l]^{\mathrm{T}}$ are the estimated horizontal position vector of AUV

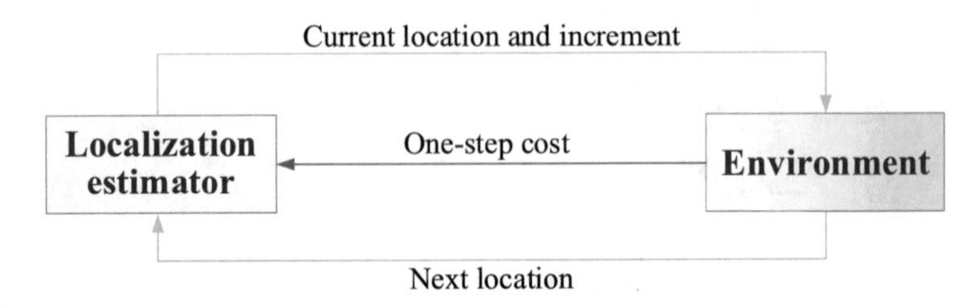

Fig. 4.4 Evaluation of RL-based localization

l, where $\hat{x}_l$ and $\hat{y}_l$ denote the estimated positions on X and Y axes, respectively. To describe the transition procedure, the discrete-time dynamical system for the localization estimator of AUV l is formulated as

$$\hat{\hat{\mathbf{P}}}_{l,k+1} = \hat{\hat{\mathbf{P}}}_{l,k} + \mathbf{u}_{l,k}, \tag{4.16}$$

where $\hat{\hat{\mathbf{P}}}_{l,k} \in \mathcal{R}^n$ denotes state vector, and $\mathbf{u}_{l,k} \in \mathcal{R}^m$ denotes increment (i.e., action) for the localization estimator of AUV l.

By using (4.6) and referring to the standard steps of RL in Lewis and Vrabie (2009), the following cost (or reward) function is given to measure the one-step cost

$$g_{\mathrm{V}}(\hat{\hat{\mathbf{P}}}_{l,k}, \mathbf{u}_{l,k}) = \mathcal{H}(\hat{x}_{l,k}, \hat{y}_{l,k}) + \mathbf{u}_{l,k}^{\mathrm{T}}\mathbf{Q}_1\mathbf{u}_{l,k}. \tag{4.17}$$

where $\mathbf{Q}_1$ denotes a positive definite matrix.

From (4.17), the total cost (or reward) function can be defined as

$$\mathcal{V}_{\mathrm{V}}(\hat{\hat{\mathbf{P}}}_{l,k}) = \sum_{i=k}^{\infty} \gamma_{\mathrm{V}}^{i-k} g_{\mathrm{V}}(\hat{\hat{\mathbf{P}}}_{l,i}, \mathbf{u}_{l,i}), \tag{4.18}$$

where $\gamma_{\mathrm{V}} \in (0, 1]$ is the discount factor for the localization of AUV $l \in \mathcal{I}_{\mathrm{V}}$.

Then, the Bellman's equation is given as

$$\mathcal{V}_{\mathrm{V}}(\hat{\hat{\mathbf{P}}}_{l,k}) = g_{\mathrm{V}}(\hat{\hat{\mathbf{P}}}_{l,k}, \mathbf{u}_{l,k}) + \gamma_{\mathrm{V}}\mathcal{V}_{\mathrm{V}}(\hat{\hat{\mathbf{P}}}_{l,i,k+1}) \tag{4.19}$$

According to Bellman optimality equation, $\mathcal{V}_{\mathrm{V}}(\hat{\hat{\mathbf{P}}}_{l,k})$ in (4.19) is interpreted as the cost (or reward) function, and the optimal value of localization estimator is to select the increment (i.e., action) that minimizes the cost (or reward) function to obtain

$$\mathcal{V}_{\mathrm{V}}^*(\hat{\hat{\mathbf{P}}}_{l,k}) = \min_{\mathbf{u}_{l,k}}(g_{\mathrm{V}}(\hat{\hat{\mathbf{P}}}_{l,k}, \mathbf{u}_{l,k}) + \gamma_{\mathrm{V}}\mathcal{V}_{\mathrm{V}}^*(\hat{\hat{\mathbf{P}}}_{l,i,k+1})). \tag{4.20}$$

Accordingly, from (4.19) and (4.20), the optimal increment is given as

$$\mathbf{u}_{l,k}^{*} = \text{argmin}_{\mathbf{u}_{l,k}} (g_{\text{V}}(\hat{\bar{\mathbf{P}}}_{l,i,k}, \mathbf{u}_{l,k}) + \gamma_{\text{V}} \mathcal{V}_{\text{V}}^{*}(\hat{\bar{\mathbf{P}}}_{l,i,k+1})). \tag{4.21}$$

With regard to (4.19), the following value iteration algorithm is developed to seek the optimal value of $\mathbf{u}_{l,k}$.

1. **Initialization**. At the beginning, the utility and cost functions are set as $\mathbf{u}_{l,k,0} = 0$ and $\mathcal{V}_{\text{V},0}(\hat{\bar{\mathbf{P}}}_{l,k}) = 0$, respectively.
2. **Value Update**. For iteration step $s = 0, 1, \ldots,$ the cost function can be updated by using the following rule

$$\mathcal{V}_{\text{V},s+1}(\hat{\bar{\mathbf{P}}}_{l,k}) = \mathcal{H}(\hat{\bar{\mathbf{P}}}_{l,k}) + \mathbf{u}_{l,k,s}^{\text{T}} \mathbf{Q}_1 \mathbf{u}_{l,k,s} + \gamma_{\text{V}} \mathcal{V}_{\text{V},s}(\hat{\bar{\mathbf{P}}}_{l,k+1}). \tag{4.22}$$

3. **Policy Improvement.** An improved increment policy can be determined by using the following rule

$$\mathbf{u}_{l,k,s+1} = \text{argmin}_{\mathbf{u}_{l,k(\cdot)}} (g_{\text{V}}(\hat{\bar{\mathbf{P}}}_{l,k}, \mathbf{u}_{l,k,s}) + \gamma_{\text{V}} \mathcal{V}_{\text{V},s+1}(\hat{\bar{\mathbf{P}}}_{l,k+1})). \tag{4.23}$$

Note that the above process relies on the accurate solutions of $\mathcal{V}_{\text{V},s}$ and $\mathbf{u}_{l,k,s}$. To smoothly approximate $\mathcal{V}_{\text{V},s}$ and $\mathbf{u}_{l,k,s}$, two neural networks are employed here. In particular, the estimation functions can be defined as

$$\mathcal{V}_{\text{V}}(\hat{\bar{\mathbf{P}}}_{l,k}) = \mathbf{W}_{\text{V}}^{\text{T}} \mathbf{\Phi}_{\text{V}}(\hat{\bar{\mathbf{P}}}_{l,k}), \ \mathbf{u}_{l,k} = \mathbf{H}_{\text{V}}^{\text{T}} \boldsymbol{\sigma}_{\text{V}}(\hat{\bar{\mathbf{P}}}_{l,k}), \tag{4.24}$$

where $\mathbf{W}_{\text{V}}$ and $\mathbf{H}_{\text{V}}$ are the weight vectors for cost and increment functions, respectively. $\mathbf{\Phi}_{\text{V}}$ and $\boldsymbol{\sigma}_{\text{V}}$ are the basis functions of $\mathbf{W}_{\text{V}}$ and $\mathbf{H}_{\text{V}}$, respectively.

Accordingly, from (4.22) and (4.24), the value update becomes

$$\mathbf{W}_{\text{V},s+1}^{\text{T}} \mathbf{\Phi}_{\text{V}}(\hat{\bar{\mathbf{P}}}_{l,k}) = g_{\text{V}}(\hat{\bar{\mathbf{P}}}_{l,k}, \mathbf{u}_{l,k}) + \gamma_{\text{V}} \mathbf{W}_{\text{V},s}^{\text{T}} \mathbf{\Phi}_{\text{V}}(\hat{\bar{\mathbf{P}}}_{l,k+1}), \tag{4.25}$$

where $\mathbf{W}_{\text{V},s+1}^{\text{T}}$ can be solved by the least squares estimator.

Similarly, the weight $\mathbf{H}_{\text{V}}$ in (4.24) can be given as

$$\begin{aligned} \mathbf{H}_{\text{V},s+1}^{q_l+1} = \mathbf{H}_{\text{V},s+1}^{q_l} - \bar{\delta}_l \boldsymbol{\sigma}_{\text{V}}(\hat{\bar{\mathbf{P}}}_{l,k}) \{ 2\mathbf{Q}_1 (\mathbf{H}_{\text{V},s+1}^{q_i})^{\text{T}} \boldsymbol{\sigma}_{\text{V}}(\hat{\bar{\mathbf{P}}}_{l,k}) \\ + [\gamma_{\text{V}} \nabla \mathbf{\Phi}_{\text{V}}^{\text{T}}(\hat{\bar{\mathbf{P}}}_{l,k}) \mathbf{W}_{\text{V},s+1}] \}^{\text{T}}, \end{aligned} \tag{4.26}$$

where $\nabla \mathbf{\Phi}_{\text{V}}(\hat{\bar{\mathbf{P}}}_l) = \partial \mathbf{\Phi}_{\text{V}}(\hat{\bar{\mathbf{P}}}_l)/\partial \hat{\bar{\mathbf{P}}}_l$. $\bar{\delta}_l > 0$ and q_l denote tuning parameter and index for gradient descent method, respectively.

If $||\mathbf{W}_{\text{V},s+1} - \mathbf{W}_{\text{V},s}|| < \varrho_{l,1}$ and $||\mathbf{H}_{\text{V},s+1} - \mathbf{H}_{\text{V},s}|| < \varrho_{l,2}$ both hold, the neural iteration process is stopped, where $\varrho_{l,1}$ and $\varrho_{l,2}$ denote positive decimals.

Therefore, the optimal increment $\mathbf{u}_{l,k}^{*}$ can be acquired. Repeating the above process, the location of AUV $l \in \mathcal{I}_{\mathrm{V}}$ can be obtained from (4.16).

In the following, we present the localization algorithm for active sensor nodes. Denote $\hat{\hat{\mathbf{P}}}_{\mathrm{A},i} = [\hat{x}_{\mathrm{A},i}, \hat{y}_{\mathrm{A},i}]^{\mathrm{T}}$ as the estimated horizontal position vector of active sensor node i, where $\hat{x}_{\mathrm{A},i}$ and $\hat{y}_{\mathrm{A},i}$ denote the estimated positions on X and Y axes, respectively. Then, the following discrete-time dynamical system can be constructed to describe the transition procedure, i.e.,

$$\hat{\hat{\mathbf{P}}}_{\mathrm{A},i,k+1} = \hat{\hat{\mathbf{P}}}_{\mathrm{A},i,k} + \mathbf{u}_{\mathrm{A},k}, \tag{4.27}$$

where $\hat{\hat{\mathbf{P}}}_{\mathrm{A},i,k} \in \mathcal{R}^{n}$ denotes state vector of active sensor node i, and $\mathbf{u}_{\mathrm{A},k} \in \mathcal{R}^{m}$ denotes its increment (i.e., action) for localization.

The cost (reward) function for one-step cost for the localization estimator of active sensor node i can be defined as

$$g_{\mathrm{A}}(\hat{\hat{\mathbf{P}}}_{\mathrm{A},i,k}, \mathbf{u}_{\mathrm{A},k}) = \mathcal{H}_{1}(\hat{x}_{\mathrm{A},i,k}, \hat{y}_{\mathrm{A},i,k}) + \mathbf{u}_{\mathrm{A},k}^{\mathrm{T}}\mathbf{Q}_{2}\mathbf{u}_{\mathrm{A},k}. \tag{4.28}$$

where Q_2 denotes a positive definite matrix.

Therefore, the total cost (or reward) function can be expressed as

$$\mathcal{V}_{\mathrm{A}}(\hat{\hat{\mathbf{P}}}_{\mathrm{A},i,k}) = \sum\nolimits_{i=k}^{\infty} \gamma_{\mathrm{A}}^{i-k} g_{\mathrm{A}}(\hat{\hat{\mathbf{P}}}_{\mathrm{A},i,i}, \mathbf{u}_{\mathrm{A},i}), \tag{4.29}$$

where $\gamma_{\mathrm{A}} \in (0, 1]$ is the discount factor for the localization of active sensor node $i \in \mathcal{I}_{\mathrm{A}}$.

The Bellman's equation is given as

$$\mathcal{V}_{\mathrm{A}}(\hat{\hat{\mathbf{P}}}_{\mathrm{A},i,k}) = g_{\mathrm{A}}(\hat{\hat{\mathbf{P}}}_{\mathrm{A},i,k}, \mathbf{u}_{\mathrm{A},k}) + \gamma_{\mathrm{A}}\mathcal{V}_{\mathrm{A}}(\hat{\hat{\mathbf{P}}}_{\mathrm{A},i,k+1}), \tag{4.30}$$

and hence, by interpreting $\mathcal{V}_{\mathrm{V}}(\hat{\hat{\mathbf{P}}}_{l,k})$ as reward function, the aim of RL-based localization algorithm is to select $\mathbf{u}_{\mathrm{A},k}$ that minimizes $\mathcal{V}_{\mathrm{A}}(\hat{\hat{\mathbf{P}}}_{\mathrm{A},i,k})$ to obtain

$$\mathcal{V}_{\mathrm{A}}^{*}(\hat{\hat{\mathbf{P}}}_{\mathrm{A},i,k}) = \min_{\mathbf{u}_{\mathrm{A},k}}(g_{\mathrm{A}}(\hat{\hat{\mathbf{P}}}_{\mathrm{A},i,k}, \mathbf{u}_{\mathrm{A},k}) + \gamma_{\mathrm{A}}\mathcal{V}_{\mathrm{A}}^{*}(\hat{\hat{\mathbf{P}}}_{\mathrm{A},i,k+1})). \tag{4.31}$$

Accordingly, from (4.30) and (4.31), the optimal increment is given as

$$\mathbf{u}_{\mathrm{A},k}^{*} = \mathrm{argmin}_{\mathbf{u}_{\mathrm{A},k}}(g_{\mathrm{A}}(\hat{\hat{\mathbf{P}}}_{\mathrm{A},i,k}, \mathbf{u}_{\mathrm{A},k}) + \gamma_{\mathrm{A}}\mathcal{V}_{\mathrm{A}}^{*}(\hat{\hat{\mathbf{P}}}_{\mathrm{A},i,k+1})). \tag{4.32}$$

Therefore, Algorithm 2 is given to find the optimal increment $\mathbf{u}_{\mathrm{A},k}^{*}$, and the value iteration process is updated as follows.

1. **Initialization.** Utility and cost functions are set as $\mathbf{u}_{\mathrm{A},k,0} = 0$ and $\mathcal{V}_{\mathrm{A},0}(\hat{\hat{\mathbf{P}}}_{\mathrm{A},i,k}) = 0$, respectively.

Algorithm 2: Localization algorithm for active sensor i

Input: The position of AUV l, i.e., $\mathbf{P}_{V,l}$, and the timestamps collected by active sensor node i

Output: The optimal increment of localization estimator

1 $s \leftarrow 0$;
2 **repeat**
3   **for** $r = 1 : L$ **do**
4   Generate an increment $\mathbf{u}_{A,r}$ with (4.35);
5   Calculate $g_A(\hat{\bar{\mathbf{P}}}_{A,r}, \mathbf{u}_{A,s,r})+\gamma_A \mathbf{W}_{A,s} \mathbf{\Phi}_A(\hat{\bar{\mathbf{P}}}_{A,r+1})$;
6   Collect data $g_A(\hat{\bar{\mathbf{P}}}_{A,r}, \mathbf{u}_{A,r})$, $\hat{\bar{\mathbf{P}}}_{A,r}$, $\hat{\bar{\mathbf{P}}}_{A,r+1}$, $\mathbf{u}_{A,r}$;
7   Build data set $\mathbf{\Gamma}_L$ and update $\mathbf{W}_{A,s+1}$;
8   Acquire an increment $\mathbf{u}_{A,s+1,r}$;
9   $s \leftarrow s + 1$;
10 **until** $\|\mathbf{W}_{A,s+1} - \mathbf{W}_{A,s}\| < \bar{\varrho}_1$ *and* $\|\mathbf{H}_{A,s+1} - \mathbf{H}_{A,s}\| < \bar{\varrho}_2$
11 **return** $\mathbf{u}^*_{A,k}$;
12 Acquire the estimated horizontal position with (4.27)

2. **Value Update.** The cost is updated through

$$\begin{aligned} &\mathcal{V}_{A,s+1}(\hat{\bar{\mathbf{P}}}_{A,i,k}) \\ &= \mathcal{H}_1(\hat{\bar{\mathbf{P}}}_{A,i,k}) + \mathbf{u}^T_{A,k,s}\mathbf{Q}_2\mathbf{u}_{A,k,s} + \gamma_A \mathcal{V}_{A,s}(\hat{\bar{\mathbf{P}}}_{A,i,k+1}). \end{aligned} \tag{4.33}$$

3. **Policy Improvement.** An improved increment policy is determined with the following rule

$$\mathbf{u}_{A,k,s+1} = \text{argmin}_{\mathbf{u}_{A,k(\cdot)}}(g_A(\hat{\bar{\mathbf{P}}}_{A,i,k}, \mathbf{u}_{A,k,s}) + \gamma_A \mathcal{V}_{A,s+1}(\hat{\bar{\mathbf{P}}}_{A,i,k+1})). \tag{4.34}$$

To smoothly approximate $\mathcal{V}_{A,s}$ and $\mathbf{u}_{A,k,s}$, the following estimation functions is defined, i.e.,

$$\mathcal{V}_A(\hat{\bar{\mathbf{P}}}_{A,i,k}) = \mathbf{W}^T_A \mathbf{\Phi}_A(\hat{\bar{\mathbf{P}}}_{A,i,k}), \mathbf{u}_{A,k} = \mathbf{H}^T_A \boldsymbol{\sigma}_A(\hat{\bar{\mathbf{P}}}_{A,i,k}), \tag{4.35}$$

where $\mathbf{W}_A$ and $\mathbf{H}_A$ are the weight vectors for cost and increment functions, respectively. $\mathbf{\Phi}_A$ and $\boldsymbol{\sigma}_A$ are the basis functions of $\mathbf{W}_A$ and $\mathbf{H}_A$, respectively.

From (4.33) and (4.35), the value update can be rearranged as

$$\mathbf{W}^T_{A,s+1}\mathbf{\Phi}_A(\hat{\bar{\mathbf{P}}}_{A,i,k}) = g_A(\hat{\bar{\mathbf{P}}}_{A,i,k},\mathbf{u}_{A,k}) + \gamma_A \mathbf{W}^T_{A,s}\mathbf{\Phi}_A(\hat{\bar{\mathbf{P}}}_{A,i,k+1}), \tag{4.36}$$

where $\mathbf{W}_{A,s+1}$ is the value of $\mathbf{W}_A$ for the $(k+1)$-th iteration.

A least squares method is employed to calculate $\mathbf{W}_{A,s+1}$. Let $\Gamma_{A,N} = \{g(\hat{\bar{\mathbf{P}}}_{A,i,r}, \mathbf{u}_{A,r}), \hat{\bar{\mathbf{P}}}_{A,i,r}, \mathbf{u}_{A,r}, \hat{\bar{\mathbf{P}}}_{A,i,r+1}, r = 1, 2, ..., L\}$ be the measured data set with size L. Based on this, by using (4.33) and (4.36), the residual error ε_r can be defined as

$$\begin{aligned}\varepsilon_r(\hat{\hat{\mathbf{P}}}_{\mathrm{A},i}, \mathbf{u}_{\mathrm{A}}) = \mathbf{W}^{\mathrm{T}}_{\mathrm{A},s+1}\boldsymbol{\Phi}_{\mathrm{A}}(\hat{\hat{\mathbf{P}}}_{\mathrm{A},i,r}) - \gamma_{\mathrm{A}}\mathbf{W}^{\mathrm{T}}_{\mathrm{A},s}\boldsymbol{\Phi}_{\mathrm{A}}(\hat{\hat{\mathbf{P}}}_{\mathrm{A},i,r+1}) \\ - g(\hat{\hat{\mathbf{P}}}_{\mathrm{A},i,r}, \mathbf{u}_{\mathrm{A},s,r}).\end{aligned} \tag{4.37}$$

In view of (4.37), the iteration of least square method is to minimize the sum of residual errors, i.e.,

$$\min \sum\nolimits_{r=1}^{L} \varepsilon_r^2, \tag{4.38}$$

and hence, by using (4.36)–(4.38), the iterative least squares solution $\mathbf{W}_{\mathrm{A},s+1}$ can be given as

$$\mathbf{W}_{\mathrm{A},s+1} = [\boldsymbol{\ell}^{\mathrm{T}}\boldsymbol{\ell}]^{-1}\boldsymbol{\ell}^{\mathrm{T}}\boldsymbol{\eta}, \tag{4.39}$$

where $\boldsymbol{\ell} = [\boldsymbol{\Phi}_{\mathrm{A},1}(\hat{\hat{\mathbf{P}}}_{\mathrm{A},i}), \ldots, \boldsymbol{\Phi}_{\mathrm{A},L}(\hat{\hat{\mathbf{P}}}_{\mathrm{A},i})]^{\mathrm{T}}$. Especially, $\boldsymbol{\Phi}_{\mathrm{A},r}(\hat{\hat{\mathbf{P}}}_{\mathrm{A},i})$ is the value of $\boldsymbol{\Phi}_{\mathrm{A}}(\hat{\hat{\mathbf{P}}}_{\mathrm{A},i})$ for the r-th subset, and $\boldsymbol{\eta} = [\eta_1, \eta_2, ..., \eta_L]^{\mathrm{T}}$ is with $\eta_r = g(\hat{\hat{\mathbf{P}}}_{\mathrm{A},i,r}, \mathbf{u}_{\mathrm{A},s,r}) + \mathbf{W}_{\mathrm{A},s}\boldsymbol{\Phi}_{\mathrm{A}}(\hat{\hat{\mathbf{P}}}_{\mathrm{A},i,r+1})$ for $r = 1, 2, ..., L$.

The weight $\mathbf{H}_{\mathrm{A}}$ in (4.35) can be estimated by the gradient descent method. Therefore, we have

$$\begin{aligned}\mathbf{H}^{q_{\mathrm{A}}+1}_{\mathrm{A},s+1} &= \mathbf{H}^{q_{\mathrm{A}}}_{\mathrm{A},s+1} - \bar{\delta}_{\mathrm{A}}\frac{\partial(g_{\mathrm{A}}(\hat{\hat{\mathbf{P}}}_{\mathrm{A},i,k}, \mathbf{u}_{\mathrm{A},k}) + \gamma_{\mathrm{A}}\mathcal{V}_{\mathrm{A},s+1}(\hat{\hat{\mathbf{P}}}_{\mathrm{A},i,k+1}))}{\partial \mathbf{H}_{\mathrm{A}}} \\ &= \mathbf{H}^{q_{\mathrm{A}}}_{\mathrm{A},s+1} - \bar{\delta}_{\mathrm{A}}\boldsymbol{\sigma}_{\mathrm{A}}(\hat{\hat{\mathbf{P}}}_{\mathrm{A},i,k})\{2\mathbf{Q}_2(\mathbf{H}^{q_{\mathrm{A}}}_{\mathrm{A},s+1})^{\mathrm{T}}\boldsymbol{\sigma}_{\mathrm{A}}(\hat{\hat{\mathbf{P}}}_{\mathrm{A},i,k}) \\ &\quad + [\gamma_{\mathrm{A}}\nabla\boldsymbol{\Phi}^{\mathrm{T}}_{\mathrm{A}}(\hat{\hat{\mathbf{P}}}_{\mathrm{A},i,k+1})\mathbf{W}_{\mathrm{A},s+1}]\}^{\mathrm{T}},\end{aligned} \tag{4.40}$$

where $\nabla\boldsymbol{\Phi}_{\mathrm{A}}(\hat{\hat{\mathbf{P}}}_{\mathrm{A},i}) = \partial\boldsymbol{\Phi}_{\mathrm{A}}(\hat{\hat{\mathbf{P}}}_{\mathrm{A},i})/\partial\hat{\hat{\mathbf{P}}}_{\mathrm{A},i}$. Specifically, $\bar{\delta}_{\mathrm{A}} > 0$ and q_{A} denote the tuning parameter and the index for gradient descent method, respectively. When the conditions $||\mathbf{W}_{\mathrm{A},s+1} - \mathbf{W}_{\mathrm{A},s}|| < \bar{\varrho}_1$ and $||\mathbf{H}_{\mathrm{A},s+1} - \mathbf{H}_{\mathrm{A},s}|| < \bar{\varrho}_2$ both hold, the iteration process is ended, where $\bar{\varrho}_1$ and $\bar{\varrho}_2$ denote positive decimals. Then, the optimal action $\mathbf{u}^*_{\mathrm{A},k}$ can be acquired. Repeating the above process, the position of active sensor node i can be ultimately acquired from (4.27).

Finally, we provide the localization algorithm for passive sensor nodes. It is noted that, the unknown variables in (4.13) include the position and clock, which are more complex than the problems in (4.6) and (4.10). In view of this, the localization process for passive sensor node j consists of three major steps, i.e., *rough position estimation, rough clock estimation, and exact position update*, as described in Fig. 4.5. It is worth mentioning that, the RL-based optimization algorithm in each step is similar to the one in Algorithm 2, and hence it is omitted here. The above three steps are summarized as follows.

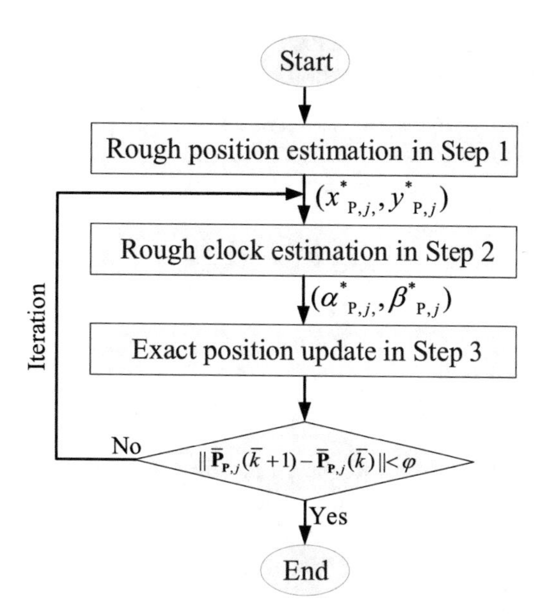

Fig. 4.5 Work flow of localization for passive sensor node j

Step 1. Rough Position Estimation
It is assumed that the clock is synchronous, i.e., $\alpha_{P,j} = 1$ and $\beta_{P,j} = 0$. Then, the optimization problem in (4.13) can be rearranged as

$$(x^*_{P,j}, y^*_{P,j}) = \text{argmin}_{(x_{P,j}, y_{P,j})} \mathcal{H}^1_2(\hat{x}_{P,j}, \hat{y}_{P,j}), \tag{4.41}$$

with

$$\mathcal{H}^1_2(\hat{x}_{P,j}, \hat{y}_{P,j}) = \frac{1}{2\sigma^2_{\text{mea}}} \sum\nolimits_{l=1}^{K} (T^{\text{P}}_{l,j} - ((T^*_{l,l} - \beta_{V,l})/\alpha_{V,l} + \tau^{\#}_{l,j}))^2. \tag{4.42}$$

Then, an RL-based localization algorithm is proposed to solve the problem in (4.41), through which a rough position vector $(x^*_{P,j}, y^*_{P,j})$ can be estimated.

Step 2. Rough Clock Estimation
With rough position $(x^*_{P,j}, y^*_{P,j})$, the optimization problem in (4.13) can be updated as

$$(\alpha^*_{P,j}, \beta^*_{P,j}) = \text{argmin}_{(\alpha_{P,j}, \beta_{P,j})} \mathcal{H}^2_2(\hat{\alpha}_{P,j}, \hat{\beta}_{P,j}), \tag{4.43}$$

with

$$\mathcal{H}^2_2(\hat{\alpha}_{P,j}, \hat{\beta}_{P,j}) = \frac{1}{2\sigma^2_{\text{mea}}} \sum\nolimits_{l=1}^{K} (T^{\text{P}}_{l,j} - \hat{\alpha}_{P,j}(\frac{T^*_{l,l} - \beta_{V,l}}{\alpha_{V,l}} + \tau^{\#}_{l,j}) - \hat{\beta}_{P,j})^2. \tag{4.44}$$

Then, a rough clock vector $(\alpha^*_{\mathrm{P},j}, \beta^*_{\mathrm{P},j})$ can be estimated via an RL-based localization algorithm.

Step 3. Exact Position Update
With the rough clock vector $(\alpha^*_{\mathrm{P},j}, \beta^*_{\mathrm{P},j})$ in Step 2, the following optimization problem can be carried out to seek the accurate position, i.e.,

$$(x^*_{\mathrm{P},j}, y^*_{\mathrm{P},j}) = \operatorname{argmin}_{(x_{\mathrm{P},j}, y_{\mathrm{P},j})} \mathcal{H}^3_2(\hat{x}_{\mathrm{P},j}, \hat{y}_{\mathrm{P},j}), \tag{4.45}$$

with

$$\mathcal{H}^3_2(\hat{x}_{\mathrm{P},j}, \hat{y}_{\mathrm{P},j}) = \frac{1}{2\sigma^2_{\mathrm{mea}}} \sum\nolimits_{l=1}^{K} (T^{\mathrm{P}}_{l,j} - \alpha^*_{\mathrm{P},j}(\frac{T^*_{l,l} - \beta_{\mathrm{V},l}}{\alpha_{\mathrm{V},l}} + \tau^{\#}_{l,j}) - \beta^*_{\mathrm{P},j})^2. \tag{4.46}$$

Thereby, an RL-based localization algorithm is developed to seek $(x^*_{\mathrm{P},j}, y^*_{\mathrm{P},j})$. In order to form a closed-loop update procedure, the estimated position vector $(x^*_{\mathrm{P},j}, y^*_{\mathrm{P},j})$ in Step 3 acts as input to Step 2 to replace the rough position vector in next round of iteration. The above iteration is ended when $\left\|\bar{\mathbf{P}}_{\mathrm{P},j}(\bar{k}+1) - \bar{\mathbf{P}}_{\mathrm{P},j}(\bar{k})\right\| < \varphi$ holds. Especially, $\bar{k}$ denotes the number of iterations and φ denotes the threshold.

4.3.3 Performance Analysis

(1) Cramér-Rao Lower Bound
For active sensor node i, the log-likelihood function is expressed as $\ln \Phi(\bar{\mathbf{P}}_{\mathrm{A},i})$. Based on this, the FIM can be constructed as

$$\begin{aligned} &\chi_{\mathrm{A}}(\bar{\mathbf{P}}_{\mathrm{A},i}) \\ &= \mathbf{E}\{[\nabla_{\bar{\mathbf{P}}_{\mathrm{A},i}} \ln \Phi(\bar{\mathbf{P}}_{\mathrm{A},i})][\nabla_{\bar{\mathbf{P}}_{\mathrm{A},i}} \ln \Phi(\bar{\mathbf{P}}_{\mathrm{A},i})]^{\mathrm{T}}\}|_{\bar{\mathbf{P}}_{\mathrm{A},i}=\tilde{\mathbf{P}}_{\mathrm{A},i}} \\ &= -\mathbf{E}\{\nabla_{\bar{\mathbf{P}}_{\mathrm{A},i}} \nabla^{\mathrm{T}}_{\bar{\mathbf{P}}_{\mathrm{A},i}} \ln \Phi(\bar{\mathbf{P}}_{\mathrm{A},i})\}|_{\bar{\mathbf{P}}_{\mathrm{A},i}=\tilde{\mathbf{P}}_{\mathrm{A},i}}, \end{aligned} \tag{4.47}$$

where $\tilde{\mathbf{P}}_{\mathrm{A},i}$ denotes ground truth and $\mathbf{E}\{\cdot\}$ denotes the expectation.

The log-likelihood function for active sensor i can be defined as

$$\begin{aligned} &\ln \Phi(\bar{\mathbf{P}}_{\mathrm{A},i}) \\ &= \frac{1}{6\sigma^2_{\mathrm{mea}}} \sum_{\bar{l}=2}^{K} \sum_{m=1}^{\bar{l}-1} [\Delta\Gamma_{m,\bar{l}} - \alpha_{\mathrm{V},\bar{l}}(\bar{\tau}_{m,\bar{l}} - \tau^*_{i,\bar{l}} + \tau^*_{i,m})]^2. \end{aligned} \tag{4.48}$$

With straightforward derivation of the horizontal position, the following results can be obtained from (4.48)

$$
\begin{aligned}
\chi_{\mathrm{A},1,1} &= \frac{1}{3\sigma_{\mathrm{mea}}^2}\sum_{\bar{l}=2}^{K}\sum_{m=1}^{\bar{l}-1}\left[-\alpha_{\mathrm{V},\bar{l}}\left(\frac{\partial\bar{\tau}_{m,\bar{l}}}{\partial x_{\mathrm{A},i}} - \frac{\partial\tau^*_{i,\bar{l}}}{\partial x_{\mathrm{A},i}} + \frac{\partial\tau^*_{i,m}}{\partial x_{\mathrm{A},i}}\right)\right]^2,\\
\chi_{\mathrm{A},2,2} &= \frac{1}{3\sigma_{\mathrm{mea}}^2}\sum_{\bar{l}=2}^{K}\sum_{m=1}^{\bar{l}-1}\left[-\alpha_{\mathrm{V},\bar{l}}\left(\frac{\partial\bar{\tau}_{m,\bar{l}}}{\partial y_{\mathrm{A},i}} - \frac{\partial\tau^*_{i,\bar{l}}}{\partial y_{\mathrm{A},i}} + \frac{\partial\tau^*_{i,m}}{\partial y_{\mathrm{A},i}}\right)\right]^2,\\
\chi_{\mathrm{A},1,2} = \chi_{\mathrm{A},2,1} &= \frac{1}{3\sigma_{\mathrm{mea}}^2}\sum_{\bar{l}=2}^{K}\sum_{m=1}^{\bar{l}-1}\left[\alpha_{\mathrm{V},\bar{l}}\left(\frac{\partial\bar{\tau}_{m,\bar{l}}}{\partial x_{\mathrm{A},i}} - \frac{\partial\tau^*_{i,\bar{l}}}{\partial x_{\mathrm{A},i}}\right.\right.\\
&\quad\left.\left. + \frac{\partial\tau^*_{i,m}}{\partial x_{\mathrm{A},i}}\right) \times \alpha_{\mathrm{V},\bar{l}}\left(\frac{\partial\bar{\tau}_{m,\bar{l}}}{\partial y_{\mathrm{A},i}} - \frac{\partial\tau^*_{i,\bar{l}}}{\partial y_{\mathrm{A},i}} + \frac{\partial\tau^*_{i,m}}{\partial y_{\mathrm{A},i}}\right)\right],
\end{aligned}
\tag{4.49}
$$

where $\chi_{\mathrm{A},i,j}$ denotes the (i, j)th element of χ_{A} for $i, j \in \{1, 2\}$.

As a result, the CRLB for $\bar{\mathbf{P}}_{\mathrm{A},i}$ is given as

$$
\begin{aligned}
\mathrm{CRLB}_{\mathrm{A}} &= [\chi_{\mathrm{A}}^{-1}(\bar{\mathbf{P}}_{\mathrm{A},i})]_{1,1} + [\chi_{\mathrm{A}}^{-1}(\bar{\mathbf{P}}_{\mathrm{A},i})]_{2,2}\\
&= \mathrm{tr}\{\chi_{\mathrm{A}}^{-1}(\bar{\mathbf{P}}_{\mathrm{A},i})\}|_{\bar{\mathbf{P}}_{\mathrm{A},i}=\tilde{\mathbf{P}}_{\mathrm{A},i}}.
\end{aligned}
\tag{4.50}
$$

With regard to (4.50), the localization error for optimization problem (4.10) satisfies the relationship of

$$
\mathbf{E}\{||\hat{\bar{\mathbf{P}}}_{\mathrm{A},i} - \tilde{\mathbf{P}}_{\mathrm{A},i}||^2\} \geq \mathrm{tr}\{\chi_{\mathrm{A}}^{-1}(\bar{\mathbf{P}}_{\mathrm{A},i})\}|_{\bar{\mathbf{P}}_{\mathrm{A},i}=\tilde{\mathbf{P}}_{\mathrm{A},i}}. \tag{4.51}
$$

In the following, we derive the CRLB for passive sensor node j. It is denoted that $\mathbf{w} = [x_{\mathrm{P},j}, y_{\mathrm{P},j}, \alpha_{\mathrm{P},j}, \beta_{\mathrm{P},j}]^{\mathrm{T}}$ and $\bar{\mathbf{h}}(\mathbf{w}) = [\alpha_{\mathrm{P},j}((T^*_{1,1} - \beta_{\mathrm{V},1})/\alpha_{\mathrm{V},1} + \tau^{\#}_{1,j}) + \beta_{\mathrm{P},j}, ..., \alpha_{\mathrm{P},j}\ ((T^*_{K,K} - \beta_{\mathrm{V},K})/\alpha_{\mathrm{V},j} + \tau^{\#}_{K,j}) + \beta_{\mathrm{P},j}]^{\mathrm{T}}$. Especially, we put the timestamps into a vector $\mathbf{T}$, i.e., $\mathbf{T} = [T^{\mathrm{P}}_{1,j}, ..., T^{\mathrm{P}}_{K,j}]^{\mathrm{T}}$, and $\mathbf{T}$ follows the distribution of $\mathbf{T} \sim \mathcal{N}(\bar{\mathbf{h}}(\mathbf{w}), \mathbf{R})$, where $\mathbf{R} = (\alpha_{\mathrm{P},j}\sigma_{\mathrm{mea}})^2\mathbf{I}$ and $\mathbf{I}$ is the identity matrix. Thus, $\mathcal{H}_2(\hat{x}_{\mathrm{P},j}, \hat{y}_{\mathrm{P},j}, \hat{\alpha}_{\mathrm{P},j}, \hat{\beta}_{\mathrm{P},j})$ in optimization problem (4.13) is given as

$$
\mathcal{H}_2(\mathbf{w}) = \frac{1}{2\sigma_{\mathrm{mea}}^2}||(\mathbf{T} - \bar{\mathbf{h}}(\mathbf{w})||^2. \tag{4.52}
$$

Referring to (4.47), the FIM for passive sensor node j is denoted as $\chi_{\mathrm{P}}(\mathbf{w})$, and its (p, q)th element is expressed as

$$
[\chi_{\mathrm{P}}(\mathbf{w})]_{p,q} = [\frac{\partial\bar{\mathbf{h}}(\mathbf{w})}{\partial[\mathbf{w}]_p}]^{\mathrm{T}}\mathbf{R}^{-1}[\frac{\partial\bar{\mathbf{h}}(\mathbf{w})}{\partial[\mathbf{w}]_q}] + \tfrac{1}{2}\mathrm{tr}[\mathbf{R}^{-1}\frac{\partial\mathbf{R}}{\partial[\mathbf{w}]_p}\mathbf{R}^{-1}\frac{\partial\mathbf{R}}{\partial[\mathbf{w}]_q}], \tag{4.53}
$$

where $[\cdot]_p$ denotes the pth element of a vector. $\frac{\partial \bar{\mathbf{h}}(\mathbf{w})}{\partial [\mathbf{w}]_p}$ is denoted as: $\frac{\partial \bar{\mathbf{h}}(\mathbf{w})}{\partial [\mathbf{w}]_p} = \alpha_{P,j} \frac{\partial \tau^{\#}_{l,j}}{\partial x_{P,j}}$ if $p = 1$; $\frac{\partial \bar{\mathbf{h}}(\mathbf{w})}{\partial [\mathbf{w}]_p} = \alpha_{P,j} \frac{\partial \tau^{\#}_{l,j}}{\partial y_{P,j}}$ if $p = 2$; $\frac{\partial \bar{\mathbf{h}}(\mathbf{w})}{\partial [\mathbf{w}]_p} = (T^{*}_{l,l} - \beta_{V,l})/\alpha_{V,l} + \tau^{\#}_{l,j}$ if $p = 3$; $\frac{\partial \bar{\mathbf{h}}(\mathbf{w})}{\partial [\mathbf{w}]_p} = 1$ if $p = 4$, where $l = 1, ..., K$. In addition, $\frac{\partial \mathbf{R}}{\partial [\mathbf{w}]_p}$ can be presented as: $\frac{\partial \mathbf{R}}{\partial [\mathbf{w}]_p} = (2\alpha_{P,j}\sigma^2_{mea})\mathbf{I}$ if $p = 3$; $\frac{\partial \mathbf{R}}{\partial [\mathbf{w}]_p} = 0$, otherwise.

In (4.53), the derivatives of propagation delay to horizontal position between AUV l and passive sensor node j are given as

$$\begin{aligned} \frac{\partial \tau^{\#}_{l,j}}{\partial x_{P,j}} &= -\frac{1}{a}\left(\frac{1}{\cos\theta^{*}_{l}}\frac{\partial \theta^{*}_{l}}{\partial r^{*}_{l,j}} - \frac{1}{\cos\theta^{P}_{l,j}}\frac{\partial \theta^{P}_{l,j}}{\partial r^{*}_{l,j}}\right)\frac{x_l - x_{P,j}}{r^{*}_{l,j}}, \\ \frac{\partial \tau^{\#}_{l,j}}{\partial y_{P,j}} &= -\frac{1}{a}\left(\frac{1}{\cos\theta^{*}_{l}}\frac{\partial \theta^{*}_{l}}{\partial r^{*}_{l,j}} - \frac{1}{\cos\theta^{P}_{l,j}}\frac{\partial \theta^{P}_{l,j}}{\partial r^{*}_{l,j}}\right)\frac{y_l - y_{P,j}}{r^{*}_{l,j}}, \end{aligned} \tag{4.54}$$

where $r^{*}_{l,j} = \sqrt{(x_l - x_{P,j})^2 + (y_l - y_{P,j})^2}$ denotes the horizontal distance between AUV l and passive sensor node j. In addition, $\theta^{*}_{l} = \phi_{l,j} - \psi_{l,j}$ and $\theta^{P}_{l,j} = \phi_{l,j} + \psi_{l,j}$ are the ray angles at AUV l and passive sensor node j, respectively. Especially, $\phi_{l,j} = \arctan \frac{z_l - z_{P,j}}{r^{*}_{l,j}}$ denotes the angle of the straight line between AUV l and passive sensor node j with regard to the horizontal axis. $\psi_{l,j} = \arctan \frac{0.5ar^{*}_{l,j}}{b+0.5a(z_l+z_{P,j})}$ denotes the angle at which the ray deviates from the straight line. z_l and $z_{P,j}$ represent locations of AUV l and passive sensor node j on Z axis, respectively.

The partial derivatives of the lay angles at AUV l and passive sensor node j can be calculated as follows

$$\begin{aligned} \frac{\partial \theta^{*}_{l}}{\partial r^{*}_{l,j}} &= -\frac{\kappa_1\kappa_2 \sin\theta^{P}_{l,j}(\sin\theta^{*}_{l} - \sin\theta^{P}_{l,j})^2}{[1-\cos(\theta^{*}_{l} - \theta^{P}_{l,j})](\kappa_1 \sin\theta^{P}_{l,j} + \sin\theta^{*}_{l})}, \\ \frac{\partial \theta^{P}_{l,j}}{\partial r^{*}_{l,j}} &= -\frac{\kappa_2(\sin\theta^{*}_{l} - \sin\theta^{P}_{l,j})^2 \sin\theta^{*}_{l}}{[1-\cos(\theta^{*}_{l} - \theta^{P}_{l,j})](\kappa_1 \sin\theta^{P}_{l,j} + \sin\theta^{*}_{l})}, \end{aligned} \tag{4.55}$$

where $\kappa_1 = (b + az_l)/(b + az_{P,j})$ and $\kappa_2 = (z_l - z_{P,j})/r^{*2}_{l,j}$.

As a result, the CRLB for $\bar{\mathbf{P}}_{P,j}$ can be expressed as

$$\mathrm{CRLB_P} = [\boldsymbol{\chi}^{-1}_{P}(\bar{\mathbf{P}}_{P,j})]_{1,1} + [\boldsymbol{\chi}^{-1}_{P}(\bar{\mathbf{P}}_{P,j})]_{2,2}, \tag{4.56}$$

where $\tilde{\mathbf{P}}_{P,j}$ is ground truth.

In view of (4.56), the localization error for optimization problem (4.13) satisfies the relationship of

$$\begin{aligned} &\mathbf{E}\{||\hat{\bar{\mathbf{P}}}_{\mathrm{P},j} - \bar{\mathbf{P}}_{\mathrm{P},j}||^2\} \geq [\chi_{\mathrm{P}}^{-1}(\mathbf{w})]_{1,1} + [\chi_{\mathrm{P}}^{-1}(\mathbf{w})]_{2,2}, \\ &\mathbf{E}\{||\hat{\alpha}_{\mathrm{P},j} - \alpha_{\mathrm{P},j}||^2\} \geq [\chi_{\mathrm{P}}^{-1}(\mathbf{w})]_{3,3}, \\ &\mathbf{E}\{||\hat{\beta}_{\mathrm{P},j} - \beta_{\mathrm{P},j}||^2\} \geq [\chi_{\mathrm{P}}^{-1}(\mathbf{w})]_{4,4}. \end{aligned} \tag{4.57}$$

(2) Convergence Analysis

Taking the active sensor nodes i as an example, the convergence analysis for RL-based localization algorithm is presented. Specifically, we aim to prove $\mathcal{V}_{\mathrm{A},s} \to \mathcal{V}_{\mathrm{A}}^*$ and $\mathbf{u}_{\mathrm{A},s} \to \mathbf{u}_{\mathrm{A}}^*$ as $s \to \infty$. Motivated by this, we denote $\boldsymbol{\mu}_s$ as an arbitrary increment of localization estimator. Then, we have

$$\Lambda_{s+1}(\hat{\bar{\mathbf{P}}}_{\mathrm{A},i,k}) = g_{\mathrm{A}}(\hat{\bar{\mathbf{P}}}_{\mathrm{A},i,k}, \boldsymbol{\mu}_{s,k}) + \gamma_{\mathrm{A}}\Lambda_s(\hat{\bar{\mathbf{P}}}_{\mathrm{A},i,k+1}). \tag{4.58}$$

Of note, $\mathcal{V}_{\mathrm{A},s}$ is minimized by $\mathbf{u}_{\mathrm{A},s,k}$ with respect to the increment $\mathbf{u}$. Afterwards, the inequality $\mathcal{V}_{\mathrm{A},s}(\hat{\bar{\mathbf{P}}}_{\mathrm{A},i,k}) \leq \Lambda_s(\hat{\bar{\mathbf{P}}}_{\mathrm{A},i,k})$ holds when $\mathcal{V}_{\mathrm{A},0}(\hat{\bar{\mathbf{P}}}_{\mathrm{A},i,k}) = \Lambda_0(\hat{\bar{\mathbf{P}}}_{\mathrm{A},i,k}) = 0$. Based on this, the following corollary can be provided.

Corollary 4.1 *For the position iteration (4.27) with cost function $\mathcal{V}_{\mathrm{A},s}$, there exists a least upper bound $Y(\hat{\bar{\mathbf{P}}}_{\mathrm{A},i,k})$, and it satisfies the condition of*

$$0 \leq \mathcal{V}_{\mathrm{A},s}(\hat{\bar{\mathbf{P}}}_{\mathrm{A},i,k}) \leq \mathcal{V}_{\mathrm{A}}^*(\hat{\bar{\mathbf{P}}}_{\mathrm{A},i,k}) \leq Y(\hat{\bar{\mathbf{P}}}_{\mathrm{A},i,k}). \tag{4.59}$$

Proof Denote $\boldsymbol{\varrho}_k$ as an arbitrary and admissible increment. In addition, set $\mathcal{V}_{\mathrm{A},0}(\hat{\bar{\mathbf{P}}}_{\mathrm{A},i,k}) = Z_0(\hat{\bar{\mathbf{P}}}_{\mathrm{A},i,k}) = 0$. Then, Z_s can be expressed as

$$Z_{s+1}(\hat{\bar{\mathbf{P}}}_{\mathrm{A},i,k}) = g_{\mathrm{A}}(\hat{\bar{\mathbf{P}}}_{\mathrm{A},i,k}, \boldsymbol{\varrho}_k) + \gamma_{\mathrm{A}} Z_s(\hat{\bar{\mathbf{P}}}_{\mathrm{A},i,k+1}). \tag{4.60}$$

By using (4.60) and adopting the induction, we have

$$\begin{aligned} &Z_{s+1}(\hat{\bar{\mathbf{P}}}_{\mathrm{A},i,k}) - Z_s(\hat{\bar{\mathbf{P}}}_{\mathrm{A},i,k}) \\ &= \gamma_{\mathrm{A}}(Z_s(\hat{\bar{\mathbf{P}}}_{\mathrm{A},i,k+1}) - Z_{s-1}(\hat{\bar{\mathbf{P}}}_{\mathrm{A},i,k+1})) \\ &= \gamma_{\mathrm{A}}^s(Z_1(\hat{\bar{\mathbf{P}}}_{\mathrm{A},i,k+s}) - Z_0(\hat{\bar{\mathbf{P}}}_{\mathrm{A},i,k+s})). \end{aligned} \tag{4.61}$$

It is noted that $Z_0(\hat{\bar{\mathbf{P}}}_{\mathrm{A},i,k}) = 0$, and thus, from (4.61), one has

$$\begin{aligned} Z_{s+1}(\hat{\bar{\mathbf{P}}}_{\mathrm{A},i,k}) &= \gamma_{\mathrm{A}}^s Z_1(\hat{\bar{\mathbf{P}}}_{\mathrm{A},i,k+s}) + Z_s(\hat{\bar{\mathbf{P}}}_{\mathrm{A},i,k}) \\ &= \textstyle\sum_{n=0}^{s} \gamma_{\mathrm{A}}^n Z_1(\hat{\bar{\mathbf{P}}}_{\mathrm{A},i,k+n}) \\ &= \textstyle\sum_{n=0}^{s} \gamma_{\mathrm{A}}^n g_{\mathrm{A}}(\hat{\bar{\mathbf{P}}}_{\mathrm{A},i,k+n}, \boldsymbol{\varrho}_{k+n}) \\ &\leq \textstyle\sum_{n=0}^{\infty} \gamma_{\mathrm{A}}^n g_{\mathrm{A}}(\hat{\bar{\mathbf{P}}}_{\mathrm{A},i,k+n}, \boldsymbol{\varrho}_{k+n}). \end{aligned} \tag{4.62}$$

Because ϱ_k is an admissible and stable increment, one has $\hat{\bar{\mathbf{P}}}_{A,i,k} \rightarrow \bar{\mathbf{P}}_{A,i,k}$ as $k \rightarrow \infty$, which yields

$$Z_{s+1}(\hat{\bar{\mathbf{P}}}_{A,i,k}) \leq \sum_{s=0}^{\infty} \gamma^s Z_1(\hat{\bar{\mathbf{P}}}_{A,i,k+s}) = Y(\hat{\bar{\mathbf{P}}}_{A,i,k}). \tag{4.63}$$

From (4.58), we define that $\boldsymbol{\mu}_{s,k} = \varrho_k$ and $\Lambda_s(\hat{\bar{\mathbf{P}}}_{A,i,k}) = Z_s(\hat{\bar{\mathbf{P}}}_{A,i,k})$. Based on the above definition, one can easily obtain $\mathcal{V}_{A,s}(\hat{\bar{\mathbf{P}}}_{A,i,k}) \leq Z_s(\hat{\bar{\mathbf{P}}}_{A,i,k}) \leq Y(\hat{\bar{\mathbf{P}}}_{A,i,k})$ for $\forall s$. To be specific, if $\varrho_k = \mathbf{u}^*_{A,k}$, one can further obtain $\sum_{n=0}^{\infty} \gamma_A^n g_A(\hat{\bar{\mathbf{P}}}_{A,i,k+n}, \mathbf{u}^*_{A,k+n}) \leq \sum_{n=0}^{\infty} \gamma_A^n g_A(\hat{\bar{\mathbf{P}}}_{A,i,k+n}, \varrho_{k+n})$, which means $0 \leq \mathcal{V}_{A,s}(\hat{\bar{\mathbf{P}}}_{A,i,k}) \leq \mathcal{V}_A^*(\hat{\bar{\mathbf{P}}}_{A,i,k}) \leq Y(\hat{\bar{\mathbf{P}}}_{A,i,k})$. That completes the proof. □

Based on Corollary 4.1, the following theorem can be provided.

Theorem 4.1 *Consider the position iteration (4.27) with $\mathcal{V}_{A,s}$ and $\mathbf{u}_{A,s}$ can be defined by (4.33) and (4.34), respectively. It is assumed that $\mathcal{V}_{A,0}(\hat{\bar{\mathbf{P}}}_{A,i,k}) = 0$, then one has $\mathcal{V}_{A,s+1}(\hat{\bar{\mathbf{P}}}_{A,i,k}) \geq \mathcal{V}_{A,s}(\hat{\bar{\mathbf{P}}}_{A,i,k})$, $\mathcal{V}_{A,s} \rightarrow \mathcal{V}_A^*$ and $\mathbf{u}_{A,s} \rightarrow \mathbf{u}_A^*$ as $s \rightarrow \infty$, i.e., $\mathcal{V}_{A,s}$ and $\mathbf{u}_{A,s}$ can converge to the optimal values.*

Proof In view of (4.58), one knows that $\Lambda_s(\hat{\bar{\mathbf{P}}}_{A,i,k})$ is the value function with an arbitrary sequence of increments $\boldsymbol{\mu}_s$. Thus, we set that $\mathcal{V}_{A,0}(\hat{\bar{\mathbf{P}}}_{A,i,k}) = \Lambda_0(\hat{\bar{\mathbf{P}}}_{A,i,k}) = 0$, then one has $\mathcal{V}_{A,s}(\hat{\bar{\mathbf{P}}}_{A,i,k}) \leq \Lambda_s(\hat{\bar{\mathbf{P}}}_{A,i,k})$. Besides that, it is assumed that $\boldsymbol{\mu}_{s,k} = \mathbf{u}_{A,s+1,k}$, then $\Lambda_{s+1}(\hat{\bar{\mathbf{P}}}_{A,i,k})$ can be denoted as

$$\begin{aligned}
&\Lambda_{s+1}(\hat{\bar{\mathbf{P}}}_{A,i,k}) \\
&= g_A(\hat{\bar{\mathbf{P}}}_{A,i,k}, \boldsymbol{\mu}_{s,k}) + \gamma_A \Lambda_s(\hat{\bar{\mathbf{P}}}_{A,i,k+1}, \boldsymbol{\mu}_{s,k}) \\
&= g_A(\hat{\bar{\mathbf{P}}}_{A,i,k}, \mathbf{u}_{A,s+1,k}) + \gamma_A \Lambda_s(\hat{\bar{\mathbf{P}}}_{A,i,k+1}, \mathbf{u}_{A,s+1,k}).
\end{aligned} \tag{4.64}$$

Because $\mathcal{V}_{A,0}(\hat{\bar{\mathbf{P}}}_{A,i,k}) = \Lambda_0(\hat{\bar{\mathbf{P}}}_{A,i,k}) = 0$, we employ the induction to update $\mathcal{V}_{A,s}(\hat{\bar{\mathbf{P}}}_{A,i,k})$ and $\Lambda_s(\hat{\bar{\mathbf{P}}}_{A,i,k})$. Therefore, the following relationship can be constructed, i.e.,

$$\begin{aligned}
&\mathcal{V}_{A,1}(\hat{\bar{\mathbf{P}}}_{A,i,k}) - \Lambda_0(\hat{\bar{\mathbf{P}}}_{A,i,k}) = \mathcal{H}_1(\hat{x}_{A,i,k}, \hat{y}_{A,i,k}) \geq 0, \\
&\mathcal{V}_{A,2}(\hat{\bar{\mathbf{P}}}_{A,i,k}) - \Lambda_1(\hat{\bar{\mathbf{P}}}_{A,i,k}) = \gamma_A \mathcal{V}_{A,1}(\hat{\bar{\mathbf{P}}}_{A,i,k+1}) \geq 0, \\
&\vdots \\
&\mathcal{V}_{A,s+1}(\hat{\bar{\mathbf{P}}}_{A,i,k}) - \Lambda_s(\hat{\bar{\mathbf{P}}}_{A,i,k}) = \gamma_A^s \mathcal{V}_{A,1}(\hat{\bar{\mathbf{P}}}_{A,i,k+s}) \geq 0.
\end{aligned} \tag{4.65}$$

From (4.65), it is obvious that $\Lambda_s(\hat{\hat{\mathbf{P}}}_{A,i,k}) \leq \mathcal{V}_{A,s+1}(\hat{\hat{\mathbf{P}}}_{A,i,k})$. In addition, according to (4.58), we have $\mathcal{V}_{A,s}(\hat{\hat{\mathbf{P}}}_{A,i,k}) \leq \Lambda_s(\hat{\hat{\mathbf{P}}}_{A,i,k})$. Based on this, one knows $\mathcal{V}_{A,s}(\hat{\hat{\mathbf{P}}}_{A,i,k}) \leq \mathcal{V}_{A,s+1}(\hat{\hat{\mathbf{P}}}_{A,i,k})$. With Corollary 4.1, we have $\mathcal{V}_{A,\infty}(\hat{\hat{\mathbf{P}}}_{A,i,k}) \leq \mathcal{V}_A^*(\hat{\hat{\mathbf{P}}}_{A,i,k})$. Besides that, for $s \to \infty$, one obtains

$$\mathcal{V}_{A,\infty}(\hat{\hat{\mathbf{P}}}_{A,i,k})-\gamma_A\mathcal{V}_{A,\infty}(\hat{\hat{\mathbf{P}}}_{A,i,k+1})=g_A(\hat{\hat{\mathbf{P}}}_{A,i,k}, \mathbf{u}_{A,\infty,k}). \tag{4.66}$$

According to (4.66), one knows $\mathcal{V}_{A,\infty}(\hat{\hat{\mathbf{P}}}_{A,i,k})$ is the value function for a stable and admissible increment $\mathbf{u}_{A,\infty,k} = \boldsymbol{\varrho}_k$. It is noted that, $\mathcal{V}_{A,s}(\hat{\hat{\mathbf{P}}}_{A,i,k})$ is a nondecreasing sequence and $\mathcal{V}_A^*(\hat{\hat{\mathbf{P}}}_{A,i,k}) \leq Y(\hat{\hat{\mathbf{P}}}_{A,i,k})$ from Corollary 4.1. Based on this, one has $\mathcal{V}_{A,\infty}(\hat{\hat{\mathbf{P}}}_{A,i,k}) = Y(\hat{\hat{\mathbf{P}}}_{A,i,k}) \geq \mathcal{V}_A^*(\hat{\hat{\mathbf{P}}}_{A,i,k})$. This yields the conclusion of $\mathcal{V}_A^*(\hat{\hat{\mathbf{P}}}_{A,i,k}) \leq \mathcal{V}_{A,\infty}(\hat{\hat{\mathbf{P}}}_{A,i,k}) \leq \mathcal{V}_A^*(\hat{\hat{\mathbf{P}}}_{A,i,k})$, i.e., $\mathcal{V}_{A,s} \to \mathcal{V}_A^*$ and $\mathbf{u}_{A,s} \to \mathbf{u}_A^*$ as $s \to \infty$. That completes the proof. □

(3) Complexity Analysis

In this section, the computational complexity of the proposed algorithm is analyzed. As demonstrated in Arasaratnam and Haykin (2008), an effective way to evaluate the computational complexity is to count the floating point operations (flops). Specifically, an addition, subtraction, multiplication, division, or square root operation in the real domain is computed by one flop. For the proposed localization solution to active sensor i, the related computation is decided by the calculations of $\mathbf{W}_{A,s+1}$ and $\mathbf{H}_{A,s+1}$, whose goals are to approximate the cost function $\boldsymbol{\Phi}_A(\hat{\hat{\mathbf{P}}}_{A,i,k})$ and the increment $\boldsymbol{\sigma}_A(\hat{\hat{\mathbf{P}}}_{A,i,k})$, respectively. Inspired by this, the basis functions for weights $\mathbf{W}_{A,s+1}$ and $\mathbf{H}_{A,s+1}$ in (4.35) can be rearranged as $\boldsymbol{\Phi}_A(\hat{\hat{\mathbf{P}}}_{A,i,k}) = [A_i, \hat{x}_{A,i,k}, \hat{y}_{A,i,k}, \hat{x}_{A,i,k}^2, \hat{y}_{A,i,k}^2, \hat{x}_{A,i,k}\hat{y}_{A,i,k}]^T$ and $\boldsymbol{\sigma}_A(\hat{\hat{\mathbf{P}}}_{A,i,k}) = [A_i, \hat{x}_{A,i,k}, \hat{y}_{A,i,k}]^T$, respectively. Besides that, A_i denotes an adjustable arbitrary constant, while $\hat{x}_{A,i,k}$ and $\hat{y}_{A,i,k}$ are the values of $\hat{x}_{A,i}$ and $\hat{y}_{A,i}$ for the k-th iteration, respectively. As a result, the computational complexity for the calculation of $\mathbf{W}_{A,s+1}$ can be calculated as

$$\begin{aligned}\Omega_1 = &\underbrace{3L}_{\text{cost of computing } \boldsymbol{\ell}} + \underbrace{112L + \tfrac{5}{2}LK(K-1)}_{\text{cost of computing } \boldsymbol{\eta}} \\ &+ \underbrace{72L - 36}_{\text{cost of computing } \boldsymbol{\ell}^T\boldsymbol{\ell}} + \underbrace{112L + \tfrac{5}{2}LK(K-1)}_{\text{cost of computing } (\boldsymbol{\ell}^T\boldsymbol{\ell})^{-1}} \\ &+ \underbrace{66L}_{\text{cost of computing } (\boldsymbol{\ell}^T\boldsymbol{\ell})^{-1}\boldsymbol{\ell}^T} + \underbrace{12L - 6}_{\text{cost of computing } (\boldsymbol{\ell}^T\boldsymbol{\ell})^{-1}\boldsymbol{\ell}^T\boldsymbol{\eta}}\end{aligned} \tag{4.67}$$

Similarly, the complexity for calculation of $\mathbf{H}_{\mathrm{A},s+1}$ is

$$
\begin{aligned}
\Omega_2 = & \underbrace{}_{\text{cost of } 2(\mathbf{H}^{q_\mathrm{A}}_{\mathrm{A},s+1})^{\mathrm{T}}\boldsymbol{\sigma}_\mathrm{A}(\hat{\bar{\mathbf{P}}}_{\mathrm{A},i,k})}^{12L} + \underbrace{}_{\text{cost of } \nabla\boldsymbol{\Phi}_\mathrm{A}(\hat{\bar{\mathbf{P}}}_{\mathrm{A},i})}^{12L} \\
& + \underbrace{}_{\text{cost of } \gamma_\mathrm{A}\nabla\boldsymbol{\Phi}^{\mathrm{T}}_\mathrm{A}(\hat{\bar{\mathbf{P}}}_{\mathrm{A},i,k+1})\mathbf{W}_{\mathrm{A},s+1}}^{24L} + \underbrace{}_{\text{cost of } \bar{\delta}_\mathrm{A}\boldsymbol{\sigma}_\mathrm{A}\{\Xi\}^{\mathrm{T}}}^{14L} \\
& + \underbrace{}_{\text{cost of } \mathbf{H}^{q_\mathrm{A}}_{\mathrm{A},s+1} - \bar{\delta}_\mathrm{A}\boldsymbol{\sigma}_\mathrm{A}\{\Xi\}^{\mathrm{T}}}^{6L},
\end{aligned}
\tag{4.68}
$$

where $\Xi = 2(\mathbf{H}^{q_\mathrm{A}}_{\mathrm{A},s+1})^{\mathrm{T}}\boldsymbol{\sigma}_\mathrm{A}(\hat{\bar{\mathbf{P}}}_{\mathrm{A},i,k}) + [\gamma_\mathrm{A}\nabla\boldsymbol{\Phi}^{\mathrm{T}}_\mathrm{A}(\hat{\bar{\mathbf{P}}}_{\mathrm{A},i,k+1})\mathbf{W}_{\mathrm{A},s+1}]$.

Similarly, the computational complexity for passive sensor node j can be calculated as $\Omega_3 = 658 + 4K + 634L$. Extend this result to the other active and passive sensor nodes, and therefore, the computational complexity for the total weight calculation can be computed as $\Omega = M(\Omega_1 + \Omega_2) + N\Omega_3$. Obviously, Ω is related to the number of active sensor nodes, the number of AUVs, and the number of passive sensor nodes.

4.4 Simulation and Experimental Results

4.4.1 Simulation Results

The simulation results are performed with MATLAB 2017b. Especially, three AUVs, two active sensor nodes and four passive sensor nodes are stationed in an area of 200 m×200 m×100 m. At the beginning, surface buoys obtain their position estimations and time references via GPS. When receiving localization request from active sensor node, AUVs enter into the monitoring area, and provide localization services for active and passive sensor nodes. Some parameters used in the simulation are listed in Table 4.1. Specially, the following current model (Liu et al. 2014; Beerens et al. 1994) is applied to describe the passive horizontal motion of underwater nodes, i.e.,

$$
\begin{aligned}
\mu_\mathrm{x} &= k_1\lambda v \sin(k_2 x)\cos(k_3 y) + k_1\lambda\cos(2k_1 t) + k_4, \\
\mu_\mathrm{y} &= -\lambda v \sin(k_3 x)\cos(k_2 y) + k_5,
\end{aligned}
\tag{4.69}
$$

where μ_x and μ_y are the velocities on X and Y axes, respectively. In addition, k_1, k_2, k_3, λ and v denote the variables that are related to the underwater environment factors. Besides that, k_4 and k_5 denotes random variables representing random factors in simulation.

Table 4.1 Parameters used in the simulation

Parameter	Value	Parameter	Value
a	0.017	b	1473
σ_{mea}	0.0003	υ	$\mathcal{N}(-1, -0.1)$
k_1	$\mathcal{N}(\pi, 0.1\pi)$	k_2	$\mathcal{N}(\pi, 0.1\pi)$
k_3	$\mathcal{N}(2\pi, 0.2\pi)$	k_4	$\mathcal{N}(1, 0.1)$
k_5	$\mathcal{N}(1, 0.1)$	λ	$\mathcal{N}(3, 0.3)$
γ_{V}	0.8	γ_{A}	0.8
$\varrho_{l,1}$	0.03	$\varrho_{l,2}$	10^{-5}
$\bar{\varrho}_1$	0.05	$\bar{\varrho}_2$	10^{-4}

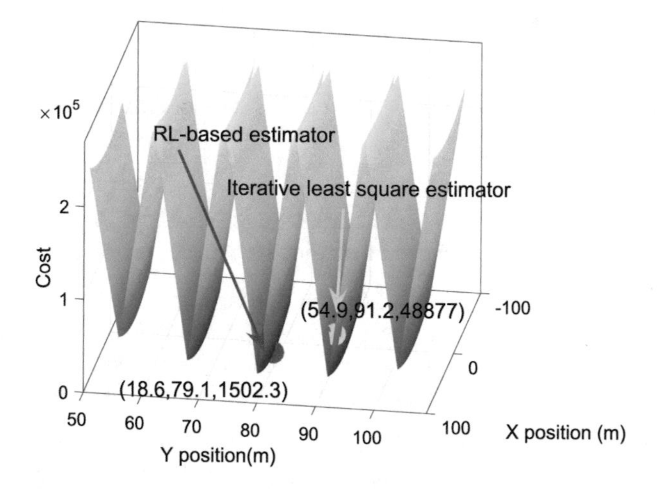

Fig. 4.6 Advantage of the RL-based localization strategy

(1) Advantage of RL-Based Localization Strategy

To start with, we verify the advantage of RL-based localization strategy. As mentioned above, the traditional least squares-based estimators, e.g., Liu et al. (2016), Zhang et al. (2017), and Mortazavi et al. (2017), can easily fall into local minimum, however the RL-based localization solution in this chapter can avoid this problem. To support this judgement, we assume that the localization procedure for active sensor node $i \in \mathcal{I}_{\text{A}}$ is disturbed by periodic noise, and thus the noise can be given as $\digamma_{\#} = 0.05\cos^2(0.25\hat{x}_{\text{A},i} - 0.025)$. Based on this, the localization optimization problem for active sensor node $i \in \mathcal{I}_{\text{A}}$ can be rearranged as $(x^*_{\text{A},i}, y^*_{\text{A},i}) = \text{argmin}_{(x_{\text{A},i}, y_{\text{A},i})}\left[\mathcal{H}_1(\hat{x}_{\text{A},i}, \hat{y}_{\text{A},i}) + \digamma_{\#}\right]$. In the following, we employ the least squares-based estimator to localize active sensor node $i \in \mathcal{I}_{\text{A}}$, where the initial estimation of $(x^*_{\text{A},i}, y^*_{\text{A},i})$ is set as (77, 12). By using least squares-based estimator, one obtains $(x^*_{\text{A},i}, y^*_{\text{A},i}) = (54.9, 91.2)$ with its optimization cost 48,877, as shown in Fig. 4.6. Meanwhile, RL strategy is also employed, through which one has $(x^*_{\text{A},i}, y^*_{\text{A},i}) = (18.6, 79.1)$ with its cost 1502.3. Clearly, the RL-based strategy can obtain the global minimum (i.e., $(x^*_{\text{A},i}, y^*_{\text{A},i}) = (18.6, 79.1)$), while the least squares-based strategy falls into local minimum (i.e., $(x^*_{\text{A},i}, y^*_{\text{A},i}) =$

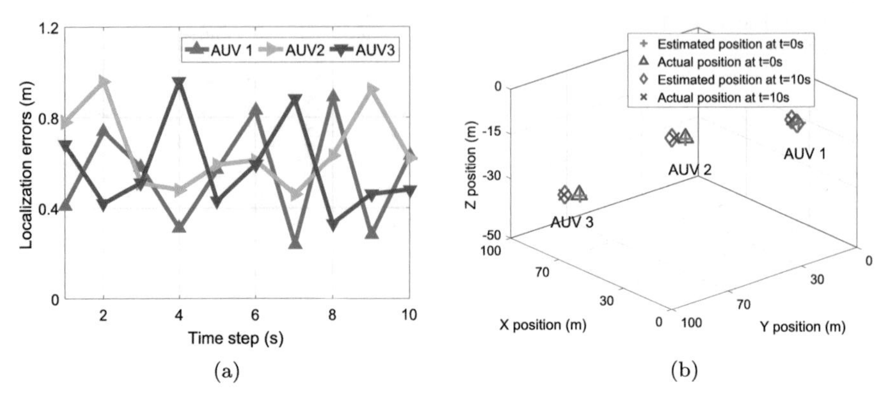

Fig. 4.7 Localization results for AUVs from 0 to 10 s. (**a**) Localization errors of AUVs. (**b**) Estimated positions of AUVs

(54.9, 91.2)). This result verifies the advantage of RL-based localization strategy, i.e., the global optimal solution can be obtained even with periodic noise.

(2) Localization of AUVs, Active and Passive Sensor Nodes

We verify the effectiveness of the proposed localization algorithm for AUVs. It is worth mentioning that, AUVs are equipped with depth units. With this assumption in hand, the localization error for AUV l can be defined as $e_{\mathrm{V},l} = \sqrt{(x_l^* - x_l)^2 + (y_l^* - y_l)^2}$ for $l \in \{1, 2, 3\}$. By using RL-based localization algorithm, the localization errors during $t \in [0,\ 10\ \mathrm{s}]$ are presented in Fig. 4.7a. Specially, the estimated positions of AUVs at $t = 0$ and $t = 10\,\mathrm{s}$ are shown by Fig. 4.7b. Clearly, the localization task for AUVs can be achieved, since the localization errors are in an acceptable range.

Next, we give the localization results for active and passive sensor nodes. The localization error for active sensor node i is defined as $e_{\mathrm{A},i} = \sqrt{(x_{\mathrm{A},i}^* - x_{\mathrm{A},i})^2 + (y_{\mathrm{A},i}^* - y_{\mathrm{A},i})^2}$ for $i \in \{1, 2\}$, while the localization error for passive sensor node j is $e_{\mathrm{P},j} = \sqrt{(x_{\mathrm{P},j}^* - x_{\mathrm{P},i})^2 + (y_{\mathrm{P},j}^* - y_{\mathrm{P},j})^2}$ for $j \in \{1, 2, 3, 4\}$. Based on this, the localization errors for active sensor nodes during $t \in [0,\ 10\ \mathrm{s}]$ are presented in Fig. 4.8a, while the estimated positions of active sensor nodes at $t = 0$ s and $t = 10$ s are shown by Fig. 4.8b. Similarly, the localization errors for passive sensor nodes during $t \in [0,\ 10\ \mathrm{s}]$ are presented in Fig. 4.8c, while the estimated positions of passive sensor nodes at $t = 0$ and $t = 10\,\mathrm{s}$ are shown by Fig. 4.8d. From Fig. 4.8a, d, one knows that the localization tasks for active and passive sensor nodes can be achieved. Meanwhile, the localization errors for sensor nodes are close to the CRLBs. From Fig. 4.8c, we also notice that the location errors of passive sensors 2, 3, and 4 at $t = 10$ s are lower than the errors at $t = 0\,\mathrm{s}$. The reason associated this phenomenon can be given as follows: due to the stochastic noises and the mobility of sensor nodes, the localization quality at every

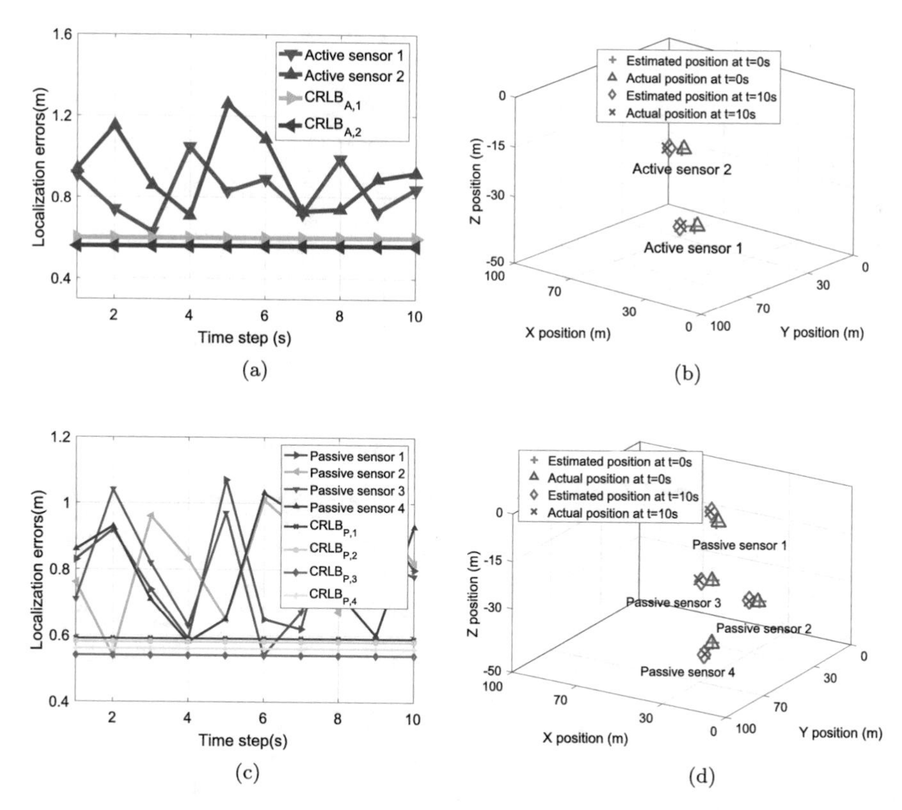

Fig. 4.8 Localization results for active and passive sensor nodes from 0 to 10 s. (**a**) Localization errors of active sensors. (**b**) Estimated positions of active sensors. (**c**) Localization errors of passive sensors. (**d**) Estimated positions of passive sensors

moment is dynamically changed, i.e., the change of location errors cannot present a convergence procedure.

(3) RL-Based Localization Process for a Single Node

In order to describe more clearly, we give the RL-based localization process for a single node. Taking AUV 1 as an example, the RL-based localization iteration process is presented in Fig. 4.9a. Clearly, the estimated value can approximate to the global optimum solution. Correspondingly, the weight vectors for value and increment functions, i.e., $\mathbf{W}_{\text{V}}$ and $\mathbf{H}_{\text{V}}$, are shown in Fig. 4.9b, c, respectively. It is clear that $\mathbf{W}_{\text{V}}$ and $\mathbf{H}_{\text{V}}$ can converge to the optimization, which reflect the effectiveness of RL-based localization algorithm for AUVs. For AUV $l \in \mathcal{I}_{\text{V}}$, the relationship between time clock and location can be represented as $\mathbf{B}_l = \mathbf{A}_l\mathbf{C}_l + \mathbf{W}_l$, where $\mathbf{B}_l = [T_{l,l}, T_{l,l}, T_{l,l}]^{\text{T}}$, $\mathbf{A}_l = [t_{l,1} - \tau_{l,1}, 1; t_{l,2} - \tau_{l,2}, 1; t_{l,3} - \tau_{l,3}, 1]$, $\mathbf{C}_l = [\alpha_{\text{V},l}, \beta_{\text{V},l}]^{\text{T}}$, and $\mathbf{W}_l$ denotes the measurement noise. By employing the least square method, the clock parameters of AUV l can be estimated as

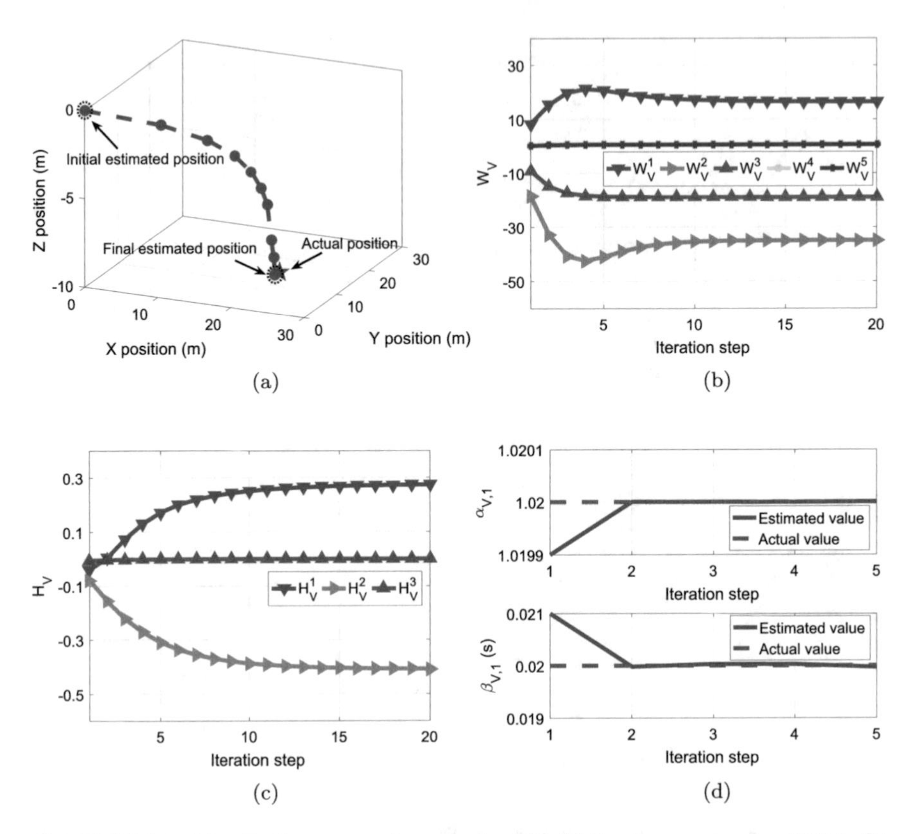

Fig. 4.9 RL-based localization process for a single AUV. (**a**) Iteration process for AUV 1. (**b**) Convergence of $\mathbf{W}_{\rm V}$. (**c**) Convergence of $\mathbf{H}_{\rm V}$. (**d**) Clock for AUV 1

$\mathbf{C}_l=(\mathbf{A}_l^{\rm T}\mathbf{A}_l)^{-1}\mathbf{A}_l^{\rm T}\mathbf{B}_l$. Based on this, the estimated clock information for AUV 1 is shown in Fig. 4.9d. It is clear that the estimated clock information can converge to the actual value, i.e., 1.02 and 0.02 s.

With respect to active sensor nodes, we give the localization iteration process for active sensor node 1, as shown in Fig. 4.10a. The weight vectors for cost and increment functions, i.e., $\mathbf{W}_{\rm A}$ and $\mathbf{H}_{\rm A}$, are presented in Fig. 4.10b, c, respectively. These results reflect the effectiveness of RL-based localization algorithm for active sensor nodes. Similar to AUVs, the time clocks of active sensor nodes can also be estimated through the position information. Inspired by this, the relationship between time clock and position for active sensor node i is expressed as $\mathbf{B}_{{\rm A},i} = \mathbf{A}_{{\rm A},i}\mathbf{C}_{{\rm A},i} + \mathbf{W}_{{\rm A},i}$, where $\mathbf{B}_{{\rm A},i} = [T_{i,0}^{\rm A}, T_{i,0}^{\rm A}, T_{i,0}^{\rm A}]^{\rm T}$, $\mathbf{A}_{{\rm A},i} = [(T_{i,1}^{\rm V} - \beta_{{\rm V},1})/\alpha_{{\rm V},1} - \tau_{i,1}^*, 1; (T_{i,2}^{\rm V} - \beta_{{\rm V},2})/\alpha_{{\rm V},2} - \tau_{i,2}^*, 1; (T_{i,3}^{\rm V} - \beta_{{\rm V},3})/\alpha_{{\rm V},3} - \tau_{i,3}^*, 1]$, $\mathbf{C}_{{\rm A},i} = [\alpha_{{\rm A},i}, \beta_{{\rm A},i}]^{\rm T}$, and $\mathbf{W}_{{\rm A},i}$ is the measurement noise. Then, the clock skew and offset for active sensor node $i \in \mathcal{I}_{\rm A}$ can be estimated as $\mathbf{C}_{{\rm A},i}=(\mathbf{A}_{{\rm A},i}^{\rm T}\mathbf{A}_{{\rm A},i})^{-1}\mathbf{A}_{{\rm A},i}^{\rm T}\mathbf{B}_{{\rm A},i}$. The estimated clock information for active sensor node 1 is shown in Fig. 4.10d. Clearly, the clock information is well estimated.

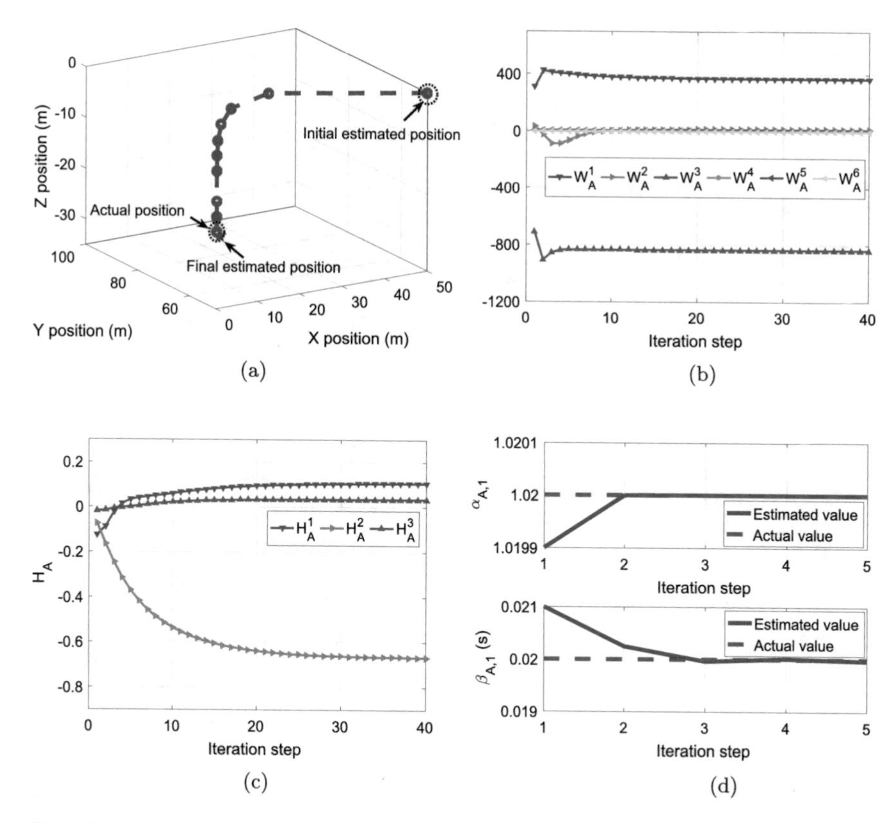

Fig. 4.10 RL-based localization process for a single active sensor node. (**a**) Iteration process for active sensor 1. (**b**) Convergence of $\mathbf{W}_{\mathrm{A}}$. (**c**) Convergence of $\mathbf{H}_{\mathrm{A}}$. (**d**) Clock for active sensor 1

Finally, the localization iteration process for passive sensor node 1 is presented (see Fig. 4.11a). The weight vectors for value and increment functions, i.e., $\mathbf{W}_{\mathrm{P}}$ and $\mathbf{H}_{\mathrm{P}}$, are described in Fig. 4.11b, c, respectively. Different from AUVs and active sensor nodes, the clock information of passive sensor nodes can be directly estimated by the RL-based localization algorithm. The estimated clock information for passive sensor 1 is represented in Fig. 4.11d. It is clear that the RL-based localization algorithm is also effective to the passive sensor nodes.

(4) Comparison with Asynchronous Localization Algorithm

A simplified clock model $T = t + \beta$ was considered in Carroll et al. (2014) and Yan et al. (2018), where the clock skew was ignored. As mentioned above, neglecting clock skew can reduce the localization accuracy. To judge this conclusion, the following two scenarios are considered for active sensor nodes: (1) Clock skew is ignored where $\alpha_{\mathrm{A},i} = 1$; (2) Clock skew is not ignored where $\alpha_{\mathrm{A},i} \neq 1$. Under Scenario 1, we employ the algorithm in Carroll et al. (2014) and Yan et al. (2018) to locate the active sensor nodes. Correspondingly, the localized positions

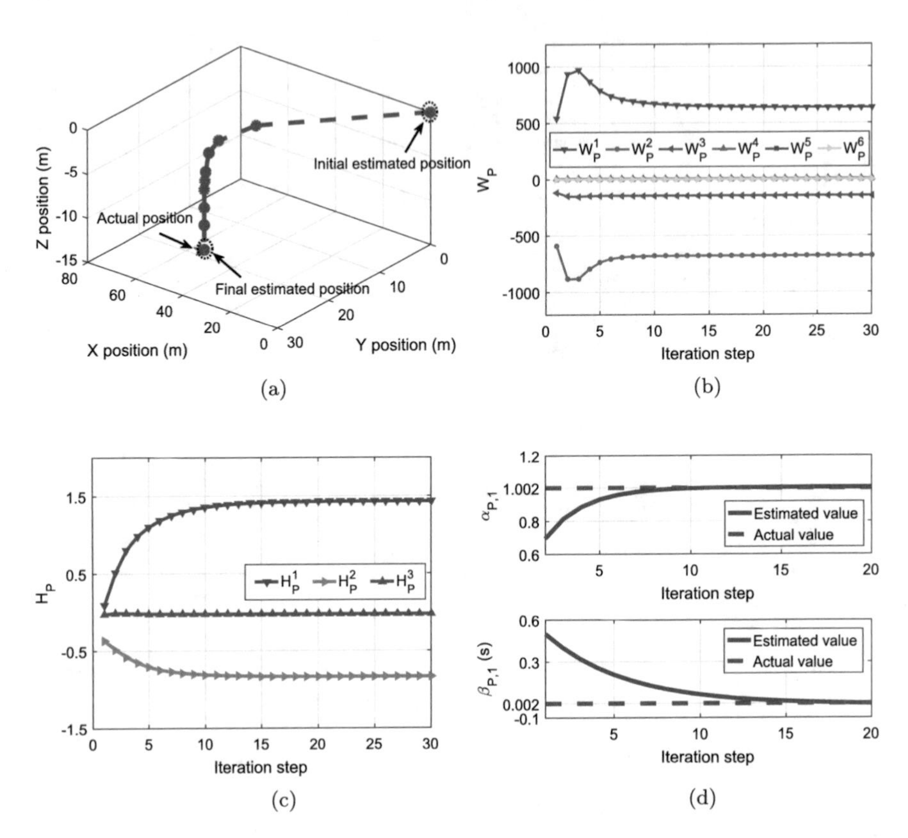

Fig. 4.11 RL-based localization process for a single passive sensor node. (**a**) Iteration process for passive sensor 1. (**b**) Convergence of $\mathbf{W}_{\mathrm{P}}$. (**c**) Convergence of $\mathbf{H}_{\mathrm{P}}$. (**d**) Clock for passive sensor 1

of active sensor nodes are presented in Fig. 4.12a, and the localization errors are shown in Fig. 4.12b. Clearly, the localization mission can be accomplished, since the localization results are within an acceptable range. Subsequently, we consider Scenario 2 where clock skew is set to be $\alpha_{\mathrm{A},i} = 1.002$. With the algorithm in Carroll et al. (2014) and Yan et al. (2018), the localized positions of active sensor nodes are presented in Fig. 4.12c, whose localization errors are given in Fig. 4.12d. Compared with the results in Fig. 4.10a, one knows the localization mission cannot be well accomplished by using the algorithm in Carroll et al. (2014) and Yan et al. (2018). By comparisons, we know the neglecting of clock skew can reduce the localization accuracy, while the clock model considered is meaningful.

In Gong et al. (2018), the clock skew and offset were both considered for the localization task, however the stratification effect was ignored. Of note, the ray bending can reduce the localization accuracy, and it is necessary to consider the stratification effect. To verify this judgement, we apply the algorithm in Gong et al. (2018) to locate active sensor nodes, and the acoustic speed is fixed as 1500 m/s. Then, the estimated positions and errors are shown in Fig. 4.12e, f, respectively.

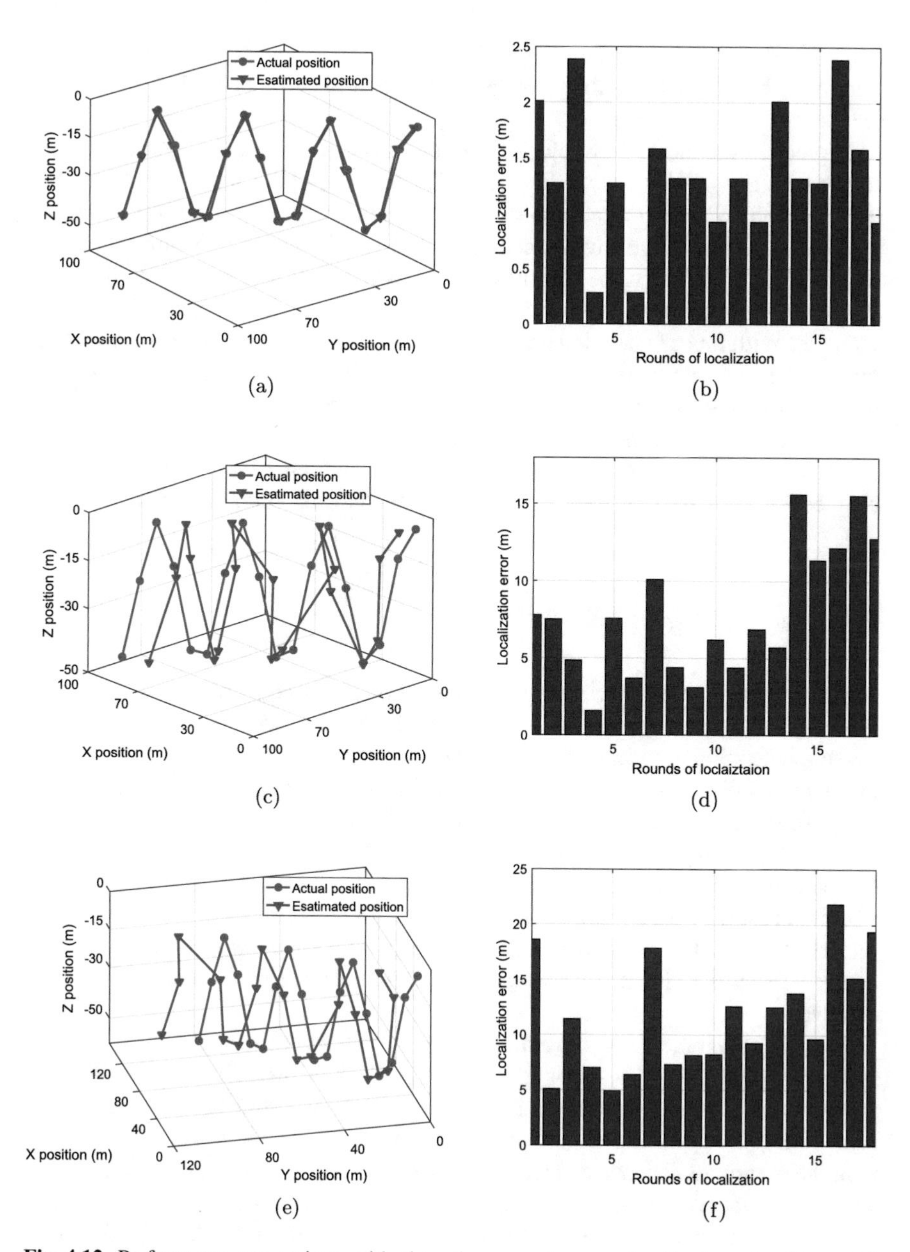

Fig. 4.12 Performance comparison with the other asynchronous localization algorithms. (**a**) Estimated positions without clock skew. (**b**) Localization errors without clock skew. (**c**) Estimated positions with clock skew. (**d**) Localization errors with clock skew. (**e**) Positions when ignoring stratification effect. (**f**) Errors when ignoring stratification effect

From Fig. 4.12e, f, one obtains that the localization accuracy is significantly reduced due to ray bending. Then, the consideration of stratification effect is necessary for USNs.

Moreover, the clock skew, clock offset and stratification effect were considered together in our previous work Yan et al. (2019b). However, the algorithm in Yan et al. (2019b) was performed in a static environment, whose localization accuracy can be poor for a mobile environment. For instance, the located positions for active sensor nodes by adopting the algorithm in Yan et al. (2019b) are shown in Fig. 4.13a. We believe that the static environment-suited algorithm can be extended to mobile scenario. With this problem in hand, we combine the motion compensation strategy proposed in this chapter (i.e., (4.12)) into the algorithm in Yan et al. (2019b), through which the located positions for active sensor nodes are shown in Fig. 4.13b, and the localization errors are shown in Fig. 4.13c. From Fig. 4.13b, c, the localization task can be accomplished, and it reflects that the motion compensation strategy in this chapter is helpful for the static environment-suited localization algorithm.

4.4.2 *Experimental Results*

Experiment results are performed to verify the RL-based localization algorithm. Due to the limited experimental conditions in our lab, we only verify the RL-based localization algorithm for AUVs. For this purpose, the experimental setup is presented in Fig. 4.14a, and the structure is depicted in Fig. 4.14b. To be specific, the hardwares are mainly composed of the following three parts: (1) *Surface Buoys*. Three surface buoys are deployed, whose locations and time references can be acquired through the ultra-wideband (UWB) system. (2) *AUV*. An AUV is recruited on the water bottom, and it can initiate the localization procedure through underwater acoustic communication with surface buoys. (3) *Control Center*. Control center acts as the signal processing unit, through which the location of AUV is estimated.

It is noted that the AUV can autonomously move. For this reason, the horizontal velocity vector of AUV is set as $\mathbf{v}_1 = [0.2, 0]^{\mathrm{T}}$. With the asynchronous localization protocol, the distance measurements between AUV and surface buoys are represented in Fig. 4.15a. The experimental data is shown in Table 4.2. It is obvious that, the measured distance can well capture the actual value. Based on an arbitrary group of distance measurements, the estimated positions of AUV at $t = 0$ s and $t = 6$ s are shown in Fig. 4.15b. It is worth mentioning that, AUV is deployed on the water bottom, thus the localization error on Z axis can be omitted. Based on this, the localization errors for AUV on X and Y axes are shown in Fig. 4.15c. From Fig. 4.15b–c, we know the localization task for an AUV can be accomplished, because the localization errors asymptotically converge to zero.

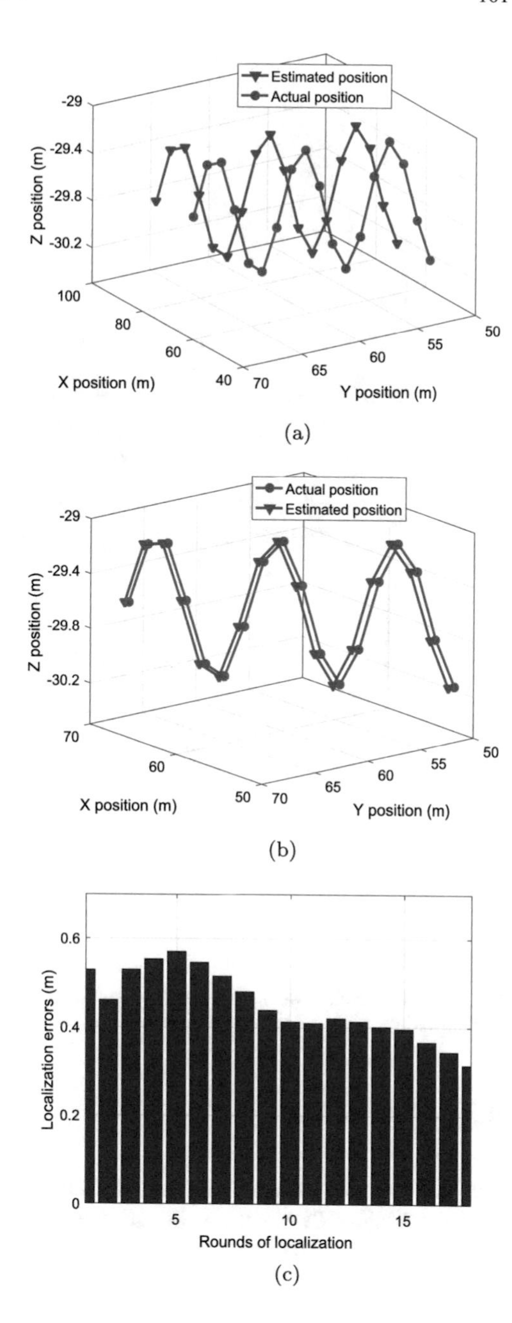

Fig. 4.13 Performance comparison with the other asynchronous localization algorithms. (**a**) Positions with Yan et al. (2019b) for mobile environment. (**b**) Positions with Yan et al. (2019b) and motion compensation. (**c**) Errors in Yan et al. (2019b) with motion compensation

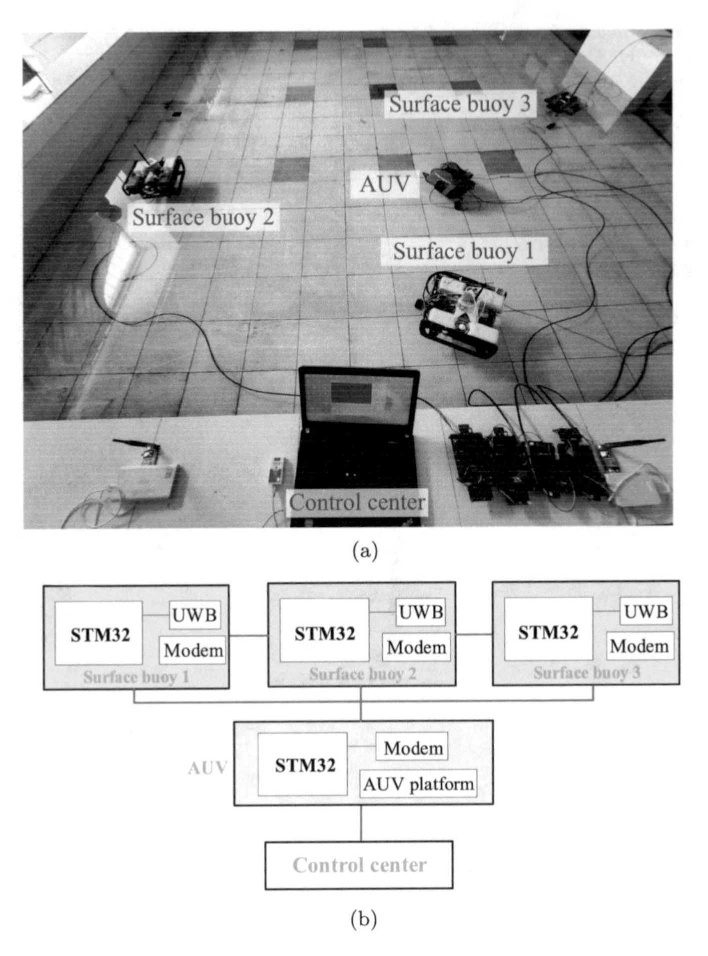

Fig. 4.14 Experiment setup for the RL-based localization algorithm. (**a**) Experimental setup in our lab. (**b**) Structure of the experiment

4.5 Conclusion

An AUV-aided asynchronous localization protocol is designed to realize localization task for USNs. To relax the linearization requirement, the localization problem is regarded as an RL problem. As a result, an RL-based localization algorithm is presented to acquired the positions of AUVs, active and passive sensor nodes. Specially, the ray compensation and mobility compensation strategies are both employed to improve the localization accuracy. Finally, simulation and experimental results are provided to verify the effectiveness of the RL-based localization method.

Fig. 4.15 Experimental results for RL-based localization algorithm. (**a**) Distance measurements between AUV and buoys. (**b**) Estimated position for AUV. (**c**) Localization errors for AUV

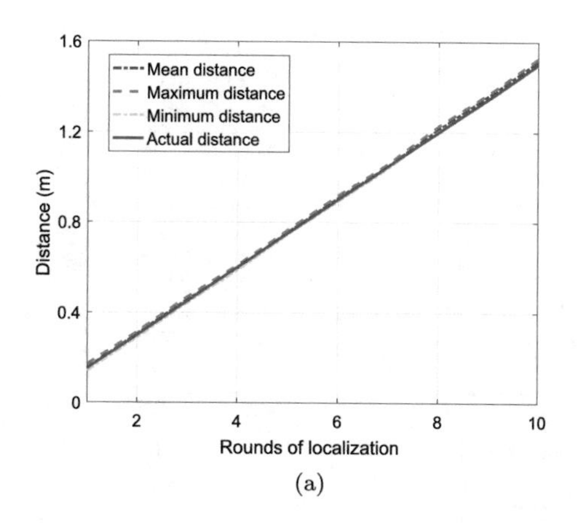

(a)

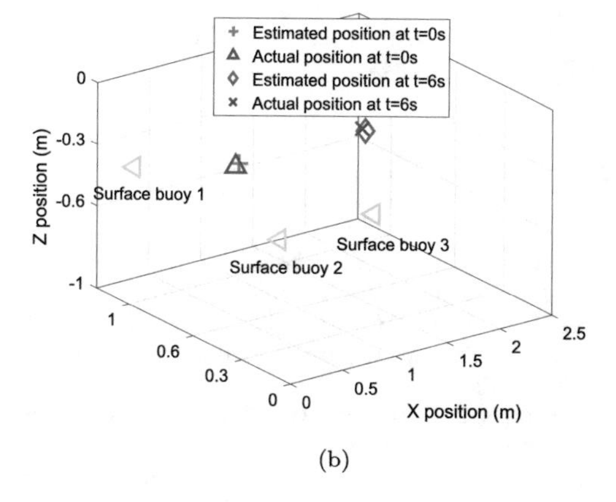

(b)

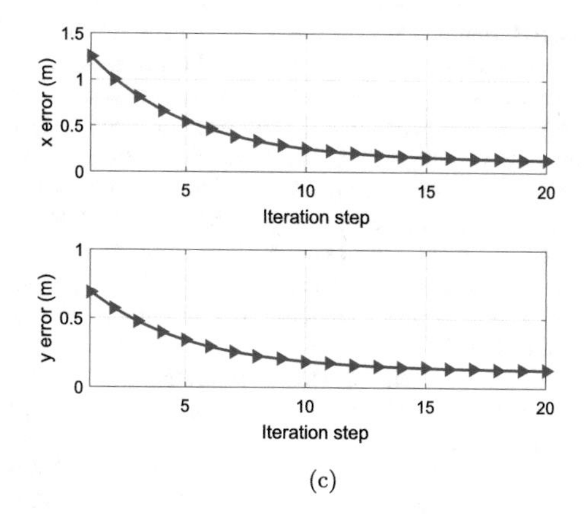

(c)

Table 4.2 Relative time difference measurements for different points (us)

	Point 1	Point 2	Point 3	Point 4	Point 5	Point 6	Point 7	Point 8	Point 9
Data 1	117	212	323	414	518	629	719	840	930
Data 2	99	213	314	405	514	629	732	832	937
Data 3	112	208	320	414	518	625	723	834	934
Data 4	108	200	322	420	514	624	719	830	947
Data 5	99	209	315	417	523	616	728	838	945
Data 6	95	213	305	405	521	625	724	847	945
Data 7	108	217	318	405	514	616	728	845	950
Data 8	112	213	314	407	527	633	732	839	941
Data 9	99	213	307	409	526	629	728	840	943
Data 10	99	209	309	413	527	627	719	830	938
Min	95	200	305	405	514	616	719	830	930
Max	117	217	323	420	527	633	732	847	947
Mean	104	210	314	410	520	625	725	837	941
Actual	103	216	310	413	516	620	723	836	940

References

Akyildiz I, Pompili D, Melodia T (2004) Challenges for efficient communication in underwater acoustic sensor networks. ACM Sigbed Rev 1(2):3–8

Aras M, Abdullah S, Shafei S, Rashid M (2012) Investigation and evaluation of low cost depth sensor system using pressure sensor for unmanned underwater vehicle. Majlesi J Electr Eng 6(2):1–12

Arasaratnam I, Haykin S (2008) Square-root quadrature Kalman filtering. IEEE Trans Signal Process 56(6):2589–2593

Beerens S, Ridderinkhof H, Zimmerman J (1994) An analytical study of chaotic stirring in tidal areas. Chaos Solitons Fractals 4(6):1011–1029

Carroll P, Mahmood K, Zhou S, Zhou H, Xu X, Cui J (2014) On-demand asynchronous localization for underwater sensor networks. IEEE Trans Signal Process 62(13):3337–3348

Diamant R, Lampe L (2013) Underwater localization with time-synchronization and propagation speed uncertainties. IEEE Trans Mob Comput 12(7):100–105

Gong Z, Li C, Jiang F (2018) AUV-aided joint localization and time synchronization for underwater acoustic sensor networks. IEEE Signal Process Lett 25(4):477–481

Hong H, Wang J, Zhou X, Xiang T, Zhang Y, Wu H, Wang Y (2017) High-accuracy positioning for indoor wireless sensor networks. In: Proc. IEEE conf commun. softw. networks, pp 311–316

Hong H, Guo S, Gui G, Yang Z, Zhang J, Sari H, Adachi F (2019) Deep learning for physical-layer 5G wireless techniques: opportunities, challenges and solutions. IEEE Wirel Commun 27(1):214–222

Houegnigan L, Safari P, Nadeu C, Schaar M, Andre M (2017) A novel approach to real-time range estimation of underwater acoustic sources using supervised machine learning. In: Proc IEEE conf oceans, pp 1–5

Isbitiren G, Akan O (2011) Three-dimensional underwater target tracking with acoustic sensor networks. IEEE Trans Veh Technol 60(8):3897–3906

Jia T, Ho K, Wang H, Shen X (2019) Effect of sensor motion on time delay and Doppler shift localization: analysis and solution. IEEE Trans Signal Process 67(22):5881–5895

Jorgensen E, Fossen T, Bryne T, Schjoberg I (2019) Underwater position and attitude estimation using acoustic, inertial, and depth measurements. IEEE J Oceanic Eng. https://doi.org/10.1109/JOE.2019.2933883

Kiumarsi B, Vamvoudakis K, Modares H, Lewis F (2018) Optimal and autonomous control using reinforcement learning: a survey. IEEE Trans Neural Netw Learn Syst 29(6):2042–2062

Lewis F, Vrabie D (2009) Reinforcement learning and adaptive dynamic programming for feedback control. IEEE Circuits Syst Mag 9(3):32–50

Liu J, Wang Z, Peng Z, Cui J, Fiondella L (2014) Suave: swarm underwater autonomous vehicle localization. In: Proc. IEEE conf. comput. commun., pp 64–72

Liu J, Wang Z, Cui J, Zhou S, Yang B (2016) A joint time synchronization and localization design for mobile underwater sensor networks. IEEE Trans Mobile Comput 15(3):530–543

Liu Y, Fang G, Chen H, Xie L, Fan R, Su X (2018) Error analysis of a distributed node positioning algorithm in underwater acoustic sensor networks. In: Proc. wirel. commun. signal proc., pp 1–6

Mortazavi E, Javidan R, Dehghani M, Kavoosi V (2017) A robust method for underwater wireless sensor joint localization and synchronization. Ocean Eng 137(1):276–286

Parras J, Zazo S, Alvarez I, Gonzalez J (2019) Model free localization with deep neural architectures by means of an underwater WSN. Sensors 19(16):1–16

Phoemphon S, So-In C, Leelathakul N (2018) Fuzzy weighted centroid localization with virtual node approximation in wireless sensor networks. IEEE Internet Things J 5(6):4728–4752

Ramezani H, Jamali-Rad H, Leus G (2013) Target localization and tracking for an isogradient sound speed profile. IEEE Trans Signal Process 61(1):1434–1446

Rauchenstein L, Vishnu A, Li X, Deng Z (2018) Improving underwater localization accuracy with machine learning. Rev Sci Instrum 89(7):1–12

Shi X, Wu J (2018) To hide private position information in localization using time difference of arrival. IEEE Trans Signal Process 66(18):4946–4956

Soares C, Gomes J, Ferreira B, Costeira J (2017) LocDyn: robust distributed localization for mobile underwater networks. IEEE J Oceanic Eng 42(4):1063–1074

Sorbelli F, Pinotti C, Ravelomanana V (2019) Range-free localization algorithm using a customary drone: towards a realistic scenario. Pervasive Mob Comput 54(1):1–15

Torrieri DJ (1984) Statistical theory of passive location systems. IEEE Trans Aerosp Electron Syst AES-20(2):183–198

Wen G, Chen P, Ge S, Yang H, Liu X (2019) Optimized adaptive nonlinear tracking control using actor-critic reinforcement learning strategy. IEEE Trans Ind Inf 15(9):4969–4977

Yan J, Zhang X, Luo X, Wang Y, Chen C, Guan X (2018) Asynchronous localization with mobility prediction for underwater acoustic sensor networks. IEEE Trans Veh Technol 67(3):2543–2556

Yan J, Zhao H, Wang Y, Luo X, Guan X (2019) Asynchronous localization for UASNs: an unscented transform-based method. IEEE Signal Process Lett 26(4):602–606

Yan J, Zhao H, Luo X, Wang Y, Chen C, Guan X (2019b) Asynchronous localization of underwater target using consensus-based unscented Kalman filtering. IEEE J Ocean Eng. https://doi.org/10.1109/JOE.2019.2923826

Zhang B, Wang H, Zheng L, Wu J, Zhuang Z (2017) Joint synchronization and localization for underwater sensor networks considering stratification effect. IEEE Access 5(1):26932–26943

Zhou Z, Peng Z, Cui J, Shi Z, Bagtzoglou A (2011) Scalable localization with mobility prediction for underwater sensor networks. IEEE Trans Mob Comput 10(3):335–348

Chapter 5
Privacy Preserving Asynchronous Localization of USNs

Abstract Under the constraint of the asynchronous clock, security attack and node mobility, a privacy-preserving asynchronous localization issue is studied. To be specific, an asynchronous localization protocol is developed, and then two privacy-preserving localization algorithms are proposed to locate the position of active and ordinary sensor nodes. Note that the proposed localization algorithms reveal disguised positions to the network, while they do not adopt any homomorphic encryption technique. Besides that, the performance analyses of the proposed algorithms are also given. Finally, simulation and experiment results show that the proposed localization algorithms can avoid the leakage of location information, while the localization accuracy can be significantly enhanced by comparing with the other works.

Keywords Localization · Mobility · Asynchronous clock · Privacy preservation · Underwater sensor networks (USNs)

5.1 Introduction

To achieve effective localization for USNs, some asynchronous localization algorithms (Carroll et al. 2014; Yan et al. 2018; Tsai et al. 2017; Alexandri et al. 2020; Haddad et al. 2016; Diamant and Lampe 2013) and joint localization and synchronization algorithms (Liu et al. 2016; Mortazavi et al. 2017; Zhang et al. 2017; Yan et al. 2019a,b) have been proposed. These algorithms mainly involve the following process. (1) *Anchor discovery*: several non-collinear anchor nodes are deployed in the network to provide localization reference for sensor nodes; (2) *Distance measurement*: the distance measurements are derived by multiplying the time (or time difference) with the transmission speed; (3) *Location estimation*: with the positions of anchor nodes and the measured distances, optimal or suboptimal estimators are designed to calculate the locations of sensor nodes. In such a process, the issue of the privacy preservation is not taken into consideration, because of the positions of anchor nodes are directly revealed to the networks. Nevertheless, the

J. Yan et al., *Localization in Underwater Sensor Networks*, Wireless Networks,
https://doi.org/10.1007/978-981-16-4831-1_5

USNs are usually deployed in harsh or even insecure environment. Therefore, the security threats cannot be avoided.

Ignoring the impact of privacy preservation can lead to privacy leakage or even failure of the localization (Li et al. 2015). For instance, malicious nodes are easy to attack the anchor nodes and destroy the whole localization system if they obtain the locations of anchor nodes. Inspired by this, many privacy-preserving localization schemes have been proposed. For example, the encryption technique was applied in Konstantinidis et al. (2015) and Shu et al. (2015), through which privacy-preserving approaches were designed for indoor localization. Other encryption technique-based privacy preservation were proposed in Jolfaei et al. (2016), Ostovari et al. (2016), and Aiash et al. (2015). Although the encryption technique can provide strong privacy preservation performance, its communication and computational overheads are high. Therefore, the encryption technique is not applicable to underwater localization (Li et al. 2015), due to the limited bandwidth and energy of USNs. To handle this issue, some researchers try to use the signal processing solutions to preserve privacy data, whose main idea is to add noises to the privacy data. For instance, the privacy preserving summation (PPS) strategy was adopted to hide private location information of anchor nodes in Shi and Wu (2018) and Wang et al. (2018a). In Wang et al. (2018b), a privacy preserving mechanism was developed for the position estimation, and a differential privacy based privacy-preserving indoor localization scheme was presented in Wang et al. (2018c). However, these localization schemes depend on the assumption of synchronous clock. Besides that, the mobility characteristic of nodes was not considered. As mentioned above, the clocks in USNs are always asynchronous, wherein the nodes often have passive motions. Ignoring the above characteristics can increase ranging errors and reduce localization accuracy. Per knowledge of the authors, how to design a localization algorithm that jointly considers the asynchronous clock, mobility and privacy preservation is not well studied.

In this chapter, a privacy preserving solution for the asynchronous localization of USNs is developed, which is ignored in the above chapters. Firstly, a hybrid network architecture including four types of nodes is proposed. In order to eliminate the influence of asynchronous clock and mobility, an asynchronous localization protocol is proposed, such that PPS and privacy-preserving diagonal product (PPDP) based localization algorithms are developed to hide privacy information. Main contributions lie in two aspects:

1. **Asynchronous localization protocol.** An asynchronous localization protocol is designed to eliminate the impact of asynchronous clock and mobility, through which the relationship between propagation delay and position is established. Different from the existing works (Konstantinidis et al. 2015; Shu et al. 2015; Shi and Wu 2018; Wang et al. 2018a), the clocks between anchors and sensors are not required to be synchronized. Besides that, the localization protocol in this chapter can compensate both the effect of clock skew and offset as compared with the works (Carroll et al. 2014; Yan et al. 2018; Cheng et al. 2008).

2. **Asynchronous localization algorithm**. PPS and PPDP based asynchronous localization algorithms are developed for USNs to hide the private location information. Compared to the existing works (Liu et al. 2016; Mortazavi et al. 2017), the position information of anchor nodes in this chapter does not require to be revealed. Per knowledge of the authors, this is the first work that applies privacy preservation strategy into the asynchronous localization of USNs.

5.2 Network Architecture and the Asynchronous Localization Protocol

5.2.1 Network Architecture

To achieve privacy-preserving asynchronous localization for USNs, a network architecture with four types of nodes is provided, as shown in Fig. 5.1.

- **Surface Buoys**. Surface buoys are installed with GPS to obtain their accurate time references and positions through electromagnetic communication. The function of surface buoys is to provide self-localization and clock synchronization services for anchor nodes.
- **Anchor Nodes**. Anchor nodes are powerful fixed nodes, and they communicates with surface buoys directly. Similar to the assumption in Zhou et al. (2011) and Liu et al. (2014), it is assumed that the time clocks of anchor nodes are synchronized and the positions are pre-known by using some existing technologies (Bechaz and Thomas 2000).
- **Active Sensor Node**. Active sensor nodes initiate the whole localization process by broadcasting timestamps to the networks. Due to the influence of

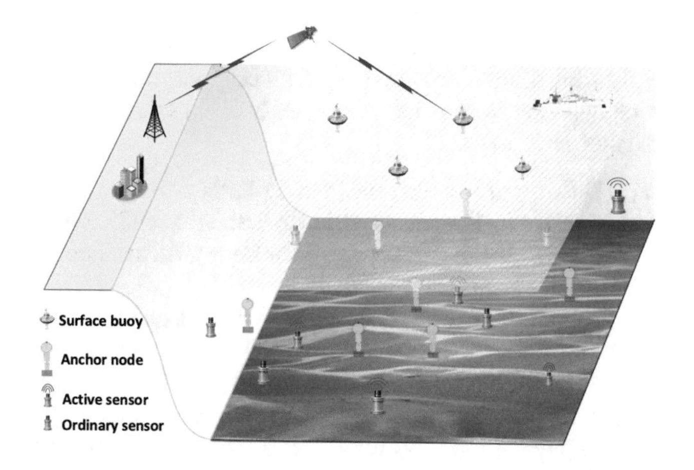

Fig. 5.1 Network architecture of USNs

water current, active sensor nodes can move passively, whose velocities can be accurately measured by Doppler velocity log (DVL) or fiber optic gyroscope (FOG). Particularly, the positions of active sensor nodes are required to be estimated and protected. It is worth emphasizing that the clocks of active sensor nodes are not well synchronized with the real time.

- **Ordinary Sensor Nodes**. Ordinary sensor nodes are low-complexity nodes, and they cannot initiatively start the whole localization process, that is, they just passively listen to the networks and then send state noises to anchor nodes. The positions of ordinary sensor nodes are required to be estimated and protected. Similar to active sensor nodes, ordinary sensor nodes can move passively, whose velocities can be accurately measured. Besides, the clocks of ordinary sensor nodes are asynchronous.

5.2.2 Asynchronous Localization Protocol

With the above network architecture, the localization process can be divided into two subprocesses: (a) active sensor localization; (b) ordinary sensor localization. Without loss of generality, one active sensor node and one ordinary sensor node are considered in this chapter, while the method for single node can be easily extended to the other nodes. In addition, each node can hear message only from its neighboring nodes, that is, each node cannot hear everybody else because of its limited sensing range. Thereby, it is assumed that m anchor nodes are deployed in the sensing range of active and ordinary nodes, where $m \geq 4$. Besides that, the IDs of neighboring anchor nodes are pre-known to active and ordinary sensor nodes.

At the beginning, active sensor node sends out an initiator message to its neighboring nodes. After receiving the initiator message, anchor nodes record and reply the initiator message. Meantime, ordinary sensor node passively listens to the messages from active sensor node and anchor nodes. Since the exchange process can be quickly completed, it assumed that the locations of active and ordinary sensor nodes are fixed during the timestamp exchange process. The timestamp transmission process can be depicted in Fig. 5.2. Accordingly, the asynchronous localization protocol is detailed as follows.

1. At time $T_{\mathrm{a,a}}$, active sensor node sends out an initiator packet to the networks, whose local length is t_{r}. The packet includes t_{r} and the sending order of anchor nodes, i.e., $i \in \{1, ..., m\}$. After, active sensor node switches into waiting mode for the replies from the other nodes.
2. At time $t_{\mathrm{a},i}$, anchor node i receives the initiator packet, and then it switches into waiting mode. At time $t_{j,i}$, anchor node i receives message from anchor node $j \in \{1, ..., i-1\}$. After the message from anchor node $i-1$ has been received, anchor node i sends out its message at time $t_{i,i}$. The length of this message is $t_{\mathrm{r},i}$, this message includes $t_{\mathrm{a},i}$, $\{t_{j,i}\}_{\forall j}$ for $j < i$, $t_{i,i}$, $t_{\mathrm{r},i}$, and the *disguised* states of anchor node i. Note that the *disguised* states can be obtained by adding noises to the real states (see Sect. 5.3).

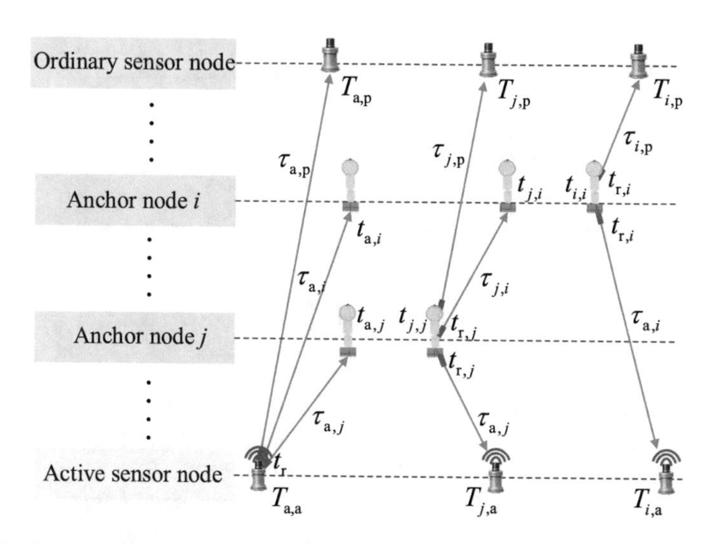

Fig. 5.2 Description of the timestamp transmission process

3. At time $T_{i,\mathrm{a}}$, active sensor node receives the reply from anchor node i. After receiving replies from all anchor nodes, active sensor node ends its waiting mode. Accordingly, active sensor node has the following measurements

$$T_{\mathrm{a,a}},\ \{t_{\mathrm{a},i}, t_{i,i}, T_{i,\mathrm{a}}, t_{\mathrm{r},i}\}_{i=1}^{m},\ \{t_{j,i}\}_{i=2,\ j=1}^{m,\ i-1},\ t_{\mathrm{r}}. \tag{5.1}$$

4. Ordinary sensor node listens to the networks passively, and it sends out state noises to anchor nodes as required. Accordingly, ordinary sensor node has the following timestamp measurements from anchor nodes

$$T_{\mathrm{a,a}},\ T_{\mathrm{a,p}},\ \{t_{\mathrm{a},i}, t_{i,i}, T_{i,\mathrm{p}}, t_{\mathrm{r},i}\}_{i=1}^{m},\ \{t_{j,i}\}_{i=2,\ j=1}^{m,\ i-1},\ t_{\mathrm{r}}. \tag{5.2}$$

5. With the collected timestamps in (5.1) and (5.2), two asynchronous localization algorithms are given to estimate locations of active and ordinary sensor nodes at timestamps $T_{\mathrm{a,a}}$ and $T_{\mathrm{a,p}}$, as shown in Sect. 5.3.

Different from the model in Carroll et al. (2014), Yan et al. (2018), and Cheng et al. (2008), the following clock model is considered

$$T_{\mathrm{a}} = \alpha_{\mathrm{a}} t + \beta_{\mathrm{a}},\ T_{\mathrm{p}} = \alpha_{\mathrm{p}} t + \beta_{\mathrm{p}}, \tag{5.3}$$

where T_{a} and T_{p} denote the local clocks of active and ordinary sensor nodes, respectively. Besides, α_{a} represents the clock skew between active sensor node and real time t, while α_{p} represents the clock skew between ordinary sensor node and real time t. Furthermore, β_{a} and β_{p} are the clock offsets of active sensor node and ordinary sensor node, respectively. According to Gong et al. (2018), one knows α_{a} and α_{p} lie in $[1 - 2\mathrm{e}^{-4}, 1 + 2\mathrm{e}^{-4}]$.

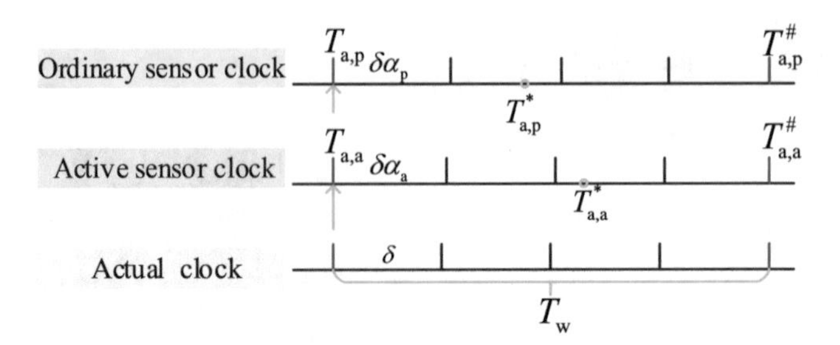

Fig. 5.3 Example for the structure of measurement window

Due to the mobility characteristic, the above localization protocol should be enforced to update the location information periodically. Nevertheless, this implementation can lead to the waste of communication costs. In order to solve this problem, we propose a mobility compensation strategy, whose aim is to balance the tradeoff between communication cost and localization accuracy. Then, the time axis is divided into multiple measurement windows with length set to T_{w}, as shown in Fig. 5.3. The position vectors of active sensor node at timestamps $T_{\mathrm{a,a}}$ and $T^{\#}_{\mathrm{a,a}}$ are denoted as $\mathbf{p}_{\mathrm{a}}$ and $\mathbf{p}^{\#}_{\mathrm{a}}$, respectively. Specifically, the position at timestamps $T_{\mathrm{a,a}}$ and $T^{\#}_{\mathrm{a,a}}$ can be updated via the proposed localization protocol. At timestamp $T^{*}_{\mathrm{a,a}} \in (T_{\mathrm{a,a}}, T^{\#}_{\mathrm{a,a}}]$, the position vector of active sensor node can be updated as

$$\mathbf{p}_{T^{*}_{\mathrm{a,a}}} = \begin{cases} \mathbf{p}_{\mathrm{a}}, & \text{if } (T^{*}_{\mathrm{a,a}} - T_{\mathrm{a,a}}) < \delta\alpha_{\mathrm{a}} \\ \mathbf{p}_{\mathrm{a}} + \delta\alpha_{\mathrm{a}} \sum_{m_{\mathrm{a}}=0}^{n_{\mathrm{a}}} \mathbf{v}_{\mathrm{a}}(\mathbf{T}_{\mathrm{a,a}}), & \text{otherwise} \end{cases} \tag{5.4}$$

where $\delta \in \mathcal{R}^{+}$ denotes the actual sampling interval. $\mathbf{T}_{\mathrm{a,a}} = T_{\mathrm{a,a}} + m_{\mathrm{a}}\delta\alpha_{\mathrm{a}}$, where $m_{\mathrm{a}} = 0, \ldots, n_{\mathrm{a}}$ and $n_{\mathrm{a}} = \lfloor (T^{*}_{\mathrm{a,a}} - T_{\mathrm{a,a}})/(\delta\alpha_{\mathrm{a}}) \rfloor$. In addition, $\mathbf{v}_{\mathrm{a}}(\mathbf{T}_{\mathrm{a,a}})$ is the velocity vector of active sensor node at timestamp $\mathbf{T}_{\mathrm{a,a}}$, which can be measured by DVL or FOG. Note that the position information of active sensor node at timestamps $T_{\mathrm{a,a}}$ and $T^{\#}_{\mathrm{a,a}}$ can be estimated with the algorithm in Sect. 5.3.1. Therefore, the clock skew α_{a} can be deduced by an exhaustive search with relationship of $\mathbf{p}^{\#}_{\mathrm{a}} - \mathbf{p}_{\mathrm{a}} = \delta\alpha_{\mathrm{a}} \sum_{m_{\mathrm{a}}=0}^{n^{\#}_{\mathrm{a}}} \mathbf{v}_{\mathrm{a}}(\mathbf{T}_{\mathrm{a,a}})$ where $n^{\#}_{\mathrm{a}} = \lfloor (T^{\#}_{\mathrm{a,a}} - T_{\mathrm{a,a}})/(\delta\alpha_{\mathrm{a}}) \rfloor$. Of note, $\lfloor \cdot \rfloor$ is the floor function.

Similar to the active sensor node, the position vectors of ordinary sensor node at timestamps $T_{\mathrm{a,p}}$ and $T^{\#}_{\mathrm{a,p}}$ can be expressed as $\mathbf{p}_{\mathrm{p}}$ and $\mathbf{p}^{\#}_{\mathrm{p}}$, respectively. The position at timestamps $T_{\mathrm{a,p}}$ and $T^{\#}_{\mathrm{a,p}}$ can be updated by the localization protocol in Sect. 5.3.2. At timestamp $T^{*}_{\mathrm{a,p}} \in (T_{\mathrm{a,p}}, T^{\#}_{\mathrm{a,p}}]$, the position of ordinary sensor node can be updated as

$$\mathbf{p}_{T^{*}_{\mathrm{a,p}}} = \begin{cases} \mathbf{p}_{\mathrm{p}}, & \text{if } (T^{*}_{\mathrm{a,p}} - T_{\mathrm{a,p}}) < \delta\alpha_{\mathrm{p}} \\ \mathbf{p}_{\mathrm{p}} + \delta\alpha_{\mathrm{p}} \sum_{m_{\mathrm{p}}=0}^{n_{\mathrm{p}}} \mathbf{v}_{\mathrm{p}}(\mathbf{T}_{\mathrm{a,p}}), & \text{otherwise} \end{cases} \tag{5.5}$$

where $\mathbf{T}_{\mathrm{a,p}} = T_{\mathrm{a,p}} + m_{\mathrm{p}}\delta\alpha_{\mathrm{p}}$ for $m_{\mathrm{p}} = 0, \ldots, n_{\mathrm{p}}$ and $n_{\mathrm{p}} = \left\lfloor (T^{*}_{\mathrm{a,p}} - T_{\mathrm{a,p}})/(\delta\alpha_{\mathrm{p}}) \right\rfloor$. Besides, $\mathbf{v}_{\mathrm{p}}(\mathbf{T}_{\mathrm{a,p}})$ represents the velocity vector of ordinary sensor node at timestamp $\mathbf{T}_{\mathrm{a,p}}$, and α_{p} can be deduced by the relationship of $\mathbf{p}^{\#}_{\mathrm{p}} - \mathbf{p}_{\mathrm{p}} = \delta\alpha_{\mathrm{p}} \sum_{m_{\mathrm{p}}=0}^{n^{\#}_{\mathrm{p}}} \mathbf{v}_{\mathrm{p}}(\mathbf{T}_{\mathrm{a,p}})$ where $n^{\#}_{\mathrm{p}} = \left\lfloor (T^{\#}_{\mathrm{a,p}} - T_{\mathrm{a,p}})/(\delta\alpha_{\mathrm{p}}) \right\rfloor$.

Problem 5.1 (Privacy-Preserving Asynchronous Localization for Active Sensor Node) The clocks are asynchronous and the position information is required to be protected in underwater environment. Inspired by this, we try to design a privacy-preserving asynchronous localization algorithm for active sensor node, where active sensor node initiates the localization process by broadcasting message to the networks. This problem can be reduced to the estimation of $\mathbf{p}_{\mathrm{a}}$ with the collected timestamps in (5.1).

Problem 5.2 (Privacy-Preserving Asynchronous Localization for Ordinary Sensor Node) Different from the active sensor node, the ordinary sensor node cannot initiatively broadcast timestamps to the networks. Thus, the localization method developed for active sensor node is not suitable for ordinary sensor node. Based on this, a privacy-preserving asynchronous localization algorithm for ordinary sensor node is designed. This problem is reduced to the estimation of $\mathbf{p}_{\mathrm{p}}$ with the collected timestamps in (5.2).

5.3 Asynchronous Localization Algorithms

5.3.1 PPS-Based Localization for Active Sensor

To remove the effect of asynchronous clock (i.e., time offset and skew), the following time difference is defined, i.e.,

$$\Delta T_{i,1} = t_{\mathrm{a},i} - t_{\mathrm{a},1}, \; i \in \{2, \ldots, m\}. \tag{5.6}$$

Without loss of generality, we assume that all nodes including anchor nodes, active and ordinary sensor nodes have the same measurement quality. To be specific, the measurement noise of each local measurement is a random variable with zero mean and variance σ^2_{mea}, which is decided by the underlying signal processing in the presence of multipath propagation and ambient noise. In viwe of (5.6), the relationship between time differences and propagation delays can be constructed as

$$\begin{aligned} \Delta T_{i,1} &= (t_{\mathrm{a,a}} + \frac{t_{\mathrm{r}}}{\alpha_{\mathrm{a}}} + \tau_{\mathrm{a},i} + \varpi_{\mathrm{a},i}) - (t_{\mathrm{a,a}} + \frac{t_{\mathrm{r}}}{\alpha_{\mathrm{a}}} + \tau_{\mathrm{a},1} + \varpi_{\mathrm{a},1}) \\ &= \tau_{\mathrm{a},i} - \tau_{\mathrm{a},1} + \varpi_{i,1}, \end{aligned} \tag{5.7}$$

where $\tau_{\mathrm{a},i}$ denotes the one-way propagation delay between active sensor node and anchor node i, $\tau_{\mathrm{a},1}$ denotes the one-way propagation delay between active sensor node and anchor node 1, $t_{\mathrm{a,a}}$ denotes the real time when the initiator message is sent to the network. In addition, $\varpi_{\mathrm{a},i}$ and $\varpi_{\mathrm{a},1}$ are the measurement noises, which satisfy the distributions of $\varpi_{\mathrm{a},i} \sim \mathcal{N}(0, \sigma_{\mathrm{mea}}^2)$ and $\varpi_{\mathrm{a},1} \sim \mathcal{N}(0, \sigma_{\mathrm{mea}}^2)$, respectively. Based on this, the noise $\varpi_{i,1}$ satisfies the distribution of $\varpi_{i,1} \sim \mathcal{N}(0, 2\sigma_{\mathrm{mea}}^2)$.

With (5.7), we have the following distance difference

$$\Upsilon_{i,1} = d_{\mathrm{a},i} - d_{\mathrm{a},1} + e_{i,1},\ i \in \{2, ..., m\} \tag{5.8}$$

where $\Upsilon_{i,1} = \mathrm{c}\Delta\mathcal{T}_{i,1}$ denotes the measured distance difference for anchor node i, $d_{\mathrm{a},i} = \mathrm{c}\tau_{\mathrm{a},i}$ denotes the relative distance between active sensor node and anchor node i, while $d_{\mathrm{a},1} = \mathrm{c}\tau_{\mathrm{a},1}$ denotes the relative distance between active sensor node and anchor node 1. Furthermore, $e_{i,1} = \mathrm{c}\varpi_{i,1}$ represents the noise of distance measurement, satisfying the distribution of $e_{i,1} \sim \mathcal{N}(0, 2\mathrm{c}^2\sigma_{\mathrm{mea}}^2)$.

It is denfined that $\mathbf{x}_\mathrm{a} = [\mathbf{p}_\mathrm{a}^\mathrm{T}, d_{\mathrm{a},1}]^\mathrm{T}$. In view of (5.8), we formulate the following least squares (LS) problem, i.e.,

$$\min_{\mathbf{x}_\mathrm{a}} \sum_{i=2}^{m} \left((\Upsilon_{i,1} + d_{\mathrm{a},1})^2 - d_{\mathrm{a},i}^2 \right)^2. \tag{5.9}$$

The position vector of anchor node $i \in \{1, ..., m\}$ is expressed as $\mathbf{p}_i \in \mathcal{R}^{3\times1}$. Then, the nonlinear equation in (5.9) can be transformed into the following linear LS problem, i.e.,

$$\min_{\mathbf{x}_\mathrm{a}} \|2\mathbf{A}\mathbf{x}_\mathrm{a} - \mathbf{B}\|^2 \tag{5.10}$$

where $\mathbf{A} = [\mathbf{a}_1, ..., \mathbf{a}_{m-1}]^\mathrm{T}$ and $\mathbf{B} = [\mathbf{b}_1, ..., \mathbf{b}_{m-1}]^\mathrm{T}$. Particularly, the j-th element of $\mathbf{A}$ is defined as $\mathbf{a}_j = [(\mathbf{p}_{j+1} - \mathbf{p}_1)^\mathrm{T}, \Upsilon_{j+1,1}]$ for $j \in \{1, ..., m-1\}$. Similarly, the j-th element of $\mathbf{B}$ is given by $\mathbf{b}_j = \mathbf{p}_{j+1}^\mathrm{T}\mathbf{p}_{j+1} - \mathbf{p}_1^\mathrm{T}\mathbf{p}_1 - \Upsilon_{j+1,1}^2$.

By using the traditional LS estimator, the direct estimation of $\mathbf{x}_\mathrm{a}$ can be given as $\hat{\mathbf{x}}_{\mathrm{a,direct}}$, i.e.,

$$\hat{\mathbf{x}}_{\mathrm{a,direct}} = \frac{1}{2}(\mathbf{A}^\mathrm{T}\mathbf{A})^{-1}\mathbf{A}^\mathrm{T}\mathbf{B}. \tag{5.11}$$

However, the direct calculation of $\frac{1}{2}(\mathbf{A}^\mathrm{T}\mathbf{A})^{-1}\mathbf{A}^\mathrm{T}\mathbf{B}$ can result in privacy leakage, since matrices $\mathbf{A}$ and $\mathbf{B}$ contain the location information of anchor nodes. To prevent privacy leakage, a PPS strategy is developed to calculate $\mathbf{A}^\mathrm{T}\mathbf{A}$ and $\mathbf{A}^\mathrm{T}\mathbf{B}$, through which $\frac{1}{2}(\mathbf{A}^\mathrm{T}\mathbf{A})^{-1}\mathbf{A}^\mathrm{T}\mathbf{B}$ can be indirectly calculated. Especially, the indirect estimation of $\mathbf{x}_\mathrm{a}$ can be expressed as $\hat{\mathbf{x}}_{\mathrm{a,indirect}}$, which can be divided into two parts, i.e., $\mathbf{A} = \mathbf{A}^\mathrm{T}\mathbf{A}$ and $\mathbf{B} = \mathbf{A}^\mathrm{T}\mathbf{B}$. Based on this and in view of (5.11), $\hat{\mathbf{x}}_{\mathrm{a,indirect}}$ can be defined as

$$\hat{\mathbf{x}}_{\mathrm{a,indirect}} = \frac{1}{2}\mathbf{A}^{-1}\mathbf{B}. \tag{5.12}$$

With regard to (5.12), the main process of PPS-based asynchronous localization algorithm is detailed as follows.

Step 1: Matrix Construction
For the convenience of computation, it is defined that $\mathbf{A} = [\mathbf{A}_{11}, \mathbf{A}_{12}; \mathbf{A}_{21}, A_{22}]$ and $\mathbf{B} = [\mathbf{B}_{11} + \mathbf{B}_{12}; B_{21} + B_{22}]$. Referring to the definitions of $\mathbf{A}$ and $\mathbf{B}$, the elements of $\mathbf{A}$ and $\mathbf{B}$ can be described as

$$
\begin{aligned}
\mathbf{A}_{11} &= \sum_{i=2}^{m} (\mathbf{p}_i - \mathbf{p}_1)(\mathbf{p}_i - \mathbf{p}_1)^{\mathrm{T}} \\
&= (m-1)\mathbf{p}_1\mathbf{p}_1^{\mathrm{T}} + \sum_{i=2}^{m} \mathbf{p}_i\mathbf{p}_i^{\mathrm{T}} - \left(\sum_{i=2}^{m} \mathbf{p}_i\right)\mathbf{p}_1^{\mathrm{T}} - \mathbf{p}_1\left(\sum_{i=2}^{m} \mathbf{p}_i^{\mathrm{T}}\right), \\
\mathbf{A}_{12} &= \sum_{i=2}^{m} (\mathbf{p}_i - \mathbf{p}_1)\Upsilon_{i,1} \\
&= \sum_{i=2}^{m} \mathbf{p}_i\Upsilon_{i,1} - \mathbf{p}_1\left(\sum_{i=2}^{m} \Upsilon_{i,1}\right), \\
\mathbf{A}_{21} &= \sum_{i=2}^{m} \Upsilon_{i,1}(\mathbf{p}_i - \mathbf{p}_1)^{\mathrm{T}} = \mathbf{A}_{12}^{\mathrm{T}}, \\
A_{22} &= \sum_{i=2}^{m} \Upsilon_{i,1}^2, \\
\mathbf{B}_{11} &= \sum_{i=2}^{m} (\mathbf{p}_i - \mathbf{p}_1)(\mathbf{p}_i^{\mathrm{T}}\mathbf{p}_i - \mathbf{p}_1^{\mathrm{T}}\mathbf{p}_1) \\
&= (m-1)\mathbf{p}_1\mathbf{p}_1^{\mathrm{T}}\mathbf{p}_1 + \sum_{i=2}^{m} \mathbf{p}_i\mathbf{p}_i^{\mathrm{T}}\mathbf{p}_i - \left(\sum_{i=2}^{m} \mathbf{p}_i\right)\mathbf{p}_1^{\mathrm{T}}\mathbf{p}_1 - \mathbf{p}_1\left(\sum_{i=2}^{m} \mathbf{p}_i^{\mathrm{T}}\mathbf{p}_i\right), \\
\mathbf{B}_{12} &= \sum_{i=2}^{m} (\mathbf{p}_i - \mathbf{p}_1)(-\Upsilon_{i,1}^2) \\
&= -\sum_{i=2}^{m} \mathbf{p}_i\Upsilon_{i,1}^2 + \mathbf{p}_1\left(\sum_{i=2}^{m} \Upsilon_{i,1}^2\right),
\end{aligned}
\tag{5.13}
$$

$$
\begin{aligned}
B_{21} &= \sum_{i=2}^{m} \Upsilon_{i,1}(\mathbf{p}_i^{\mathrm{T}}\mathbf{p}_i - \mathbf{p}_1^{\mathrm{T}}\mathbf{p}_1) \\
&= \sum_{i=2}^{m} \mathbf{p}_i^{\mathrm{T}}\mathbf{p}_i\Upsilon_{i,1} - \mathbf{p}_1^{\mathrm{T}}\mathbf{p}_1\left(\sum_{i=2}^{m} \Upsilon_{i,1}\right), \\
B_{22} &= \sum_{i=2}^{m} \Upsilon_{i,1}(-\Upsilon_{i,1}^2) = -\sum_{i=2}^{m} \Upsilon_{i,1}^3.
\end{aligned}
\tag{5.14}
$$

Thereby, the estimation of $\mathbf{x}_{\mathrm{a}}$ in (5.12) can be rewritten as

$$
\hat{\mathbf{x}}_{\mathrm{a,indirect}} = \frac{1}{2}\begin{bmatrix} \mathbf{A}_{11} & \mathbf{A}_{12} \\ \mathbf{A}_{21} & A_{22} \end{bmatrix}^{-1} \begin{bmatrix} \mathbf{B}_{11} + \mathbf{B}_{12} \\ B_{21} + B_{22} \end{bmatrix}.
\tag{5.15}
$$

Step 2: PPS-Based State Calculation
As described in Sect. 5.2.2, anchor nodes send disguised states to the networks. The real states of anchor node 1 are denoted as $\mathbf{Y}_1 = [(m-1)\mathbf{p}_1\mathbf{p}_1^{\mathrm{T}} - (\sum_{i=2}^{m}\mathbf{p}_i)\mathbf{p}_1^{\mathrm{T}} - \mathbf{p}_1(\sum_{i=2}^{m}\mathbf{p}_i^{\mathrm{T}}),\ (m-1)\mathbf{p}_1\mathbf{p}_1^{\mathrm{T}}\mathbf{p}_1 - (\sum_{i=2}^{m}\mathbf{p}_i)\mathbf{p}_1^{\mathrm{T}}\mathbf{p}_1 - \mathbf{p}_1(\sum_{i=2}^{m}\mathbf{p}_i^{\mathrm{T}}\mathbf{p}_i)]$ and $\mathbf{Z}_1 = [-\mathbf{p}_1,\ \mathbf{p}_1,\ (0;\ -\mathbf{p}_1^{\mathrm{T}}\mathbf{p}_1;\ 0)]$. The real states of anchor node $i \in \{2, ..., m\}$ are given as $\mathbf{X}_i = [\mathbf{p}_i,\ (0;\ \mathbf{p}_i^{\mathrm{T}}\mathbf{p}_i;\ 0)]$, $\mathbf{Y}_i = [\mathbf{p}_i\mathbf{p}_i^{\mathrm{T}}, \mathbf{p}_i\mathbf{p}_i^{\mathrm{T}}\mathbf{p}_i]$ and $\mathbf{Z}_i = [\mathbf{p}_i, -\mathbf{p}_i, (0;\ \mathbf{p}_i^{\mathrm{T}}\mathbf{p}_i;\ 0)]$. To calculate $\mathbf{A}$ and $\mathbf{B}$, the summations of these states, i.e., $\sum_{i=2}^{m}\mathbf{X}_i$, $\sum_{i=1}^{m}\mathbf{Y}_i$, $\mathbf{Z}_1\sum_{i=2}^{m}\mathrm{diag}\{\Upsilon_{i,1}, \Upsilon_{i,1}^2, \Upsilon_{i,1}\} + \sum_{i=2}^{m}\mathbf{Z}_i\ \mathrm{diag}\{\Upsilon_{i,1}, \Upsilon_{i,1}^2, \Upsilon_{i,1}\}$, are required to be known. In the following, we give a PPS-based strategy (Shi and Wu 2018; Wang et al. 2018a) to calculate these summations.

Anchor node $i \in \{1, ..., m\}$ randomly generates matrices $\mathbf{S}_{2,ij}$ and $\mathbf{S}_{3,ij}$, where $\sum_{j=1}^{m} \mathbf{S}_{2,ij} = \mathbf{0}$ and $\sum_{j=1}^{m} \mathbf{S}_{3,ij} = \mathbf{0}$. Then, anchor node i keeps $\mathbf{S}_{2,ii}$ and $\mathbf{S}_{3,ii}$ in memory, and sends the rest to the other $m-1$ anchor nodes. Meantime, anchor node i receives the matrices from the other $m-1$ anchor nodes. By adding the received matrices to $\mathbf{S}_{2,ii}$ and $\mathbf{S}_{3,ii}$, two random matrices $\boldsymbol{\delta}_{2i}$ and $\boldsymbol{\delta}_{3i}$ can be obtained. According, the disguised state of $\mathbf{Y}_i$ can be denoted as $\tilde{\mathbf{Y}}_i = \mathbf{Y}_i + \boldsymbol{\delta}_{2i}$ for $i \in \{1, ..., m\}$. Denote the k-th column of $\boldsymbol{\delta}_{3i}$ as $\boldsymbol{\delta}_{3i,k}$ for $k \in \{1, 2, 3\}$. Based on this, the disguised state of $\mathbf{Z}_i$ is given as $\tilde{\mathbf{Z}}_1 = \mathbf{Z}_1 + \boldsymbol{\delta}_{31}$ or $\tilde{\mathbf{Z}}_i = \mathbf{Z}_i + \mathbf{W}_i$, where $\mathbf{W}_i = [\frac{\boldsymbol{\delta}_{3i,1}}{\gamma_i}, \frac{\boldsymbol{\delta}_{3i,2}}{\gamma_{2i}}, \frac{\boldsymbol{\delta}_{3i,3}}{\gamma_i}]$, $\gamma_i = \frac{\Upsilon_{i,1}}{\sum_{i=2}^{m} \Upsilon_{i,1}}$, and $\gamma_{2i} = \frac{\Upsilon_{i,1}^2}{\sum_{i=2}^{m} \Upsilon_{i,1}^2}$ for $i \in \{2, ..., m\}$. For clear description, the calculation process of $\sum_{i=1}^{m} \mathbf{Y}_i$ is depicted in Fig. 5.4a.

With the similar strategy, the state $\mathbf{X}_i$ can also be disguised. For example, anchor node $i \in \{2, ..., m\}$ randomly generates matrix $\mathbf{S}_{1,ij}$ in the same size as $\mathbf{X}_i$ where $j \in \{2, ..., m\}$, such that $\sum_{j=2}^{m} \mathbf{S}_{1,ij} = \mathbf{0}$. Through state summation, the disguised state of $\mathbf{X}_i$ can be denoted as $\tilde{\mathbf{X}}_i = \mathbf{X}_i + \boldsymbol{\delta}_{1i}$, where $\boldsymbol{\delta}_{1i}$ is a random matrix with $i \in \{2, ..., m\}$. Particularly, the summation process of $\sum_{i=2}^{m} \mathbf{X}_i$ can be depicted by Fig. 5.4b.

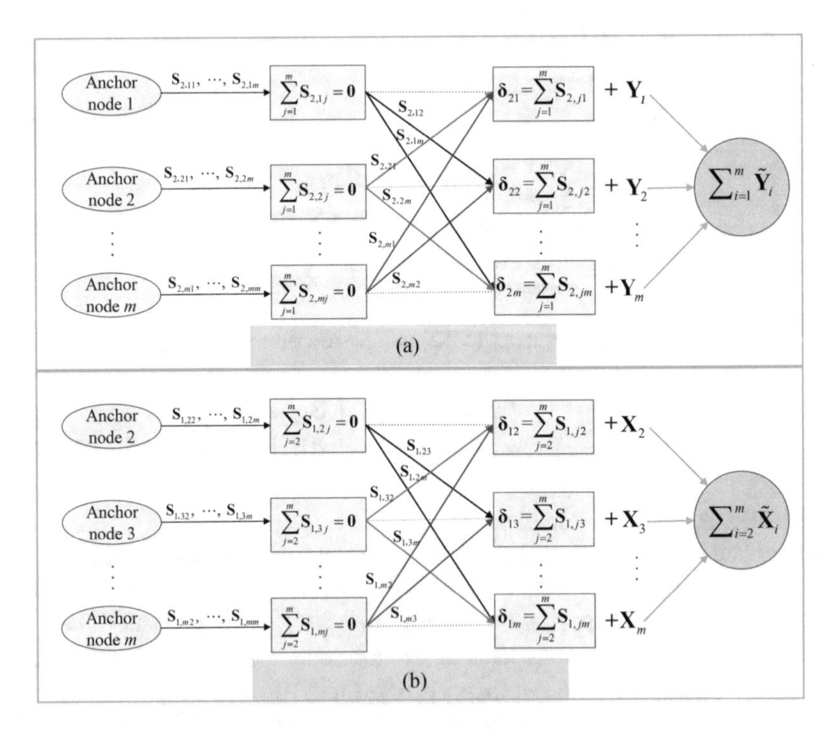

Fig. 5.4 Description of the PPS-based state calculation. (**a**) Calculation process of $\sum_{i=1}^{m} \mathbf{Y}_i$. (**b**) Calculation process of $\sum_{i=1}^{m} \mathbf{X}_i$

Afterwards, anchor node 1 sends disguised states $\tilde{\mathbf{Y}}_1$ and $\tilde{\mathbf{Z}}_1$ to the networks, while anchor node $i \in \{2, ..., m\}$ sends disguised states $\tilde{\mathbf{X}}_i$, $\tilde{\mathbf{Y}}_i$ and $\tilde{\mathbf{Z}}_i$ to the networks. Obviously, the real states of anchor nodes are not revealed.

Step 3: Calculation of **A** *and* **B**
It is noted that $\sum_{i=2}^{m} \tilde{\mathbf{X}}_i = \sum_{i=2}^{m} (\mathbf{X}_i + \delta_{1i}) = \sum_{i=2}^{m} \mathbf{X}_i$. Applying this property to all the elements, the values of $\mathbf{A}_{11}$ and $\mathbf{B}_{11}$ can be acquired with $\tilde{\mathbf{X}}_i$ and $\tilde{\mathbf{Y}}_i$. Meantime, the values of $\mathbf{A}_{12}$, $\mathbf{A}_{21}$, $\mathbf{B}_{12}$ and B_{21} can be acquired with $\tilde{\mathbf{Z}}_i$. Of note, the values of A_{22} and B_{22} can be directly acquired, because $\Upsilon_{i,1}^2$ and $\Upsilon_{i,1}^3$ are pre-known. Accordingly, the values of **A** and **B** are indirectly calculated, where the privacy information of anchor nodes is not leaked.

Step 4: Position Estimation
With the results in Step 3, the estimation of $\mathbf{x}_a$ in (5.15) can be easily calculated. Accordingly, the clock skew α_a can be obtained by an exhaustive search, and hence, one has

$$\begin{aligned} \frac{T_{a,a} - \beta_a}{\alpha_a} + \frac{t_r}{\alpha_a} + \tau_{a,i} &= t_{a,i}, \\ t_{i,i} + t_{r,i} + \tau_{a,i} &= \frac{T_{i,a} - \beta_a}{\alpha_a}. \end{aligned} \tag{5.16}$$

According to (5.16), one has

$$\tau_{a,i} = \frac{(T_{i,a} - T_{a,a}) - \alpha_a (t_{i,i} - t_{a,i} + \frac{t_r}{\alpha_a} + t_{r,i})}{2\alpha_a}. \tag{5.17}$$

Aa a result, by substituting (5.17) into (5.16), the clock offset β_a can be obtained as $\beta_a = T_{a,a} - \alpha_a (t_{a,i} - \frac{t_r}{\alpha_a} - \tau_{a,i})$.

5.3.2 PPS and PPDP Based Localization for Ordinary Sensor

The active sensor node and anchor node $i \in \{1, ..., m\}$ provide localization services for ordinary sensor node in this section. To remove the effect of clock offset (i.e., β_p), the following time difference can be defined, i.e.,

$$\Delta T_{i,a,p} = T_{i,p} - T_{a,p}, \; i \in \{1, ..., m\}. \tag{5.18}$$

In view of (5.18), the following relationship can be built, i.e.,

$$\begin{aligned} \Delta T_{i,a,p} &= [\alpha_p (t_{i,i} + t_{r,i} + \tau_{i,p} + \varpi_{i,p}) + \beta_p] \\ &\quad - [\alpha_p (\frac{T_{a,a} - \beta_a}{\alpha_a} + \frac{t_r}{\alpha_a} + \tau_{a,p} + \varpi_{a,p}) + \beta_p \\ &= \alpha_p [(t_{i,i} + t_{r,i} - \frac{T_{a,a} - \beta_a}{\alpha_a} - \frac{t_r}{\alpha_a}) \\ &\quad + (\tau_{i,p} - \tau_{a,p}) + \varpi_{i,a,p}], \end{aligned} \tag{5.19}$$

where $\tau_{i,\mathrm{p}}$ denotes the one-way propagation delay between anchor node i and ordinary sensor node, $\tau_{\mathrm{a,p}}$ denotes the one-way propagation delay between ordinary sensor node and active sensor node. Besides, $\varpi_{i,\mathrm{p}}$ and $\varpi_{\mathrm{a,p}}$ are the measurement noises, satisfying the distributions of $\varpi_{i,\mathrm{p}} \sim \mathcal{N}(0, \sigma_{\mathrm{mea}}^2)$ and $\varpi_{\mathrm{a,p}} \sim \mathcal{N}(0, \sigma_{\mathrm{mea}}^2)$. Based on this, the noise $\varpi_{i,\mathrm{a,p}}$ satisfies the distribution of $\varpi_{i,\mathrm{a,p}} \sim \mathcal{N}(0, 2\sigma_{\mathrm{mea}}^2)$.

With (5.19), the following distance difference can be obtained, i.e.,

$$\mathrm{c}\Delta\mathcal{T}_{i,\mathrm{a,p}} = \alpha_{\mathrm{p}}(M_i + d_{i,\mathrm{p}} - d_{\mathrm{a,p}} + e_{i,\mathrm{a,p}}), \tag{5.20}$$

where $M_i = \mathrm{c}(t_{i,i} + t_{\mathrm{r},i} - \frac{T_{\mathrm{a,a}}-\beta_{\mathrm{a}}}{\alpha_{\mathrm{a}}} - \frac{t_{\mathrm{r}}}{\alpha_{\mathrm{a}}})$. Besides that, $d_{i,\mathrm{p}} = \mathrm{c}\tau_{i,\mathrm{p}}$ denotes the relative distance between anchor node i and ordinary sensor node, while $d_{\mathrm{a,p}} = \mathrm{c}\tau_{\mathrm{a,p}}$ is the relative distance between ordinary sensor node and active sensor node. In addition, $e_{i,\mathrm{a,p}} = \mathrm{c}\varpi_{i,\mathrm{a,p}}$ denotes the noise of distance difference measurement, satisfying the distribution of $e_{i,\mathrm{a,p}} \sim \mathcal{N}(0, 2\mathrm{c}^2\sigma_{\mathrm{mea}}^2)$.

Due to the ordinary sensor node cannot initiatively send timestamps, the clock skew (i.e., α_{p}) cannot be removed by applying the relationship in (5.7). To remove the effect of clock skew, the following equation is defined, i.e.,

$$\frac{\Delta\mathcal{T}_{1,\mathrm{a,p}}}{\Delta\mathcal{T}_{i,\mathrm{a,p}}} = \frac{M_1 + d_{1,\mathrm{p}} - d_{\mathrm{a,p}} + e_{1,\mathrm{a,p}}}{M_i + d_{i,\mathrm{p}} - d_{\mathrm{a,p}} + e_{i,\mathrm{a,p}}}, \ i \in \{2, ..., m\}. \tag{5.21}$$

Rearranging (5.21), one can obtain $\frac{\Delta\mathcal{T}_{1,\mathrm{a,p}}}{\Delta\mathcal{T}_{i,\mathrm{a,p}}}(M_i + d_{i,\mathrm{p}} - d_{\mathrm{a,p}} + e_{i,\mathrm{a,p}}) = M_1 + d_{1,\mathrm{p}} - d_{\mathrm{a,p}} + e_{1,\mathrm{a,p}}$, i.e., $D_{i,1,\mathrm{p}}M_i - M_1 + (1 - D_{i,1,\mathrm{p}})d_{\mathrm{a,p}} = d_{1,\mathrm{p}} - D_{i,1,\mathrm{p}}d_{i,\mathrm{p}} - D_{i,1,\mathrm{p}}e_{i,\mathrm{a,p}} + e_{1,\mathrm{a,p}}$, where $D_{i,1,\mathrm{p}} = \frac{\Delta\mathcal{T}_{1,\mathrm{a,p}}}{\Delta\mathcal{T}_{i,\mathrm{a,p}}}$. Based on this, one has

$$Q_{i,1,\mathrm{p}} + (1 - D_{i,1,\mathrm{p}})d_{\mathrm{a,p}} = d_{1,\mathrm{p}} - D_{i,1,\mathrm{p}}d_{i,\mathrm{p}} + e_{i,1,\mathrm{p}}, \tag{5.22}$$

where $Q_{i,1,\mathrm{p}} = D_{i,1,\mathrm{p}}M_i - M_1$ and $e_{i,1,\mathrm{p}} = e_{1,\mathrm{a,p}} - D_{i,1,\mathrm{p}}e_{i,\mathrm{a,p}}$.

To minimize the measurement noise, we define $\mathbf{x}_{\mathrm{p}} = \left[\mathbf{p}_{\mathrm{p}}^{\mathrm{T}}, d_{\mathrm{a,p}}, d_{1,\mathrm{p}}\right]^{\mathrm{T}}$, and therefore the following LS estimation problem can be formulated, i.e.,

$$\min_{\mathbf{x}_{\mathrm{p}}} \sum_{i=2}^{m} \left\{\left[Q_{i,1,\mathrm{p}} + (1 - D_{i,1,\mathrm{p}})d_{\mathrm{a,p}} - d_{1,\mathrm{p}}\right]^2 - (D_{i,1,p}d_{i,\mathrm{p}})^2\right\}^2. \tag{5.23}$$

Similar to the active sensor node, the nonlinear estimation problem in (5.23) can be transformed into the following linear LS problem

$$\min_{\mathbf{x}_{\mathrm{p}}} \left\|2\mathbf{H}\mathbf{x}_{\mathrm{p}} - \mathbf{Q}\right\|^2, \tag{5.24}$$

where $\mathbf{H} = [\mathbf{h}_1, ..., \mathbf{h}_{m-1}]^{\mathrm{T}}$ and $\mathbf{Q} = [\mathbf{q}_1, ..., \mathbf{q}_{m-1}]^{\mathrm{T}}$. Especially, the j-th element of $\mathbf{H}$ is defined as $\mathbf{h}_j = [D_{j+1,1,\mathrm{p}}(\mathbf{p}_{\mathrm{a}} - \mathbf{p}_1)^{\mathrm{T}} + D_{j+1,1,\mathrm{p}}^2(\mathbf{p}_{j+1} - \mathbf{p}_{\mathrm{a}})^{\mathrm{T}}, (1 - D_{j+1,1,\mathrm{p}})Q_{j+1,1,\mathrm{p}}, -Q_{j+1,1,\mathrm{p}}]$ for $j \in \{1, ..., m-1\}$. In addition, the j-th element of $\mathbf{Q}$ is $\mathbf{q}_j = -Q^2_{j+1,1,\text{p}} - (1 - D_{j+1,1,\text{p}})^2\mathbf{p}_\text{a}^\text{T}\mathbf{p}_\text{a} - \mathbf{p}_1^\text{T}\mathbf{p}_1 + 2(1 - D_{j+1,1,\text{p}})\mathbf{p}_\text{a}^\text{T}\mathbf{p}_1 + D^2_{j+1,1,\text{p}}\mathbf{p}_{j+1}^\text{T}\mathbf{p}_{j+1}$.

Accordingly, the direct estimation of $\mathbf{x}_\text{p}$ can be obtained by applying the traditional LS estimator, which can be given as

$$\hat{\mathbf{x}}_{\text{p,direct}} = \tfrac{1}{2}(\mathbf{H}^\text{T}\mathbf{H})^{-1}\mathbf{H}^\text{T}\mathbf{Q}. \tag{5.25}$$

Similar to Sect. 5.3.1, the direct calculation of $\mathbf{x}_\text{p}$ can result in privacy leakage, because matrices $\mathbf{H}$ and $\mathbf{Q}$ contain the position information of anchor nodes. To handle this issue, a PPS and PPDP based strategy is presented to indirectly calculate $\hat{\mathbf{x}}_{\text{p,direct}}$. Based on this, we denote $\mathbf{H}^\text{T}\mathbf{H}$ and $\mathbf{H}^\text{T}\mathbf{Q}$ as $\mathbf{H}$ and $\mathbf{Q}$, respectively. Then, the indirect estimation of $\mathbf{x}_\text{p}$ can be expressed as

$$\hat{\mathbf{x}}_{\text{p,indirect}} = \tfrac{1}{2}\mathbf{H}^{-1}\mathbf{Q}. \tag{5.26}$$

With regard to (5.26), the main calculation process for ordinary sensor node is detailed as follows.

Step 1: Matrix Construction

For the convenience of computation, we define $\mathbf{H} = [\mathbf{H}_{11}, \mathbf{H}_{12}, \mathbf{H}_{13}; \mathbf{H}_{21}, H_{22}, H_{23}; \mathbf{H}_{31}, H_{32}, H_{33}] \in \mathcal{R}^{5\times5}$ and $\mathbf{Q} = [\mathbf{Q}_{11}; Q_{21}; Q_{31}] \in \mathcal{R}^5$. Referring to the definitions of $\mathbf{H}$ and $\mathbf{Q}$, the elements of $\mathbf{H}$ and $\mathbf{Q}$ are constructed as

$$\begin{aligned}
\mathbf{H}_{11} &= \sum\nolimits_{i=2}^{m}[D_{i,1,\text{p}}(\mathbf{p}_\text{a} - \mathbf{p}_1) + D^2_{i,1,\text{p}}(\mathbf{p}_i - \mathbf{p}_\text{a})] \\
&\quad \times [D_{i,1,\text{p}}(\mathbf{p}_\text{a} - \mathbf{p}_1)^\text{T} + D^2_{i,1,\text{p}}(\mathbf{p}_i - \mathbf{p}_\text{a})^\text{T}] \\
&= (\mathbf{p}_\text{a} - \mathbf{p}_1)(\mathbf{p}_\text{a} - \mathbf{p}_1)^\text{T}\sum\nolimits_{i=2}^{m} D^2_{i,1,\text{p}} + (\mathbf{p}_\text{a} \\
&\quad - \mathbf{p}_1)(-\mathbf{p}_\text{a}^\text{T}\sum\nolimits_{i=2}^{m} D^3_{i,1,\text{p}} + \sum\nolimits_{i=2}^{m} D^3_{i,1,\text{p}}\mathbf{p}_i^\text{T}) \\
&\quad + (-\mathbf{p}_\text{a}\sum\nolimits_{i=2}^{m} D^3_{i,1,\text{p}} + \sum\nolimits_{i=2}^{m} D^3_{i,1,\text{p}}\mathbf{p}_i)(\mathbf{p}_\text{a} \\
&\quad - \mathbf{p}_1)^\text{T} + \sum\nolimits_{i=2}^{m} D^4_{i,1,\text{p}}\mathbf{p}_\text{a}\mathbf{p}_\text{a}^\text{T} + \sum\nolimits_{i=2}^{m} D^4_{i,1,\text{p}}\mathbf{p}_i\mathbf{p}_i^\text{T} \\
&\quad - \sum\nolimits_{i=2}^{m} D^4_{i,1,\text{p}}\mathbf{p}_i\mathbf{p}_\text{a}^\text{T} - \mathbf{p}_\text{a}\sum\nolimits_{i=2}^{m} D^4_{i,1,\text{p}}\mathbf{p}_i^\text{T}, \\
\mathbf{H}_{12} &= \sum\nolimits_{i=2}^{m}[D_{i,1,\text{p}}(\mathbf{p}_\text{a} - \mathbf{p}_1) + D^2_{i,1,\text{p}}(\mathbf{p}_i - \mathbf{p}_\text{a})](1 - D_{i,1,\text{p}})Q_{i,1,\text{p}} \\
&= (\mathbf{p}_\text{a} - \mathbf{p}_1)\sum\nolimits_{i=2}^{m}(1 - D_{i,1,\text{p}})D_{i,1,\text{p}}Q_{i,1,\text{p}} \\
&\quad - \sum\nolimits_{i=2}^{m}(1 - D_{i,1,\text{p}})D^2_{i,1,\text{p}}Q_{i,1,\text{p}}\mathbf{p}_\text{a} \\
&\quad + \sum\nolimits_{i=2}^{m}(1 - D_{i,1,\text{p}})D^2_{i,1,\text{p}}Q_{i,1,\text{p}}\mathbf{p}_i, \\
\mathbf{H}_{13} &= \sum\nolimits_{i=2}^{m}[D_{i,1,\text{p}}(\mathbf{p}_\text{a}-\mathbf{p}_1)+D^2_{i,1,\text{p}}(\mathbf{p}_i-\mathbf{p}_\text{a})](-Q_{i,1,\text{p}}) \\
&= (\mathbf{p}_\text{a} - \mathbf{p}_1)\sum\nolimits_{i=2}^{m}(-D_{i,1,\text{p}}Q_{i,1,\text{p}}) + \sum\nolimits_{i=2}^{m} D^2_{i,1,\text{p}} \\
&\quad \times Q_{i,1,\text{p}}\mathbf{p}_\text{a} - \sum\nolimits_{i=2}^{m} D^2_{i,1,\text{p}}Q_{i,1,\text{p}}\mathbf{p}_i,
\end{aligned}$$

$$
\begin{aligned}
\mathbf{H}_{21} &= \mathbf{H}_{12}^{\mathrm{T}},\\
\mathbf{H}_{31} &= \mathbf{H}_{13}^{\mathrm{T}},\\
H_{22} &= \sum_{i=2}^{m}(1-D_{i,1,\mathrm{p}})^{2}Q_{i,1,\mathrm{p}}^{2},\\
H_{23} &= -\sum_{i=2}^{m}(1-D_{i,1,\mathrm{p}})Q_{i,1,\mathrm{p}}^{2},\\
H_{32} &= -\sum_{i=2}^{m}(1-D_{i,1,\mathrm{p}})Q_{i,1,\mathrm{p}}^{2},\\
H_{33} &= \sum_{i=2}^{m}Q_{i,1,\mathrm{p}}^{2},\\
Q_{21} &= \sum_{i=2}^{m}[(1-D_{i,1,\mathrm{p}})Q_{i,1,\mathrm{p}}](-Q_{i,1,\mathrm{p}}^{2}-(1\text{-}D_{i,1,\mathrm{p}})^{2}\mathbf{p}_{\mathrm{a}}^{\mathrm{T}}\mathbf{p}_{\mathrm{a}}\\
&\quad-\mathbf{p}_{1}^{\mathrm{T}}\mathbf{p}_{1}+2(1-D_{i,1,\mathrm{p}})\mathbf{p}_{\mathrm{a}}^{\mathrm{T}}\mathbf{p}_{1}+D_{i,1,\mathrm{p}}^{2}\mathbf{p}_{i}^{\mathrm{T}}\mathbf{p}_{i})\\
&= -\sum_{i=2}^{m}(1-D_{i,1,\mathrm{p}})Q_{i,1,\mathrm{p}}^{3}-(\mathbf{p}_{\mathrm{a}}-\mathbf{p}_{1})^{\mathrm{T}}(\mathbf{p}_{\mathrm{a}}-\mathbf{p}_{1})\sum_{i=2}^{m}(1-D_{i,1,\mathrm{p}})\\
&\quad\times Q_{i,1,\mathrm{p}}+2\mathbf{p}_{\mathrm{a}}^{\mathrm{T}}(\mathbf{p}_{\mathrm{a}}-\mathbf{p}_{1})\sum_{i=2}^{m}(1-D_{i,1,\mathrm{p}})D_{i,1,\mathrm{p}}Q_{i,1,\mathrm{p}}+(-\mathbf{p}_{\mathrm{a}}^{\mathrm{T}}\mathbf{p}_{\mathrm{a}}\\
&\quad\times\sum_{i=2}^{m}(1-D_{i,1,\mathrm{p}})D_{i,1,\mathrm{p}}^{2}Q_{i,1,\mathrm{p}}+\sum_{i=2}^{m}(1-D_{i,1,\mathrm{p}})D_{i,1,\mathrm{p}}^{2}Q_{i,1,\mathrm{p}}\mathbf{p}_{i}^{\mathrm{T}}\mathbf{p}_{i}),
\end{aligned}
\tag{5.27}
$$

$$
\begin{aligned}
\mathbf{Q}_{11} &= \sum_{i=2}^{m}[D_{i,1,\mathrm{p}}(\mathbf{p}_{\mathrm{a}}-\mathbf{p}_{1})+D_{i,1,\mathrm{p}}^{2}(\mathbf{p}_{i}-\mathbf{p}_{\mathrm{a}})](-Q_{i,1,\mathrm{p}}^{2}-(1\\
&\quad-D_{i,1,\mathrm{p}})^{2}\mathbf{p}_{\mathrm{a}}^{\mathrm{T}}\mathbf{p}_{\mathrm{a}}-\mathbf{p}_{1}^{\mathrm{T}}\mathbf{p}_{1}+2(1-D_{i,1,\mathrm{p}})\mathbf{p}_{\mathrm{a}}^{\mathrm{T}}\mathbf{p}_{1}+D_{i,1,\mathrm{p}}^{2}\mathbf{p}_{i}^{\mathrm{T}}\mathbf{p}_{i})\\
&= (\mathbf{p}_{\mathrm{a}}-\mathbf{p}_{1})\sum_{i=2}^{m}(-D_{i,1,\mathrm{p}}Q_{i,1,\mathrm{p}}^{2})+(\mathbf{p}_{\mathrm{a}}\sum_{i=2}^{m}D_{i,1,\mathrm{p}}^{2}Q_{i,1,\mathrm{p}}^{2}-\sum_{i=2}^{m}\\
&\quad D_{i,1,\mathrm{p}}^{2}Q_{i,1,\mathrm{p}}^{2}\mathbf{p}_{i})+(\mathbf{p}_{\mathrm{a}}-\mathbf{p}_{1})(\mathbf{p}_{\mathrm{a}}^{\mathrm{T}}\mathbf{p}_{\mathrm{a}}+\mathbf{p}_{1}^{\mathrm{T}}\mathbf{p}_{1})\sum_{i=2}^{m}(-D_{i,1,\mathrm{p}})\\
&\quad+2(\mathbf{p}_{\mathrm{a}}-\mathbf{p}_{1})\mathbf{p}_{\mathrm{a}}^{\mathrm{T}}\mathbf{p}_{1}\sum_{i=2}^{m}D_{i,1,\mathrm{p}}+(\mathbf{p}_{\mathrm{a}}\sum_{i=2}^{m}D_{i,1,\mathrm{p}}^{2}-\sum_{i=2}^{m}\\
&\quad D_{i,1,\mathrm{p}}^{2}\mathbf{p}_{i})(\mathbf{p}_{\mathrm{a}}^{\mathrm{T}}\mathbf{p}_{\mathrm{a}}+\mathbf{p}_{1}^{\mathrm{T}}\mathbf{p}_{1})+2(\mathbf{p}_{\mathrm{a}}-\mathbf{p}_{1})\mathbf{p}_{\mathrm{a}}^{\mathrm{T}}(\mathbf{p}_{\mathrm{a}}-\mathbf{p}_{1})\sum_{i=2}^{m}\\
&\quad D_{i,1,\mathrm{p}}^{2}+2(-\mathbf{p}_{\mathrm{a}}\sum_{i=2}^{m}D_{i,1,\mathrm{p}}^{2}+\sum_{i=2}^{m}D_{i,1,\mathrm{p}}^{2}\mathbf{p}_{i})\mathbf{p}_{\mathrm{a}}^{\mathrm{T}}\mathbf{p}_{1}+2(-\mathbf{p}_{\mathrm{a}}\\
&\quad\sum_{i=2}^{m}D_{i,1,\mathrm{p}}^{3}+\sum_{i=2}^{m}D_{i,1,\mathrm{p}}^{3}\mathbf{p}_{i})\mathbf{p}_{\mathrm{a}}^{\mathrm{T}}\mathbf{p}_{\mathrm{a}}+(\mathbf{p}_{\mathrm{a}}\text{-}\mathbf{p}_{1})(-\mathbf{p}_{\mathrm{a}}^{\mathrm{T}}\mathbf{p}_{\mathrm{a}}\sum_{i=2}^{m}\\
&\quad D_{i,1,\mathrm{p}}^{3}+\sum_{i=2}^{m}D_{i,1,\mathrm{p}}^{3}\mathbf{p}_{i}^{\mathrm{T}}\mathbf{p}_{i})+2(\mathbf{p}_{\mathrm{a}}\sum_{i=2}^{m}D_{i,1,\mathrm{p}}^{3}-\sum_{i=2}^{m}D_{i,1,\mathrm{p}}^{3}\\
&\quad\times\mathbf{p}_{i})\mathbf{p}_{\mathrm{a}}^{\mathrm{T}}\mathbf{p}_{1}+(\mathbf{p}_{\mathrm{a}}\mathbf{p}_{\mathrm{a}}^{\mathrm{T}}\mathbf{p}_{\mathrm{a}}\sum_{i=2}^{m}D_{i,1,\mathrm{p}}^{4}+\sum_{i=2}^{m}D_{i,1,\mathrm{p}}^{4}\mathbf{p}_{i}\mathbf{p}_{i}^{\mathrm{T}}\mathbf{p}_{i})\\
&\quad-\sum_{i=2}^{m}D_{i,1,\mathrm{p}}^{4}\mathbf{p}_{i}\mathbf{p}_{\mathrm{a}}^{\mathrm{T}}\mathbf{p}_{\mathrm{a}}-\sum_{i=2}^{m}D_{i,1,\mathrm{p}}^{4}\mathbf{p}_{\mathrm{a}}\mathbf{p}_{i}^{\mathrm{T}}\mathbf{p}_{i},\\
Q_{31} &= \sum_{i=2}^{m}(-Q_{i,1,\mathrm{p}})(-Q_{i,1,\mathrm{p}}^{2}-(1-D_{i,1,\mathrm{p}})^{2}\mathbf{p}_{\mathrm{a}}^{\mathrm{T}}\mathbf{p}_{\mathrm{a}}\\
&\quad-\mathbf{p}_{1}^{\mathrm{T}}\mathbf{p}_{1}+2(1-D_{i,1,\mathrm{p}})\mathbf{p}_{\mathrm{a}}^{\mathrm{T}}\mathbf{p}_{1}+D_{i,1,\mathrm{p}}^{2}\mathbf{p}_{i}^{\mathrm{T}}\mathbf{p}_{i})\\
&= \sum_{i=2}^{m}Q_{i,1,\mathrm{p}}^{3}+(\mathbf{p}_{\mathrm{a}}-\mathbf{p}_{1})^{\mathrm{T}}(\mathbf{p}_{\mathrm{a}}-\mathbf{p}_{1})\sum_{i=2}^{m}Q_{i,1,\mathrm{p}}-2\mathbf{p}_{\mathrm{a}}^{\mathrm{T}}(\mathbf{p}_{\mathrm{a}}-\mathbf{p}_{1})\sum_{i=2}^{m}\\
&\quad\times D_{i,1,\mathrm{p}}Q_{i,1,\mathrm{p}}+(\mathbf{p}_{\mathrm{a}}^{\mathrm{T}}\mathbf{p}_{\mathrm{a}}\sum_{i=2}^{m}D_{i,1,\mathrm{p}}^{2}Q_{i,1,\mathrm{p}}-\sum_{i=2}^{m}D_{i,1,\mathrm{p}}^{2}Q_{i,1,\mathrm{p}}\mathbf{p}_{i}^{\mathrm{T}}\mathbf{p}_{i}).
\end{aligned}
\tag{5.28}
$$

Therefore, the estimation of $\mathbf{x}_\mathrm{p}$ in (5.26) can be rewritten as

$$\hat{\mathbf{x}}_{\mathrm{p,indirect}} = \frac{1}{2}\begin{bmatrix} \mathbf{H}_{11} & \mathbf{H}_{12} & \mathbf{H}_{13} \\ \mathbf{H}_{21} & H_{22} & H_{23} \\ \mathbf{H}_{31} & H_{32} & H_{33} \end{bmatrix}^{-1} \begin{bmatrix} \mathbf{Q}_{11} \\ Q_{21} \\ Q_{31} \end{bmatrix}. \tag{5.29}$$

Step 2: PPS and PPDP Based State Calculation

Similar to Sect. 5.3.1, the real states of active sensor node are denoted as $\bar{\mathbf{X}}_\mathrm{a} = [-\mathbf{p}_\mathrm{a}, \mathbf{p}_\mathrm{a}\mathbf{p}_\mathrm{a}^\mathrm{T}, -\mathbf{p}_\mathrm{a}, \mathbf{p}_\mathrm{a}, \mathbf{p}_\mathrm{a}, \mathbf{p}_\mathrm{a}, (0; -\mathbf{p}_\mathrm{a}^\mathrm{T}\mathbf{p}_\mathrm{a}; 0), \mathbf{p}_\mathrm{a}\mathbf{p}_\mathrm{a}^\mathrm{T}\mathbf{p}_\mathrm{a}, (0; -\mathbf{p}_\mathrm{a}^\mathrm{T}\mathbf{p}_\mathrm{a}; 0), (0; \mathbf{p}_\mathrm{a}^\mathrm{T}\mathbf{p}_\mathrm{a}; 0)]$, $\bar{\mathbf{Y}}_\mathrm{a} = [\mathbf{p}_\mathrm{a}, (0; \mathbf{p}_\mathrm{a}^\mathrm{T}\mathbf{p}_\mathrm{a}; 0)]$, $\bar{\mathbf{Z}}_\mathrm{a} = [\mathbf{p}_\mathrm{a}^\mathrm{T}; (0, \mathbf{p}_\mathrm{a}^\mathrm{T}\mathbf{p}_\mathrm{a}, 0); \mathrm{diag}\{\mathbf{p}_\mathrm{a}\}]$ and $\bar{\mathbf{M}}_\mathrm{a} = \mathbf{p}_\mathrm{a}^\mathrm{T}$. The real states of anchor node 1 are $\bar{\mathbf{Y}}_1 = [-\mathbf{p}_1, (0; \mathbf{p}_1^\mathrm{T}\mathbf{p}_1; 0)]$ and $\bar{\mathbf{M}}_1 = \mathbf{p}_1$. The real states of anchor node $i \in \{2, ..., m\}$ are $\bar{\mathbf{X}}_i = [\mathbf{p}_i, \mathbf{p}_i\mathbf{p}_i^\mathrm{T}, \mathbf{p}_i, -\mathbf{p}_i, -\mathbf{p}_i, -\mathbf{p}_i, (0; \mathbf{p}_i^\mathrm{T}\mathbf{p}_i; 0), \mathbf{p}_i\mathbf{p}_i^\mathrm{T}\mathbf{p}_i, (0; \mathbf{p}_i^\mathrm{T}\mathbf{p}_i; 0), (0; -\mathbf{p}_i^\mathrm{T}\mathbf{p}_i; 0)]$, $\bar{\mathbf{Z}}_i = [\mathbf{p}_i, \mathbf{p}_i, \mathrm{diag}\{\mathbf{p}_i^\mathrm{T}\mathbf{p}_i; \mathbf{p}_i^\mathrm{T}\mathbf{p}_i; \mathbf{p}_i^\mathrm{T}\mathbf{p}_i\}]$, and $\bar{\mathbf{M}}_2 = \mathbf{p}_\mathrm{a} - \mathbf{p}_1$.

To protect the privacy of $\bar{\mathbf{X}}_\mathrm{a}$ and $\bar{\mathbf{X}}_i$, the random matrix $\mathbf{W}_{1,\bar{i}\bar{j}}$ is generated for node $\bar{i}$ where node $\bar{i}$ is active sensor node and anchor node i, i.e., $\bar{i} \in \{\mathrm{a}, 2, ..., m\}$ and $\bar{j} \in \{\mathrm{a}, 2, ..., m\}$. Note that $\mathbf{W}_{1,\bar{i}a} + \sum_{j=2}^{m}\mathbf{W}_{1,\bar{i}\bar{j}} = \mathbf{0}$. Subsequently, node $\bar{i}$ keeps $\mathbf{W}_{1,\bar{i}\bar{i}}$ and sends the rest to the other $m-1$ nodes. By adding the received matrices to $\mathbf{W}_{1,\bar{i}\bar{i}}$, a random matrix $\boldsymbol{\varphi}_{1\bar{i}}$ can be acquired. Denote the k-th column of $\boldsymbol{\varphi}_{1i}$ as $\boldsymbol{\varphi}_{1i,k}$ for $k \in \{1, ..., 12\}$. Based on this, the disguised state of $\bar{\mathbf{X}}_{\bar{i}}$ can be denoted as $\bar{\bar{\mathbf{X}}}_\mathrm{a} = \bar{\mathbf{X}}_\mathrm{a} + \boldsymbol{\varphi}_{1\mathrm{a}}$ or $\bar{\bar{\mathbf{X}}}_i = \bar{\mathbf{X}}_i + \mathbf{K}_i$, where $\mathbf{K}_i = [\frac{\boldsymbol{\varphi}_{1i,1}}{\vartheta_{2i}}, \frac{\boldsymbol{\varphi}_{1i,2}}{\vartheta_{3i}}, \frac{\boldsymbol{\varphi}_{1i,3}}{\vartheta_{3i}}, \frac{\boldsymbol{\varphi}_{1i,4}}{\vartheta_{3i}}, \frac{\boldsymbol{\varphi}_{1i,5}}{\vartheta_{4i}}, \frac{\boldsymbol{\varphi}_{1i,6}}{\vartheta_{5i}}, \frac{\boldsymbol{\varphi}_{1i,7}}{\vartheta_{6i}}, \frac{\boldsymbol{\varphi}_{1i,8}}{\vartheta_{i}}, \frac{\boldsymbol{\varphi}_{1i,9}}{\vartheta_{2i}}, \frac{\boldsymbol{\varphi}_{1i,10}}{\vartheta_{3i}}, \frac{\boldsymbol{\varphi}_{1i,11}}{\vartheta_{4i}}, \frac{\boldsymbol{\varphi}_{1i,12}}{\vartheta_{5i}}]$, $\vartheta_i = \frac{D_{i,1,p}^2}{\sum_{i=2}^{m}D_{i,1,p}^2}$, $\vartheta_{2i} = \frac{D_{i,1,p}^3}{\sum_{i=2}^{m}D_{i,1,p}^3}$, $\vartheta_{3i} = \frac{D_{i,1,p}^4}{\sum_{i=2}^{m}D_{i,1,p}^4}$, $\vartheta_{4i} = \frac{(1-D_{i,1,p})D_{i,1,p}^2 Q_{i,1,p}}{\sum_{i=2}^{m}(1-D_{i,1,p})D_{i,1,p}^2 Q_{i,1,p}}$, $\vartheta_{5i} = \frac{D_{i,1,p}^2 Q_{i,1,p}}{\sum_{i=2}^{m}D_{i,1,p}^2 Q_{i,1,p}}$, and $\vartheta_{6i} = \frac{D_{i,1,p}^2 Q_{i,1,p}^2}{\sum_{i=2}^{m}D_{i,1,p}^2 Q_{i,1,p}^2}$ for $i \in \{2, ..., m\}$. Then, active sensor node sends disguised state $\bar{\bar{\mathbf{X}}}_\mathrm{a}$, while anchor node $i \in \{2, ..., m\}$ sends disguised state $\bar{\bar{\mathbf{X}}}_i$ to the networks. Obviously, this process is based on the PPS strategy, while the real states of active sensor and anchor nodes cannot be revealed.

With the similar strategy, the state $\bar{\mathbf{Y}}_\mathrm{a}$ and $\bar{\mathbf{Y}}_1$ can also be disguised, i.e., $\bar{\bar{\mathbf{Y}}}_\mathrm{a} = \bar{\mathbf{Y}}_\mathrm{a} + \boldsymbol{\varphi}_{2\mathrm{a}}$ and $\bar{\bar{\mathbf{Y}}}_1 = \bar{\mathbf{Y}}_1 + \boldsymbol{\varphi}_{21}$, where $\boldsymbol{\varphi}_{2\mathrm{a}}$ and $\boldsymbol{\varphi}_{21}$ denote the random noise generated by active sensor node and anchor node 1, respectively. Subsequently, active sensor node and anchor node 1 send $\bar{\bar{\mathbf{Y}}}_\mathrm{a}$ and $\bar{\bar{\mathbf{Y}}}_1$ to the networks. To protect the privacy of $\bar{\mathbf{Z}}_i$ for $i \in \{2, ..., m\}$, the ordinary sensor node randomly generates matrix $\boldsymbol{\varphi}_{3i}$ to anchor node i. Denote the k-th column of $\boldsymbol{\varphi}_{3i}$ as $\boldsymbol{\varphi}_{3i,k}$ for $k \in \{1, 2, 3, 4, 5\}$. As a result, the disguised state of $\bar{\mathbf{Z}}_i$ is denoted as $\bar{\bar{\mathbf{Z}}}_i = \bar{\mathbf{Z}}_i + \mathbf{K}_{3i}$, where $\mathbf{K}_{3i} = [\frac{\boldsymbol{\varphi}_{3i,1}}{\vartheta_{3i}}, \frac{\boldsymbol{\varphi}_{3i,2}}{\vartheta_{3i}}, \frac{\boldsymbol{\varphi}_{3i,3}}{\vartheta_{3i}}, \frac{\boldsymbol{\varphi}_{3i,4}}{\vartheta_{3i}}, \frac{\boldsymbol{\varphi}_{3i,5}}{\vartheta_{3i}}]$ for $i \in \{2, ..., m\}$. Then, anchor node i sends $\bar{\bar{\mathbf{Z}}}_i$ to active sensor node. Active sensor node acquires $\bar{\bar{\mathbf{Z}}}_i\bar{\mathbf{Z}}_\mathrm{a}$ and sends it to the networks. For clear description, calculation process of $\bar{\mathbf{Z}}_i\bar{\mathbf{Z}}_\mathrm{a}$ is described in Fig. 5.5a.

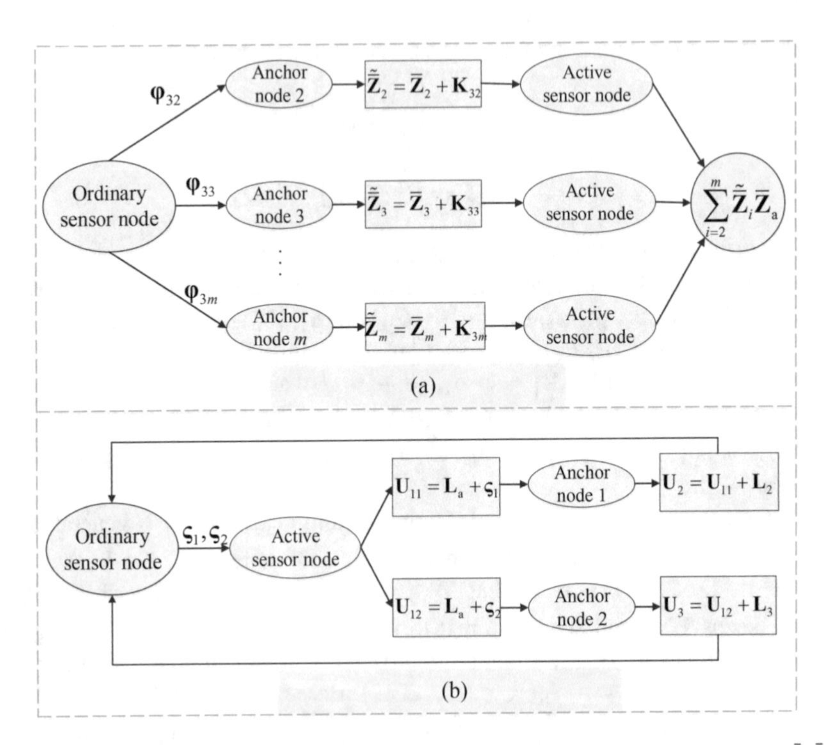

Fig. 5.5 Description of PPS and PPDP based state calculation. (**a**) Calculation process of $\bar{\mathbf{Z}}_i\bar{\mathbf{Z}}_a$. (**b**) Calculation process of $\mathbf{U}_2$ and $\mathbf{U}_3$

To protect the privacies of $\bar{\mathbf{M}}_a$, $\bar{\mathbf{M}}_1$ and $\bar{\mathbf{M}}_2$, we construct $\mathbf{L}_a$ =diag{log $\bar{\mathbf{M}}_a$}, $\mathbf{L}_1$ =diag{log $\bar{\mathbf{M}}_1$} and $\mathbf{L}_2$ =diag{log $\bar{\mathbf{M}}_2$} for active sensor node, anchor node 1, anchor node 2, respectively. This process is named as PPDP, because the adoption of multiplication. Subsequently, the ordinary sensor node generates two random matrices ς_1 and ς_2 with the same size as $\mathbf{L}_1$ and sends them to active sensor node. The active sensor node calculates $\mathbf{U}_{11} = \mathbf{L}_a + \varsigma_1$ and $\mathbf{U}_{12} = \mathbf{L}_a + \varsigma_2$, and then sends them to the anchor nodes 1 and 2, respectively. Then, anchor node 1 calculates $\mathbf{U}_2 = \mathbf{U}_{11} + \mathbf{L}_1$ and sends it to ordinary sensor node. Meantime, anchor node 2 calculates $\mathbf{U}_3 = \mathbf{U}_{12} + \mathbf{L}_2$ and sends it to ordinary sensor node. The ordinary sensor node retrieves the logarithm matrix by subtracting ς_1 and ς_2, i.e., $\mathbf{U}_2 - \varsigma_1 = \mathbf{L}_a + \mathbf{L}_1$ =diag{log $\bar{\mathbf{M}}_a\bar{\mathbf{M}}_1$} and $\mathbf{U}_3 - \varsigma_2 = \mathbf{L}_a + \mathbf{L}_2$ =diag{log $\bar{\mathbf{M}}_a\bar{\mathbf{M}}_2$}. Through the exponential transform, the values of $\bar{\mathbf{M}}_a\bar{\mathbf{M}}_1$ and $\bar{\mathbf{M}}_a\bar{\mathbf{M}}_2$ can be acquired. This process is shown in Fig. 5.5b.

Step 3: Calculation of **H** *and* **Q**
According to (5.29), the values of $\mathbf{H}_{11}$, $\mathbf{H}_{12}$, $\mathbf{H}_{13}$, $\mathbf{H}_{21}$, $\mathbf{H}_{31}$, H_{32}, $\mathbf{Q}_{11}$, Q_{21} and Q_{31} can be acquired via $\tilde{\bar{\mathbf{X}}}_a$, $\tilde{\bar{\mathbf{X}}}_i$, $\tilde{\bar{\mathbf{Y}}}_a$, $\tilde{\bar{\mathbf{Y}}}_1$, $\tilde{\bar{\mathbf{Z}}}_i\bar{\mathbf{Z}}_a$, $\mathbf{U}_2$ and $\mathbf{U}_3$. Meantime, the values of H_{22}, H_{23}, H_{32} and H_{33} can be directly obtained, because $D_{i,1,\text{p}}$ and $Q_{i,1,\text{p}}$ are pre-

known. Accordingly, the values of **H** and **Q** are calculated, and more importantly, the privacy can be protected.

Step 4: Position Estimation
According to Step 3, $\mathbf{x}_{\mathrm{p}}$ in (5.29) can be easily estimated, i.e., the position of ordinary sensor node can be estimated under the constraints of asynchronous clock and privacy preservation.

5.3.3 Consequence when There Exist Dishonest Nodes

The proposed privacy preserving algorithms need cooperation among nodes, and require each node to be honest. Nevertheless, underwater nodes are usually deployed in harsh and open environments, making them extremely vulnerable to security attack. For example, an attacker can cause nodes to have imprecise locations, which leads to the dishonesty of nodes. To deal with this issue, we give the consequence when there exist dishonest nodes.

Without loss of generality, it is assumed that all the nodes are honest in the first localization interval, i.e., $[T_{\mathrm{a,a}}, \delta\alpha_{\mathrm{a}}]$ for the active sensor node and $[T_{\mathrm{a,p}}, \delta\alpha_{\mathrm{p}}]$ for the ordinary sensor node. As a result, from (5.4), the position vector of active sensor node at timestamp $T^{\#}_{\mathrm{a,a}}$ is predicted as

$$\mathbf{p}_{T^{\#}_{\mathrm{a,a}}} = \mathbf{p}_{\mathrm{a}} + \delta\alpha_{\mathrm{a}} \sum_{m_{\mathrm{a}}=0}^{n^{\#}_{\mathrm{a}}} \mathbf{v}_{\mathrm{a}}(\mathbf{T}_{\mathrm{a,a}}). \tag{5.30}$$

By using the PPS-based localization algorithm in Sect. 5.3.1, the estimated position of active sensor node at timestamp $T^{\#}_{\mathrm{a,a}}$ can be denoted as $\tilde{\mathbf{p}}_{T^{\#}_{\mathrm{a,a}}}$. Therefore, the difference between $\mathbf{p}_{T^{\#}_{\mathrm{a,a}}}$ and $\tilde{\mathbf{p}}_{T^{\#}_{\mathrm{a,a}}}$ can be expressed as $\mathrm{e}_{\mathrm{a}}(T^{\#}_{\mathrm{a,a}}) = \left|\mathbf{p}_{T^{\#}_{\mathrm{a,a}}} - \tilde{\mathbf{p}}_{T^{\#}_{\mathrm{a,a}}}\right|$.

Similar to active sensor node, from (5.5), the position vector of ordinary sensor node at timestamp $T^{\#}_{\mathrm{a,p}}$ can be predicted as

$$\mathbf{p}_{T^{\#}_{\mathrm{a,p}}} = \mathbf{p}_{\mathrm{p}} + \delta\alpha_{\mathrm{p}} \sum_{m_{\mathrm{p}}=0}^{n^{\#}_{\mathrm{p}}} \mathbf{v}_{\mathrm{p}}(\mathbf{T}_{\mathrm{a,p}}). \tag{5.31}$$

With the PPS and PPDP based localization algorithm in Sect. 5.3.2, the estimated position of ordinary sensor node at timestamp $T^{\#}_{\mathrm{a,p}}$ can be denoted as $\tilde{\mathbf{p}}_{T^{\#}_{\mathrm{a,p}}}$. Accordingly, the difference between $\mathbf{p}_{T^{\#}_{\mathrm{a,p}}}$ and $\tilde{\mathbf{p}}_{T^{\#}_{\mathrm{a,p}}}$ can be expressed as $\mathrm{e}_{\mathrm{p}}(T^{\#}_{\mathrm{a,p}}) = \left|\mathbf{p}_{T^{\#}_{\mathrm{a,p}}} - \tilde{\mathbf{p}}_{T^{\#}_{\mathrm{a,p}}}\right|$.

In the following, let $\boldsymbol{\rho}$ denote the threshold of $\mathrm{e}_{\mathrm{a}}(T^{\#}_{\mathrm{a,a}})$, where $\boldsymbol{\rho}$ is dependent on the accuracy requirement for the localization system. Especially, when the condition of $\mathrm{e}_{\mathrm{a}}(T^{\#}_{\mathrm{a,a}}) > \boldsymbol{\rho}$ and $\mathrm{e}_{\mathrm{p}}(T^{\#}_{\mathrm{a,p}}) > \boldsymbol{\rho}$ holds, the nodes in USNs can be considered to be dishonest, otherwise the nodes in USNs are considered honest. In order to show more clearly, Fig. 5.6 is proposed to depict the consequence when there exist dishonest nodes.

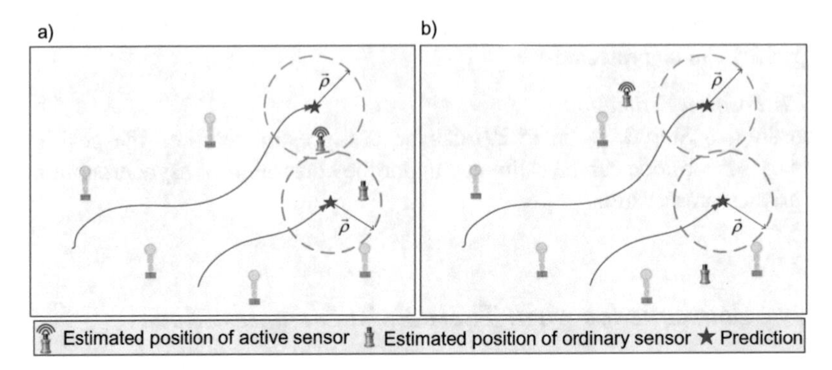

Fig. 5.6 Depiction of consequence when there exist dishonest nodes. (**a**) Nodes are honest with $e_a < \bar{\rho}$ and $e_p < \bar{\rho}$. (**b**) Nodes are dishonest with $e_a > \bar{\rho}$ and $e_p > \bar{\rho}$

5.4 Performance Analyses

5.4.1 Equivalence Analyses

At the beginning, we investigate the equivalences of estimators (5.12) and (5.26). Then, the following theorems can be given.

Theorem 5.1 *Consider the linear LS problem for active sensor node, as described in (5.10). By using the PPS-based asynchronous localization algorithm,* $\hat{\mathbf{X}}_{a,indirect}$ *in (5.12) is equivalent to the one in (5.11).*

Proof It is noted that, $\mathbf{A}_{11}$, $\mathbf{A}_{12}$, $\mathbf{A}_{21}$, $\mathbf{B}_{11}$, $\mathbf{B}_{12}$ and B_{21} can be derived from $\tilde{\mathbf{X}}_i$, $\tilde{\mathbf{Y}}_i$ and $\tilde{\mathbf{Z}}_i$. Particularly,

$$\begin{aligned}
&[\mathbf{A}_{12}, \mathbf{B}_{12}, (0;\, B_{21};\, 0)] \\
&= \tilde{\mathbf{Z}}_1 \sum_{i=2}^{m} \mathrm{diag}\{\Upsilon_{i,1}, \Upsilon_{i,1}^2, \Upsilon_{i,1}\} + \sum_{i=2}^{m} \tilde{\mathbf{Z}}_i \mathrm{diag}\{\Upsilon_{i,1}, \Upsilon_{i,1}^2, \Upsilon_{i,1}\} \\
&= \mathbf{Z}_1 \sum_{i=2}^{m} \mathrm{diag}\{\Upsilon_{i,1}, \Upsilon_{i,1}^2, \Upsilon_{i,1}\} + \sum_{i=2}^{m} \mathbf{Z}_i \mathrm{diag}\{\Upsilon_{i,1}, \Upsilon_{i,1}^2, \Upsilon_{i,1}\} \\
&\quad + \boldsymbol{\delta}_{31} \sum_{i=2}^{m} \mathrm{diag}\{\Upsilon_{i,1}, \Upsilon_{i,1}^2, \Upsilon_{i,1}\} + \sum_{i=2}^{m} \mathbf{W}_i \mathrm{diag}\{\Upsilon_{i,1}, \Upsilon_{i,1}^2, \Upsilon_{i,1}\}.
\end{aligned} \tag{5.32}$$

As $\boldsymbol{\delta}_{31} \sum_{i=2}^{m} \mathrm{diag}\{\Upsilon_{i,1}, \Upsilon_{i,1}^2, \Upsilon_{i,1}\} + \sum_{i=2}^{m} \mathbf{W}_i \mathrm{diag}\{\Upsilon_{i,1}, \Upsilon_{i,1}^2, \Upsilon_{i,1}\} = \mathbf{0}$, the value of $[\mathbf{A}_{12}, \mathbf{B}_{12}, (0;\, B_{21};\, 0)]$ is equivalent to the direct calculation, i.e., $\mathbf{A}_{12}$, $\mathbf{B}_{12}$ and B_{21} can be calculated correctly. Using this implementation to the other elements, one has $[\mathbf{A}_{11}, \mathbf{B}_{11}] = \sum_{i=1}^{m} \tilde{\mathbf{Y}}_i = \sum_{i=1}^{m} \mathbf{Y}_i + \sum_{i=1}^{m} \boldsymbol{\delta}_{2i} = \sum_{i=1}^{m} \mathbf{Y}_i$ because $\sum_{i=1}^{m} \boldsymbol{\delta}_{2i} = \mathbf{0}$. As a result, it is known that $\mathbf{A}_{11}$ and $\mathbf{B}_{11}$ can be correctly calculated.

It is noted that the values of A_{22} and B_{22} can be computed directly by $\Upsilon_{i,1}$. In addition, one has $\mathbf{A}_{21} = \mathbf{A}_{12}^{\mathrm{T}}$. Therefore, the values of $\mathbf{A}_{11}$, $\mathbf{A}_{12}$, $\mathbf{A}_{21}$, A_{22}, $\mathbf{B}_{11}$, $\mathbf{B}_{12}$, B_{21} and B_{22} are equivalent to the direct calculation. Thereby, we can know $\hat{\mathbf{X}}_{\mathrm{a,indirect}}$ is equivalent to $\hat{\mathbf{X}}_{\mathrm{a,direct}}$. □

Theorem 5.2 *Consider the linear LS problem for ordinary sensor node, as shown in (5.24). By using PPS and PPDP based asynchronous localization algorithm, estimation $\hat{\mathbf{X}}_{p,indirect}$ in (5.26) is equivalent to the one in (5.25).*

Proof Referring to the elements of $\mathbf{H}_{11}$, one knows that the matrix $\mathbf{p}_{\mathrm{a}} - \mathbf{p}_1$ can be acquired from $\bar{\bar{\mathbf{Y}}}_{\mathrm{a}} + \bar{\bar{\mathbf{Y}}}_1$. Because of $\boldsymbol{\varphi}_{2\mathrm{a}} + \boldsymbol{\varphi}_{21} = \mathbf{0}$, one has $\bar{\bar{\mathbf{Y}}}_{\mathrm{a}} + \bar{\bar{\mathbf{Y}}}_1 = \bar{\mathbf{Y}}_{\mathrm{a}} + \boldsymbol{\varphi}_{2\mathrm{a}} + \bar{\mathbf{Y}}_1 + \boldsymbol{\varphi}_{21} = \bar{\mathbf{Y}}_{\mathrm{a}} + \bar{\mathbf{Y}}_1$, i.e., $\mathbf{p}_{\mathrm{a}} - \mathbf{p}_1$ can be calculated correctly.

Then, the other elements of $\mathbf{H}_{11}$ are studied. Obviously, $-\mathbf{p}_{\mathrm{a}} \sum_{i=2}^{m} D_{i,1,\mathrm{p}}^{3} + \sum_{i=2}^{m} D_{i,1,\mathrm{p}}^{3} \mathbf{p}_i$ and $\sum_{i=2}^{m} D_{i,1,\mathrm{p}}^{4} \mathbf{p}_{\mathrm{a}} \mathbf{p}_{\mathrm{a}}^{\mathrm{T}} + \sum_{i=2}^{m} D_{i,1,\mathrm{p}}^{4} \mathbf{p}_i \mathbf{p}_i^{\mathrm{T}}$ can be acquired from $\bar{\bar{\mathbf{X}}}_{\mathrm{a}}$ and $\bar{\bar{\mathbf{X}}}_i$ for $i \in \{2, ..., m\}$. According to Step 2 in Sect. 5.3.2, one has

$$
\begin{aligned}
&\left[\sum_{i=2}^{m} D_{i,1,\mathrm{p}}^{3} \mathbf{p}_i - \mathbf{p}_{\mathrm{a}} \sum_{i=2}^{m} D_{i,1,\mathrm{p}}^{3}, \sum_{i=2}^{m} D_{i,1,\mathrm{p}}^{4} (\mathbf{p}_{\mathrm{a}} \mathbf{p}_{\mathrm{a}}^{\mathrm{T}} + \mathbf{p}_i \mathbf{p}_i^{\mathrm{T}})\right] \\
&= \bar{\bar{\mathbf{X}}}_{\mathrm{a}}(:, 1{:}4) \sum\nolimits_{i=2}^{m} \mathrm{diag}\{D_{i,1,\mathrm{p}}^{3}, D_{i,1,\mathrm{p}}^{4}, D_{i,1,\mathrm{p}}^{4}, D_{i,1,\mathrm{p}}^{4}\} \\
&\quad + \sum\nolimits_{i=2}^{m} \bar{\bar{\mathbf{X}}}_i(:, 1{:}4) \mathrm{diag}\{D_{i,1,\mathrm{p}}^{3}, D_{i,1,\mathrm{p}}^{4}, D_{i,1,\mathrm{p}}^{4}, D_{i,1,\mathrm{p}}^{4}\} \\
&= \bar{\mathbf{X}}_{\mathrm{a}}(:, 1{:}4) \sum\nolimits_{i=2}^{m} \mathrm{diag}\{D_{i,1,\mathrm{p}}^{3}, D_{i,1,\mathrm{p}}^{4}, D_{i,1,\mathrm{p}}^{4}, D_{i,1,\mathrm{p}}^{4}\} \\
&\quad + \sum\nolimits_{i=2}^{m} \bar{\mathbf{X}}_i(:, 1{:}4) \mathrm{diag}\{D_{i,1,\mathrm{p}}^{3}, D_{i,1,\mathrm{p}}^{4}, D_{i,1,\mathrm{p}}^{4}, D_{i,1,\mathrm{p}}^{4}\} \\
&\quad + \boldsymbol{\varphi}_{1\mathrm{a}}(:, 1{:}4) \sum\nolimits_{i=2}^{m} \mathrm{diag}\{D_{i,1,\mathrm{p}}^{3}, D_{i,1,\mathrm{p}}^{4}, D_{i,1,\mathrm{p}}^{4}, D_{i,1,\mathrm{p}}^{4}\} \\
&\quad + \sum\nolimits_{i=2}^{m} \mathbf{K}_i(:, 1{:}4) \mathrm{diag}\{D_{i,1,\mathrm{p}}^{3}, D_{i,1,\mathrm{p}}^{4}, D_{i,1,\mathrm{p}}^{4}, D_{i,1,\mathrm{p}}^{4}\}.
\end{aligned}
\tag{5.33}
$$

Note that $\boldsymbol{\varphi}_{1\mathrm{a}}(:, 1:4) \sum_{i=2}^{m} \mathrm{diag}\{D_{i,1,\mathrm{p}}^{3}, D_{i,1,\mathrm{p}}^{4}, D_{i,1,\mathrm{p}}^{4}, D_{i,1,\mathrm{p}}^{4}\} + \sum_{i=2}^{m} \mathbf{K}_i(:, 1:4) \mathrm{diag}\{D_{i,1,\mathrm{p}}^{3}, D_{i,1,\mathrm{p}}^{4}, D_{i,1,\mathrm{p}}^{4}, D_{i,1,\mathrm{p}}^{4}\} = \mathbf{0}$. Then, $-\mathbf{p}_{\mathrm{a}} \sum_{i=2}^{m} D_{i,1,\mathrm{p}}^{3} + \sum_{i=2}^{m} D_{i,1,\mathrm{p}}^{3} \mathbf{p}_i$ and $\sum_{i=2}^{m} D_{i,1,\mathrm{p}}^{4} \mathbf{p}_{\mathrm{a}} \mathbf{p}_{\mathrm{a}}^{\mathrm{T}} + \sum_{i=2}^{m} D_{i,1,\mathrm{p}}^{4} \mathbf{p}_i \mathbf{p}_i^{\mathrm{T}}$ can be calculated correctly. In addition, the last element of $\mathbf{H}_{11}$, i.e., $\sum_{i=2}^{m} D_{i,1,\mathrm{p}}^{4} \mathbf{p}_i \mathbf{p}_{\mathrm{a}}^{\mathrm{T}}$, can be acquired from $\bar{\bar{\mathbf{Z}}}_i \bar{\mathbf{Z}}_1$ for $i \in \{2, ..., m\}$. Because $\sum_{i=2}^{m} D_{i,1,\mathrm{p}}^{4} \mathbf{K}_{3i}(:, 1) \bar{\mathbf{Z}}_1(:, 1) = \mathbf{0}$, one has $\sum_{i=2}^{m} D_{i,1,\mathrm{p}}^{4} \mathbf{p}_i \mathbf{p}_{\mathrm{a}}^{\mathrm{T}} = \sum_{i=2}^{m} D_{i,1,\mathrm{p}}^{4} \bar{\bar{\mathbf{Z}}}_i(:, 1) \bar{\mathbf{Z}}_1(:, 1) = \sum_{i=2}^{m} D_{i,1,\mathrm{p}}^{4} \bar{\mathbf{Z}}_i(:, 1) \bar{\mathbf{Z}}_1(:, 1)$. Based on this, we know $\sum_{i=2}^{m} D_{i,1,\mathrm{p}}^{4} \mathbf{p}_i \mathbf{p}_{\mathrm{a}}^{\mathrm{T}}$ can be calculated.

Based on this, we know $\mathbf{H}_{11}$ can be calculated correctly. Using this implementation to the other parts, the values of $\mathbf{H}_{12}$, $\mathbf{H}_{13}$, $\mathbf{H}_{21}$, $\mathbf{H}_{31}$, $\mathbf{Q}_{11}$, Q_{21} and Q_{31} can be correctly obtained. As H_{22}, H_{23}, H_{32} and H_{33} are pre-known, $\hat{\mathbf{x}}_{\mathrm{p,indirect}}$ in (5.26) is equivalent to $\hat{\mathbf{x}}_{\mathrm{p,direct}}$ in (5.25). □

5.4.2 Level of Privacy Preservation

In the following, the level of privacy preservation for the proposed localization algorithms is investigated. Then, the following definition is provided.

Definition 5.1 Shi and Wu (2018) for node $\mathcal{A}$, it needs extra information to construct at least N_{p} independent equations to estimate the private position of node $\mathcal{B}$. Then, node $\mathcal{B}$ can preserve $\{N_{\mathrm{p}}\}$-*Privacy* to node $\mathcal{A}$, where $N_{\mathrm{p}} = N_{\mathrm{scal}} - N_{\mathrm{eq}}$. It is noted that, N_{scal} is the number of unknown scalar variables in the private information of node $\mathcal{B}$, and N_{eq} denotes maximum number of independent equations that node $\mathcal{A}$ can construct.

Based on Definition 5.1, the following Theorem is given.

Theorem 5.3 *With the LS estimator (5.12), the PPS-based localization algorithm for active sensor node can guarantee the following levels of privacy preservation: (1) To anchor node* 1*, the active sensor node can preserve* $\{3\}$*-Privacy while anchor node* $i \in \{2, ..., m\}$ *can preserve* $\{3m - 7\}$*-Privacy; (2) To anchor node* $i \in \{2, ..., m\}$*, the active sensor node and anchor node* $j \in \{1, ..., m\}/\{i\}$ *can preserve* $\{3m - 2\}$*-Privacy; (3) To active sensor node, if* $m > \frac{15}{2}$*, anchor node* $i \in \{1, ..., m\}$ *can preserve* $\{3m - (15 + m)\}$*-Privacy.*

Proof For anchor node 1, we have the values of $\sum_{i=2}^{m} \mathbf{p}_i$ and $\sum_{i=2}^{m} \mathbf{p}_i^{\mathrm{T}} \mathbf{p}_i$, where the number of independent equations is at most 4 and the number of unknown scalar variables is $3(m - 1)$. Especially, anchor node 1 has no information about active sensor node. With regard to this case, active sensor node can preserve $\{3\}$-*Privacy* to anchor node 1. Since $3(m - 1) > 4$ always holds, anchor node 1 is unable to know the other anchor nodes' locations. As a result, anchor node $i \in \{2, ..., m\}$ can preserve $\{3(m - 1) - 4\}$-*Privacy* to anchor node 1, which implies the statement (1) holds.

For anchor node $i \in \{2, ..., m\}$, it only has the range difference ratios γ_i and γ_{2i}, which can construct at most 2 independent equations. The number of unknown scalar variables to anchor node $i \in \{2, ..., m\}$ is $3m$. Therefore, the positions of active sensor node and anchor node $j \in \{1, ..., m\}/\{i\}$ can be protected since $3m > 2$ holds. Thus, statement (2) holds.

For active sensor node, the available information includes: range difference $\Upsilon_{i,1}$, $i \in \{2, ..., m\}$, $\mathbf{A}_{11}$, $\mathbf{A}_{12}$, $\mathbf{A}_{21}$, $\mathbf{B}_{11}$, $\mathbf{B}_{12}$ and B_{21}. Obviously, $\mathbf{A}_{11}$ is a symmetric matrix, and then active sensor node can construct at most 6 independent equations. For $\Upsilon_{i,1}$, $\mathbf{A}_{12}$, $\mathbf{B}_{11}$, $\mathbf{B}_{12}$ and B_{21}, the maximum numbers of independent equations that active sensor node can build are $m - 1$, 3, 3, 3 and 1, respectively. Accordingly, the total number of independent equations is at most $15 + m$. The number of the unknown scalars to active sensor node is $3m$. Then, the active sensor node cannot know anchor nodes' locations when the condition of $m > \frac{15}{2}$ holds. That means anchor node $i \in \{1, ..., m\}$ can preserve $\{3m - (15 + m)\}$-*Privacy*, i.e, the statement 3) holds. □

Besides that, the level of privacy preservation for the PPS and PPDP based localization algorithm is investigated.

Theorem 5.4 *Consider the LS estimator (5.26), we have the following levels of privacy preservation: (1) To active sensor node, ordinary sensor node can preserve* $\{3\}$*-Privacy while anchor node* $i \in \{1, ..., m\}$ *can preserve* $\{3m\}$*-Privacy; (2) To anchor node 1, ordinary sensor node can preserve* $\{3\}$*-Privacy while active sensor node and anchor node* $j \in \{2, ..., m\}$ *can preserve* $\{3m\}$*-Privacy; (3) To anchor node 2, ordinary sensor node, active sensor node and anchor node* $j \in \{1, ..., m\}/\{2\}$ *can preserve* $\{3(m+1)-9\}$*-Privacy; (4) To anchor node* $i \in \{3, ..., m\}$*, ordinary sensor node, active sensor node and anchor node* $j \in \{1, ..., m\}/\{i\}$ *can preserve* $\{3(m+1)-6\}$*-Privacy; (5) To ordinary sensor node, if* $m > 12$*, anchor node i can preserve* $\{3(m+1)-(2m+15)\}$*-Privacy.*

Proof For active sensor node, the position information of anchor node $i \in \{1, ..., m\}$ and ordinary sensor node is not available. Therefore, ordinary sensor node can preserve $\{3\}$-*Privacy* and anchor node $i \in \{1, ..., m\}$ can preserve $\{3m\}$-*Privacy*. Hence, statement (1) holds. The proof of statement (2) is similar to the one of statement (1), thus it is omitted.

For anchor node 2, we know the values of $\mathbf{p}_{\mathrm{a}} - \mathbf{p}_1$, ϑ_3, ϑ_{23}, ϑ_{33}, ϑ_{43}, ϑ_{53} and ϑ_{63}. Based on this, at most 9 independent equations can be constructed. It is worth mentioning that, the number of unknown scalar variables to anchor node 2 is $3(m+1)$. Therefore, the positions of ordinary sensor node, active sensor node and anchor node $j \in \{1, ..., m\}/\{2\}$ can be protected since $3(m+1) > 9$ always holds. Thus, the statement (3) holds. The proof of statement (4) is similar to the one of statement (3), and hence it is omitted.

For ordinary sensor node, the available information includes: $D_{i,1,\mathrm{p}}$, $Q_{i,1,\mathrm{p}}$, $i \in \{2, ..., m\}$, $\mathbf{H}_{11}$, $\mathbf{H}_{12}$, $\mathbf{H}_{13}$, $\mathbf{H}_{21}$, $\mathbf{H}_{31}$, $\mathbf{Q}_{11}$, Q_{21} and Q_{31}. Obviously, $\mathbf{H}_{11}$ is a symmetric matrix. Therefore, the ordinary sensor node can construct at most 6 independent equations. For $D_{i,1,\mathrm{p}}$, $Q_{i,1,\mathrm{p}}$, $\mathbf{H}_{12}$, $\mathbf{H}_{13}$, $\mathbf{Q}_{11}$, Q_{21} and Q_{31}, the maximum numbers of linear equations that ordinary sensor node can construct are given as $m-1$, $m-1$, 3, 3, 3, 1 and 1, respectively. Accordingly, the total number of independent equations is at most $2m+15$. Since the number of the unknown scalars to ordinary sensor node is $3(m+1)$, the ordinary sensor node cannot know the position of anchor nodes when the condition $m > 12$ holds. That means anchor node $i \in \{1, ..., m\}$ can preserve $\{3(m+1)-(2m+15)\}$-*Privacy*, i.e, the statement (5) holds. □

5.4.3 Collision Avoidance of Packet

Corollary 5.1 *Define* $\epsilon = (d_{a,i} - d_{a,j} - d_{j,i})/c - t_{r,j}$, $\epsilon_1 = (d_{a,p} - d_{a,j} - d_{j,p})/c - t_{r,j}$, $\epsilon_2 = (d_{j,p} - d_{j,i} - d_{i,p})/c - t_{r,i}$, *and* $\epsilon_3 = (d_{a,j} - d_{j,i} - d_{a,i})/c - t_{r,i}$*. For* $i \in \{1, ...m\}$ *and* $j \in \{1, ..., i-1\}$*, packet collision can be avoided if* $t_{j,j} - t_{a,j} > \max\{\epsilon, \epsilon_1\}$ *and* $t_{i,i} - t_{j,i} > \max\{\epsilon_2, \epsilon_2\}$ *are both satisfied.*

Proof At the beginning, the time difference when anchor node i receives packet from active sensor node and anchor node j is expressed as $\Delta_{j,i}$, the time difference when ordinary sensor node receives packet from active sensor node and anchor node j is expressed as $\Delta_{j,\mathrm{p}}$, the time difference when ordinary sensor node receives packet from anchor node j and anchor node i is expressed as $\Delta_{i,\mathrm{p}}$, while the time difference when active sensor node receives packet from anchor node j and anchor node i is expressed as $\Delta_{i,\mathrm{a}}$. According to these definitions and noting Fig. 5.2, we have the following relationship

$$\begin{aligned} &\Delta_{j,i} = t_{j,i} - t_{\mathrm{a},i},\ \Delta_{j,\mathrm{p}} = T_{j,\mathrm{p}} - T_{\mathrm{a,p}},\\ &\Delta_{i,\mathrm{p}} = T_{i,\mathrm{p}} - T_{j,\mathrm{p}},\ \Delta_{i,\mathrm{a}} = T_{i,\mathrm{a}} - T_{j,\mathrm{a}}. \end{aligned} \tag{5.34}$$

According to Fig. 5.2, it can be inferred that the conditions of preventing packet collision are $\Delta_{j,i} > 0$, $\Delta_{j,\mathrm{p}} > 0$, $\Delta_{i,\mathrm{p}} > 0$, and $\Delta_{i,\mathrm{a}} > 0$. To this end, (5.34) can be rearranged as

$$\begin{aligned} &\Delta_{j,i} = \tau_{\mathrm{a},j} + (t_{j,j} - t_{\mathrm{a},j}) + t_{\mathrm{r},j} + \tau_{j,i} - \tau_{\mathrm{a},i} > 0,\\ &\Delta_{j,\mathrm{p}} = \tau_{\mathrm{a},j} + (t_{j,j} - t_{\mathrm{a},j}) + t_{\mathrm{r},j} + \tau_{j,\mathrm{p}} - \tau_{\mathrm{a,p}} > 0,\\ &\Delta_{i,\mathrm{p}} = \tau_{j,i} + (t_{i,i} - t_{j,i}) + t_{\mathrm{r},i} + \tau_{i,\mathrm{p}} - \tau_{j,\mathrm{p}} > 0,\\ &\Delta_{i,\mathrm{a}} = \tau_{j,i} + (t_{i,i} - t_{j,i}) + t_{\mathrm{r},i} + \tau_{\mathrm{a},i} - \tau_{\mathrm{a},j} > 0, \end{aligned} \tag{5.35}$$

which yields

$$\begin{aligned} &t_{j,j} - t_{\mathrm{a},j} > (d_{\mathrm{a},i} - d_{\mathrm{a},j} - d_{j,i})/c - t_{\mathrm{r},j},\\ &t_{j,j} - t_{\mathrm{a},j} > (d_{\mathrm{a,p}} - d_{\mathrm{a},j} - d_{j,\mathrm{p}})/c - t_{\mathrm{r},j},\\ &t_{i,i} - t_{j,i} > (d_{j,\mathrm{p}} - d_{j,i} - d_{i,\mathrm{p}})/c - t_{\mathrm{r},i},\\ &t_{i,i} - t_{j,i} > (d_{\mathrm{a},j} - d_{j,i} - d_{\mathrm{a},i})/c - t_{\mathrm{r},i}. \end{aligned} \tag{5.36}$$

Obviously, if $t_{j,j} - t_{\mathrm{a},j} > \max\{\epsilon, \epsilon_1\}$ and $t_{i,i} - t_{j,i} > \max\{\epsilon_2, \epsilon_2\}$ are both satisfied, the inequality (5.36) holds. □

5.4.4 Communication Overhead

The communication overhead of the proposed localization algorithms is analyzed in this section. Referring to Shu et al. (2015), one knows the communication overhead is evaluated by the transmission of elements. In addition, it is easy to see that the communication overhead of PPS-based localization algorithm for active sensor node is dominated by the state calculation in Step 2. Similarly, the communication overhead of PPS and PPDP based localization algorithm for ordinary sensor node is also dominated by the state calculation in Step 2. Accordingly, the transmission of elements of localization algorithm for active sensor node can be calculated as

$$\begin{aligned}\Psi_{\mathrm{a}} &= \underbrace{6(m-1)(m-2)+6(m-1)}_{\text{elements of calculating } \sum_{i=2}^{m}\mathbf{X}_i} \\ &+ \underbrace{12m(m-1)+12m}_{\text{elements of calculating } \sum_{i=1}^{m}\mathbf{Y}_i} \\ &+ \underbrace{2(m-1)+9m(m-1)+9m}_{\text{elements of calculating } \mathbf{Z}_i} \\ &= 27m^2-10m+4.\end{aligned} \tag{5.37}$$

Similarly, the transmission of elements of localization algorithm for ordinary sensor node can be calculated as

$$\begin{aligned}\Psi_{\mathrm{p}} &= \underbrace{36m(m-1)+6(m-1)+36m}_{\text{elements of calculating } \bar{\mathbf{X}}_{\mathrm{a}}\text{and } \bar{\mathbf{X}}_i} \\ &+ \underbrace{6\times 2+6\times 2}_{\text{elements of calculating } \bar{\mathbf{Y}}_{\mathrm{a}} \text{ and } \bar{\mathbf{Y}}_1} \\ &+ \underbrace{15(m-1)+15(m-1)+9}_{\text{elements of calculating } \bar{\bar{\mathbf{Z}}}_i\bar{\mathbf{Z}}_{\mathrm{a}}} \\ &+ \underbrace{3\times 3\times 3}_{\text{elements of calculating } \bar{\mathbf{M}}_{\mathrm{a}}\bar{\mathbf{M}}_1} \\ &+ \underbrace{3\times 3\times 3}_{\text{elements of calculating } \bar{\mathbf{M}}_{\mathrm{a}}\bar{\mathbf{M}}_2} \\ &= 36m^2+36m+51.\end{aligned} \tag{5.38}$$

5.5 Simulation and Experiment Results

5.5.1 Simulation Studies

The simulations are implemented on MATLAB 2017b. Initially, ten active sensor nodes and ten ordinary sensor nodes are randomly deployed in an area of 600 m$\times$600 m$\times$600 m, as described in Fig. 5.7a. It is designed that anchor nodes can efficiently communicate with active and ordinary sensor nodes, i.e., data packets

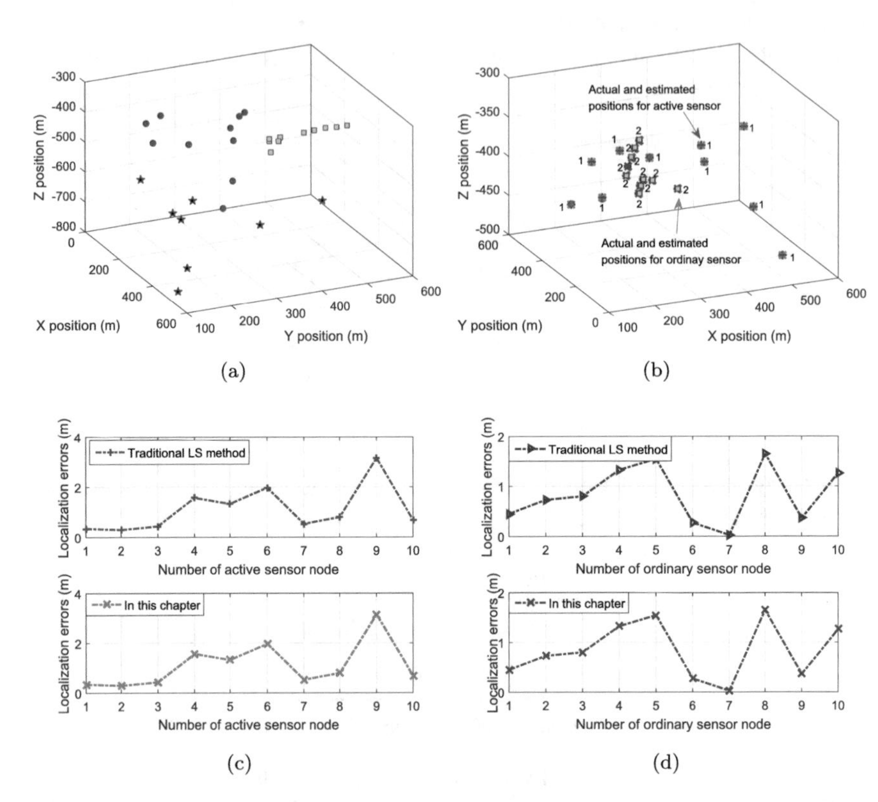

Fig. 5.7 (**a**) Deployment of anchor and sensor nodes. The circle (i.e., ◯), rectangle (i.e., □) and star (i.e., ★) denote the active sensor node, ordinary sensor and anchor node, respectively; (**b**) Positions of active and ordinary sensor. (**c**) Location errors for active sensor. (**d**) Location errors for ordinary sensor. Equivalence evaluation of positions and location errors for sensor nodes

during the communication process can be successfully received and decoded. Besides that, some parameters used for the simulation are given as: $\sigma_{\text{mea}} = 0.001$, $m = 8$, $\boldsymbol{\rho} = 20$ m, and c $= 1500$ m/s.

(1) Equivalence Between Our Method and Traditional Method

It is noted that, the privacy preservation is employed into the asynchronous localization of USNs, through which PPS and PPDP based localization algorithms are designed to hide privacy information. To verify the effectiveness of the proposed localization algorithms, the actual and localized positions of active sensor node (labelled as '1') are shown in Fig. 5.7b. In this figure, the traditional LS method refers to the estimation in (5.11), where the privacy preservation of anchor nodes is not considered. For clear description, the localization error of active sensor node is defined as $\text{err}_1 = \left\| \mathbf{p}_{\text{a}} - \hat{\mathbf{p}}_{\text{a}} \right\|$. Correspondingly, the localization errors are shown in Fig. 5.7c. From Fig. 5.7b, c, we can see the localization accuracies obtained by PPS-based asynchronous localization algorithm and traditional LS algorithm are completely the same. This result verifies the effectiveness of Theorem 5.1.

Similarly, the actual and localized positions of ordinary sensor node (labelled as '2') are also shown in Fig. 5.7b. Define the localization error of ordinary sensor node as $\mathrm{err}_2 = \left\|\mathbf{p}_\mathrm{p} - \hat{\mathbf{p}}_\mathrm{p}\right\|$, then localization errors with the two methods are presented in Fig. 5.7d. Clearly, localization accuracies obtained by PPS and PPDP-based asynchronous localization algorithm are coincided with the ones by traditional LS estimation (5.25). This result verifies the effectiveness of Theorem 5.2.

(2) Comparison with Synchronous Localization Algorithm

In Shi and Wu (2018), the authors adopted PPS into TDOA-based localization, where the target sends localization request to anchor nodes. However, the TDOA-based localization method in Shi and Wu (2018) can achieve localization task only in the case of synchronized clock, i.e., $\alpha_\mathrm{a} = 1$ ($\alpha_\mathrm{p} = 1$) and $\beta_\mathrm{a} = 0$ ($\beta_\mathrm{p} = 0$). To verify this judgement, we consider the following two scenarios for ordinary sensor node: (1) $\alpha_\mathrm{p} = 1$ and $\beta_\mathrm{p} = 0$; (2) $\alpha_\mathrm{p} \neq 1$ and $\beta_\mathrm{p} \neq 0$. Based on this, the actual and localized positions of ordinary sensor node in Scenario 1 are shown in Fig. 5.8a, and the localization errors are presented in Fig. 5.8b. Obviously, the localization task

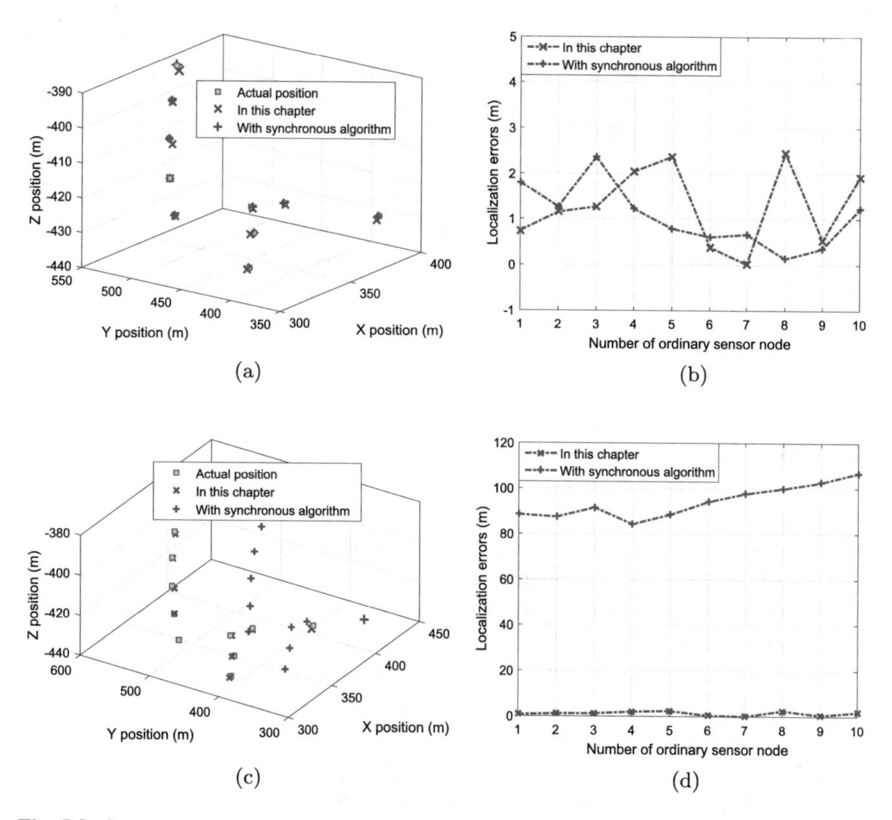

Fig. 5.8 Comparison with synchronous localization algorithm, e.g., Shi and Wu (2018). (**a**) Positions under synchronous clock. (**b**) Errors under synchronous clock. (**c**) Positions under asynchronous clock. (**d**) Errors under asynchronous clock

can be achieved for ordinary node under two methods, since the location errors are both confined into a desired bound. Under the same assumption, the clock skew is set as $\alpha_{\mathrm{p}} = 1.02$ while the clock offset is set as $\beta_{\mathrm{p}} = 0.5\mathrm{s}$ for Scenario 2. Correspondingly, the actual and localized positions of ordinary sensor node are shown in Fig. 5.8c, while the localization errors are shown in Fig. 5.8d. Clearly, the asynchronous clocks seriously affect the localization accuracy and the localization task is not achieved by the method in Shi and Wu (2018). By comparison, one obtains the consideration of asynchronous clocks is necessary in the case of sensor nodes achieving location task passively.

(3) Comparison with Asynchronous Localization Algorithm
To remove the influence of asynchronous clocks, an asynchronous localization algorithm based on iterative least squares estimators was presented in Yan et al. (2018), where the clock skew was ignored, i.e., $\alpha_{\mathrm{a}} = 1$ and $\alpha_{\mathrm{p}} = 1$. In addition, the location privacy of anchor nodes is not considered in Yan et al. (2018). In the following, we ignore the clock skew, through which the asynchronous localization algorithm in Yan et al. (2018) and the proposed algorithm are both conducted to locate sensor nodes under the same assumption. Then, the positions and localization errors of sensor nodes are presented in Fig. 5.9a, b, respectively. It is obvious that the localization task can be achieved without clock skew under two methods. In the following, we consider a general case, i.e., $\alpha_{\mathrm{a}} \neq 1$ and $\alpha_{\mathrm{p}} \neq 1$. Specially, the clock skews are set as $\alpha_{\mathrm{a}} = 1.03$ and $\alpha_{\mathrm{p}} = 1.02$, while the clock offsets are set as $\beta_{\mathrm{a}} = 0.1\mathrm{s}$ and $\beta_{\mathrm{p}} = 0.5\mathrm{s}$. Based on this, the localized positions of active sensor node are shown in Fig. 5.9c, and the localization errors are given in Fig. 5.9d. Meanwhile, the localized positions of ordinary sensor node are shown in Fig. 5.9e, whose localization errors are given in Fig. 5.9f. From Fig. 5.9c–f, one knows the clock skew reduces the localization accuracy. Alternatively, the proposed algorithm can eliminate the effect of clock offset and clock skew. Thereby, the position task can be achieved. By comparisons, we obtain that the consideration of clock skew is necessary, and the proposed asynchronous localization algorithm in this chapter can effectively eliminate the impact of the clock skew.

(4) Performance for the Mobility Compensation Strategy
As provided in Sect. 5.2.2, a mobility compensation strategy is developed to balance the tradeoff between communication cost and localization accuracy. To verify the effectiveness, the measurement window T_{w} and sampling interval δ are set as 4s and 1s, respectively. Of note, the clock skew α_{a} (or α_{p}) is estimated by the located positions at timestamps $T_{\mathrm{a,a}}$ (or $T_{\mathrm{a,p}}$) and $T^{\#}_{\mathrm{a,a}}$ (or $T^{\#}_{\mathrm{a,p}}$). Thus, the clock skew α_{a} (or α_{p}) in the first measurement window cannot be estimated, due to the lack of position information (i.e., $\mathbf{p}^{\#}_{\mathrm{a}}$ or $\mathbf{p}^{\#}_{\mathrm{p}}$). In view of this, the clock skew α_{a} (or α_{p}) in the first measurement window is set as 1. Accordingly, the localization errors of an active sensor node and an ordinary sensor node in three measurement windows are shown in Fig. 5.10a, b, respectively. Correspondingly, the estimated clock skews α_{a} and α_{p} are represented in Fig. 5.10c. It is clear that, the estimated position in the first measurement window is not very accurate. After the first measurement window, the clock skews α_{a} and α_{p} are both corrected, whose accuracy is improved.

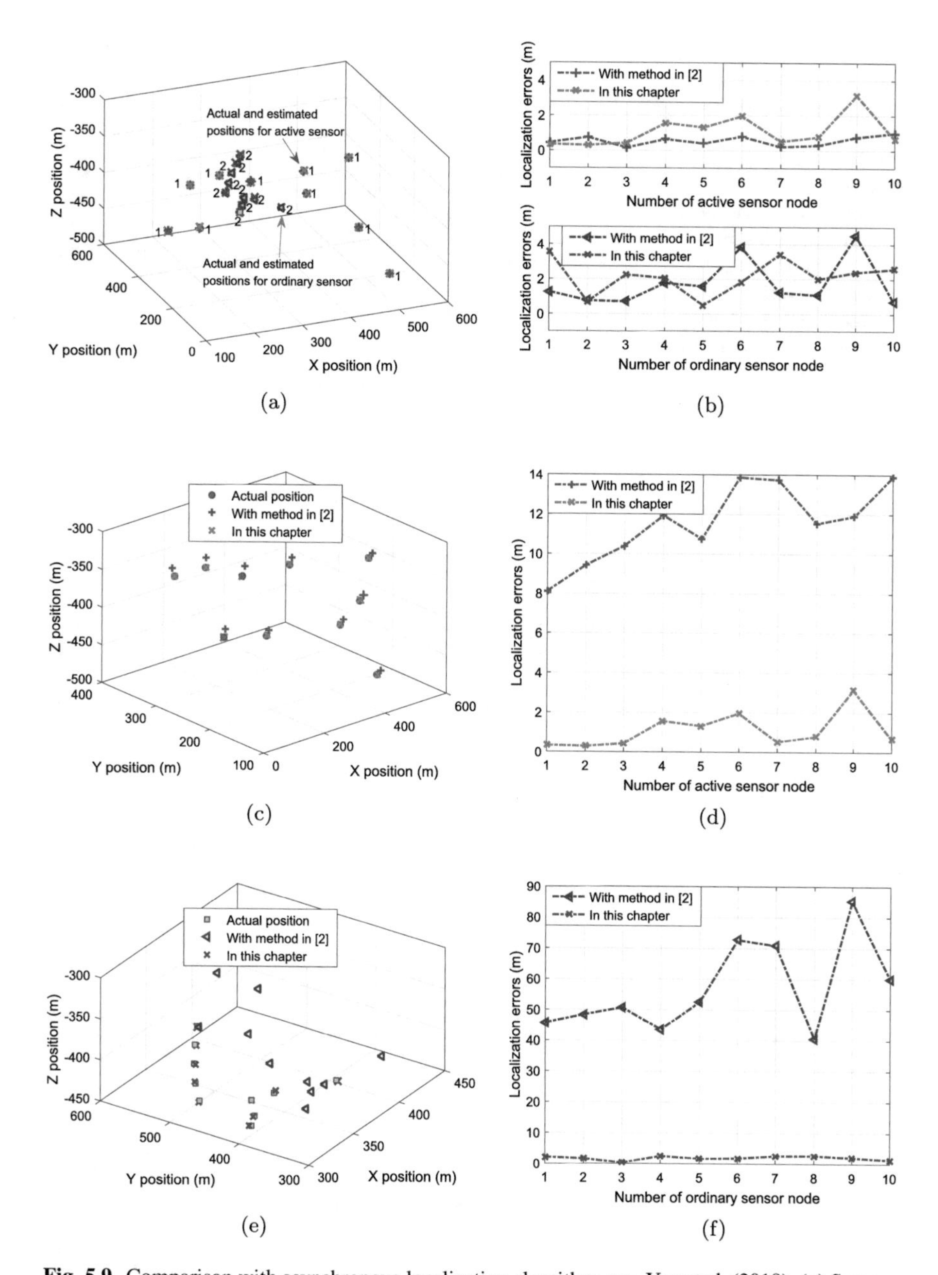

Fig. 5.9 Comparison with asynchronous localization algorithm, e.g., Yan et al. (2018). (**a**) Sensors positions without clock skew. (**b**) Errors without clock skew. (**c**) Active sensor positions with clock skew. (**d**) Errors of active sensor with skew. (**e**) Ordinary sensor positions with clock skew. (**f**) Errors of ordinary sensor with skew

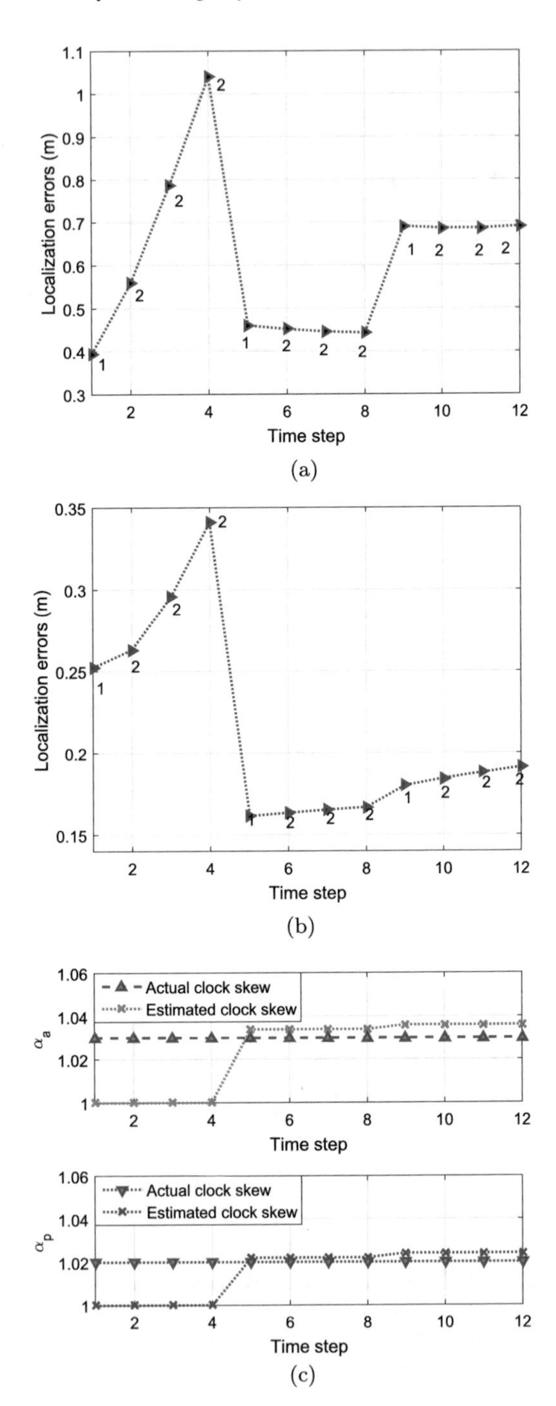

Fig. 5.10 Performance for the mobility compensation strategy in three measurement windows. Of note, '1' denotes estimated errors by the algorithm in Sect. 5.3, '2' denotes estimated errors by mobility compensation strategy. (**a**) Estimated errors of an active sensor. (**b**) Estimated errors of an ordinary sensor. (**c**) Estimated clock skews α_a and α_p

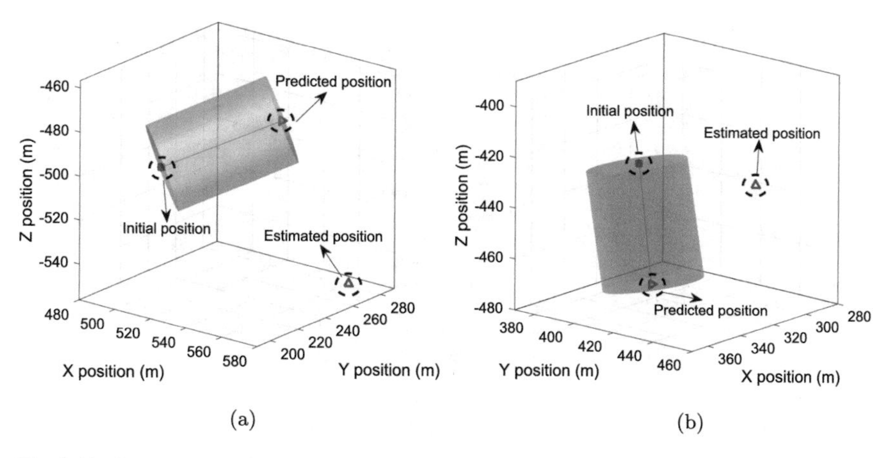

Fig. 5.11 Consequence when there exist dishonest nodes. (**a**) Active sensor node. (**b**) Ordinary sensor node

(5) Consequence when There Exist Dishonest Nodes
In the above results, the nodes are assumed to be honest. As mentioned in Sect. 5.3.3, if the condition of $\mathrm{e_a}(T^{\#}_{\mathrm{a,a}}) > \rho$ and $\mathrm{e_p}(T^{\#}_{\mathrm{a,p}}) > \rho$ holds, the nodes in USNs can be considered to be dishonest, otherwise the nodes are dishonest. To show the consequence when there exist dishonest nodes, dishonest data are added to the position and timestamps information of anchor node 4. To be specific, $\mathbf{p}_4$ is tampered with $\mathbf{p}_4 + 10\boldsymbol{\xi}_1$, the time $t_{\mathrm{a},4}$ is tampered with $t_{\mathrm{a},4} + 0.5\xi_2$, and the time $T_{4,\mathrm{p}}$ is tampered with $T_{4,\mathrm{p}} + 0.5\xi_3$, where $\boldsymbol{\xi}_1$, ξ_2 and ξ_3 denote zero-mean Gaussian noises with a variance 1. Correspondingly, the predicted trajectories and estimated positions of active sensor node are shown in Fig. 5.11a, while the predicted trajectories and estimated positions of ordinary sensor node are shown in Fig. 5.11b. Clearly, the estimated positions of active sensor node and ordinary sensor node are outside the cylinder when there exist dishonest nodes, i.e., the distance differences exceed the threshold ρ.

5.5.2 Experiment Studies

Experiment results are presented in this section, and we only check the effectiveness of the PPS-based asynchronous localization algorithm for active sensor node due to limited experiment conditions in our lab. To be specific, the experiment is carried out in a tank of 6 m ×4 m ×1 m, and the hardware in the experiment is mainly comprised of the following three parts: (1) *Base Stations*: Base stations act as surface buoys, whose role is to provide self-localization and clock synchronization services for anchor nodes. (2) *Anchor Nodes*: When anchor nodes are on the surface of the water, they employ ultra-wideband technology (Aiello and Rogerson 2003)

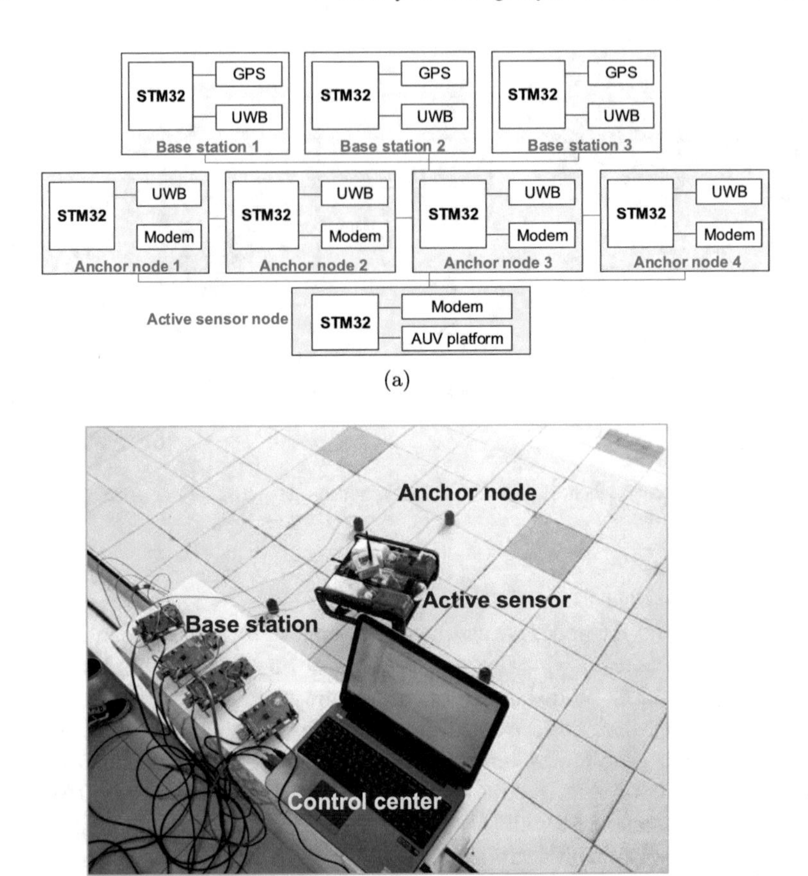

Fig. 5.12 Experiment description of the PPS-based asynchronous localization algorithm for active sensor node. (**a**) Configuration of the experiment. (**b**) Experiment setup in our lab

to acquire their positions via the cooperation with base stations. After acquiring their position information, anchor nodes sink to the bottom of the tank, and then provide localization service for the active sensor node. (3) *Active Sensor Node*: An AUV is employed to perform the role of active sensor node. It is worth mentioning that the aforementioned communication links are all wireless. The experimental configuration is shown in Fig. 5.12a, and the experiment setup is depicted in Fig. 5.12b.

In order to estimate the position of active sensor node, underwater acoustic modems are required to provide communication capability for the sensor-to-anchor wireless links. Then, an acoustic modem that consists of STM32 processor, transmitter and receiver is designed in our lab, as shown in Fig. 5.13a. To verify the effectiveness of our modem, the received analog signals are presented on the top of Fig. 5.13b. These analog signals can be converted to digital signals, as shown on

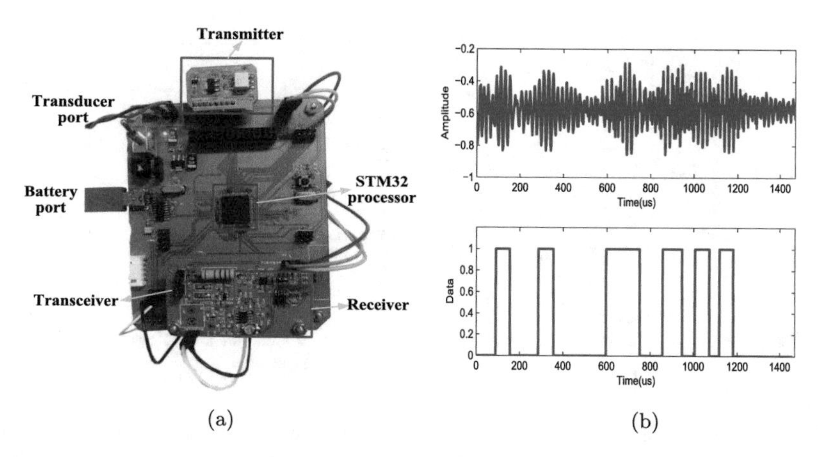

Fig. 5.13 Construction of acoustic modem module. (**a**) Acoustic modem module. (**b**) Received signals with the modem

Table 5.1 Relative timestamps measurements (μs)

	$T_{a,a}$	$t_{a,1}$	$t_{a,2}$	$t_{a,3}$	$t_{a,4}$
Point 1	64,762	65,584	65,892	65,582	65,042
Point 2	65,637	66,530	66,637	66,270	66,084
Point 3	62,384	63,104	63,233	63,104	62,949
Point 4	72,563	73,129	73,283	73,412	73,284
Point 5	84,369	84,816	85,002	85,369	85,263
Point 6	57,296	57,929	57,743	58,190	58,296
Point 7	66,437	67,257	66,717	67,257	67,567
Point 8	82,653	83,547	83,100	83,286	83,653
Point 9	54,698	55,698	55,331	55,145	55,592
Point 10	72,734	73,864	73,554	73,014	73,558
Point 11	63,647	64,927	64,447	63,847	64,667

the bottom of Fig. 5.13b. Clearly, the transmitted preamble and the guard interval in Fig. 5.13 can be identified, which indirectly verify the effectiveness of the designed acoustic modem.

With the embedded communication system, the position of active sensor node can be estimated. To this end, the position vectors of anchor nodes are set as $\mathbf{p}_1 = [-0.6, 0.6, -0.5]^{\mathrm{T}}$, $\mathbf{p}_2 = [0.9, 0.6, -0.5]^{\mathrm{T}}$, $\mathbf{p}_3 = [0.9, -0.9, -0.5]^{\mathrm{T}}$ and $\mathbf{p}_4 = [-0.6, -0.9, -0.5]^{\mathrm{T}}$. Besides, we assume that eleven different points are required to be localized, i.e., the active senor node is deployed at eleven different points. By using the localization protocol, some timestamp measurements (see Table 5.1) are conducted to estimate the position of active sensor node. Accordingly, the actual and localized positions of active sensor node are shown in Fig. 5.14a, while the localization errors are presented in Fig. 5.14b. In addition, the clock of active sensor node can be inferred as $\alpha_{\mathrm{a}} = 1.0008$ and $\beta_{\mathrm{a}} = 152.0373\,\mu\mathrm{s}$. Clearly, the PPS-based asynchronous localization algorithm in this chapter can achieve localization

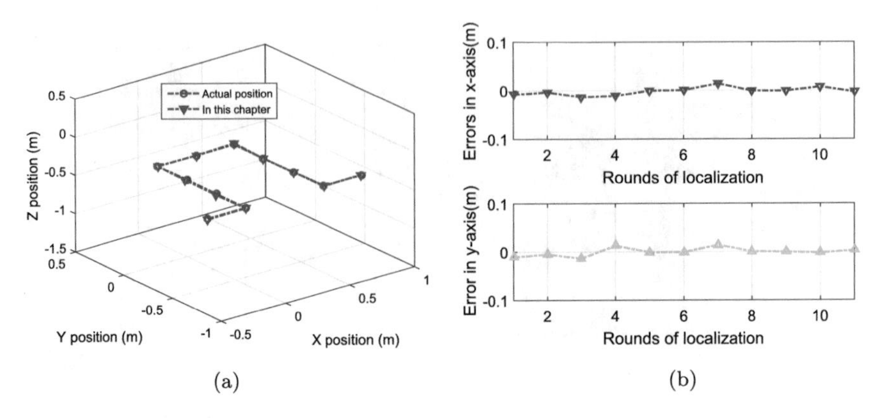

Fig. 5.14 Experiment results of the PPS-based asynchronous localization algorithm for active sensor node. (**a**) Actual and estimated positions. (**b**) Localization errors

task, since all the localization errors are close to zero. These results reflect that the asynchronous localization algorithm designed in Sect. 5.3.1 is effective and meaningful.

5.6 Conclusion

With consideration of privacy preservation, an asynchronous localization issue for USNs is studied. To eliminate the effect of asynchronous clocks (i.e., clock offset and skew), an asynchronous localization protocol including mobility compensation strategy is presented. Based on this, the PPS and PPDP based localization algorithms are designed to hide privacy information, where the performance analyses are also presented. Finally, simulation and experiment results show that the localization algorithm in this chapter can achieve localization task, while the effect of clock can be eliminated.

References

Aiello G, Rogerson G (2003) Ultra-wideband wireless systems. IEEE Microwave Mag 4(2):36–47

Alexandri T, Miller E, Spanier E, Diamant R (2020) Tracking the slipper lobster using acoustic tagging: testbed description. IEEE J Ocean Eng 45(2):577–585

Aiash M, Colson R, Kallash M (2015) Introducing a hybrid infrastructure and information-centric approach for secure cloud computing. In: Proceedings of IEEE international conference on advanced information networking and applications, Gwangiu, pp 154–159

Bechaz C, Thomas H (2000) GIB system: the underwater GPS solution. In: Proceedings of 5th ECUA, Villeurbanne, pp 613–618

Carroll P, Mahmood K, Zhou S, Zhou H, Xu X, Cui J (2014) On-demand asynchronous localization for underwater sensor networks. IEEE Trans Signal Process 62(13):3337–3348

Cheng X, Shu H, Liang Q, Du D (2008) Silent positioning in underwater acoustic sensor networks. IEEE Trans Veh Technol 57(3):1756–1766

Diamant R, Lampe L (2013) Underwater localization with time-synchronization and propagation speed uncertainties. IEEE Trans Mob Comput 12(7):1257–1269

Gong Z, Li C, Jiang F (2018) AUV-aided joint localization and time synchronization for underwater acoustic sensor networks. IEEE Signal Process Lett 25(4):477–481

Haddad D, Martins W, Costa M, Biscainho L, Nunes L, Lee B (2016) Robust acoustic self-localization of mobile devices. IEEE Trans Mob Comput 15(4):982–995

Jolfaei A, Wu X, Muthukkumarasamy V (2016) On the security of permutation-only image encryption schemes. IEEE Trans Inf Forensics Secur 11(2):235–246

Konstantinidis A, Chatzimilioudis G, Zeinalipour-Yazti D, Mpeis P, Pelekis N, Theodoridis Y (2015) Privacy-preserving indoor localization on smartphones. IEEE Trans Knowl Data Eng 27(11):3042–3055

Li H, He Y, Cheng X, Zhu H, Sun L (2015) Security and privacy in localization for underwater sensor networks. IEEE Commun Mag 53(11):56–62

Liu J, Wang Z, Peng Z, Cui J, Fiondella L (2014) Suave: swarm underwater autonomous vehicle localization. In: Proceedings of IEEE INFOCOM, Toronto, pp 64–72

Liu J, Wang Z, Cui J, Zhou S, Yang B (2016) A joint time synchronization and localization design for mobile underwater sensor networks. IEEE Trans. Mob. Comput. 15(3):530–543

Mortazavi E, Javidan R, Dehghani M, Kavoosi V (2017) A robust method for underwater wireless sensor joint localization and synchronization. Ocean Eng. 137(2):276–286

Ostovari P, Wu J, Khreishah A, Shroff N (2016) Scalable video streaming with helper nodes using random linear network coding. IEEE/ACM Trans Netw 24(3):1574–1587

Shi X, Wu J (2018) To hide private position information in localization using time difference of arrival. IEEE Trans Signal Process 66(18):4946–4956

Shu T, Chen Y, Yang J (2015) Protecting multi-lateral privacy in pervasive environments. IEEE/ACM Trans Netw 23(5):1688–1701

Tsai P, Tsai R, Wang S (2017) Hybrid localization approach for underwater sensor networks. J. Sensors 2017(2017):1–13

Wang G, He J, Shi X, Pan J, Shen S (2018a) Analyzing and evaluating efficient privacy-preserving localization for pervasive computing. IEEE Internet Things J 5(4):2993–3007

Wang S, Huang L, Nie Y, Wang P, Xu H, Yang W (2018b) PrivSet: set-valued data analyses with locale differential privacy. In: Proceedings of IEEE INFOCOM, Honolulu, pp 1088–1096

Wang Y, Huang M, Jin Q, Ma J (2018c) DP3: a differential privacy-based privacy-preserving indoor localization mechanism. IEEE Commun Lett 22(12):2547–2550

Yan, J, Zhang X, Luo X, Wang Y, Chen C, Guan X (2018) Asynchronous localization with mobility prediction for underwater acoustic sensor networks. IEEE Trans Veh Technol 67(3):2543–2556

Yan J, Ban H, Luo X, Zhao H, Guan X (2019a) Joint localization and tracking design for AUV with asynchronous clocks and state disturbances. IEEE Trans Veh Technol 68(5):4707–4720

Yan J, Zhao H, Wang Y, Luo X, Guan X (2019b) Asynchronous localization for UASNs: an unscented transform-based method. IEEE Signal Process Lett 26(4):602–606

Zhang B, Wang H, Zheng L, Wu J, Zhuang Z (2017) Joint synchronization and localization for underwater sensor networks considering stratification effect. IEEE Access 5(1):26932–26943

Zhou Z, Peng Z, Cui J, Shi Z, Bagtzoglou A (2011) Scalable localization with mobility prediction for underwater sensor networks. IEEE Trans Mob Comput 10(3):335–348

Chapter 6
Privacy Preserving Asynchronous Localization with Attack Detection and Ray Compensation

Abstract In this chapter, we are concerned with a privacy-preserving localization solution for USNs, and then asynchronous clock, stratification effect and forging attack are considered in cyber channels. In particular, a privacy-preserving asynchronous transmission protocol is developed to eliminate the influence of asynchronous clock and protect the private position information, where a RSS-based detection strategy is designed to detect the malicious anchor nodes. Based on the above, the position information of target is estimated by a least squares estimator. Finally, experiment and simulation results are represented to reveal that the developed localization solution outperforms the other existing works in terms of localization accuracy and effectiveness.

Keywords Ray compensation · Localization · Privacy-preserving · Asynchronous · Underwater sensor networks (USNs)

6.1 Introduction

In order to accomplish effective localization for USNs, some underwater localization schemes have been proposed, e.g., Carroll et al. (2014), Liu et al. (2016), Mortazavi et al. (2017), Yan et al. (2018), and Yan et al. (2019). These schemes mainly include the following two stages, i.e., *information collection* and *position estimation*. In the first stage, several underwater anchor nodes are employed to provide localization reference for the target, and hence, the position-related information such as signal strength, timestamp or distance can be measured, as depicted by Fig. 6.1. In the second stage, the position information of target is calculated by a centralized node or by itself locally. For such two stages, the position information leakage is inevitable, because the position information of anchor nodes requires to be revealed to the networks.

With regard to this question, we know that many privacy-preserving localization schemes have been developed for terrestrial networks. To be specific, a homomorphic encryption based localization solution which can provide different levels of privacy preservation was designed in Shu et al. (2015). Instead of employing the

J. Yan et al., *Localization in Underwater Sensor Networks*, Wireless Networks,
https://doi.org/10.1007/978-981-16-4831-1_6

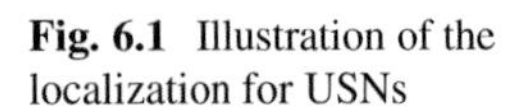

Fig. 6.1 Illustration of the localization for USNs

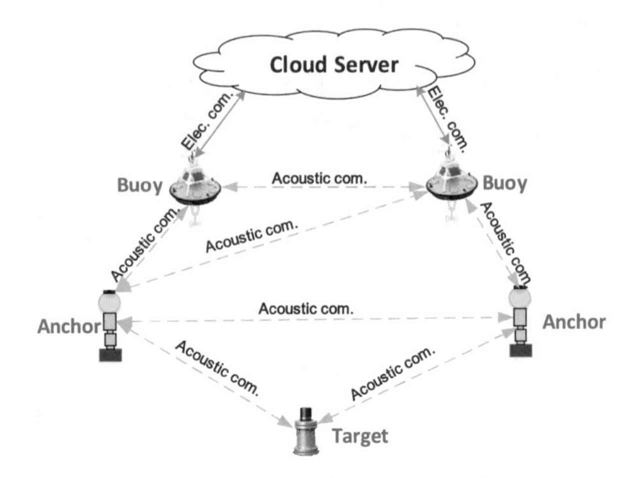

homomorphic encryption technique, Sweeney (2002) and Samarati and Sweeney (1998) employed the anonymization technique to refine the privacy model, through which a k-anonymity based localization algorithm was designed in Buccafurri and Lax (2009) to locate patients. In addition, some researchers attempt to apply the obfuscation technique to protect the privacy. In Ardagna et al. (2011) developed an obfuscation-based approach to protect the position privacy, where the measurements were perturbed. Besides that, Shi and Wu (2018) and Wang et al. (2018a) employed PPS into the localization procedure, and then two privacy-preserving localization algorithms were developed to hide the private position information. Also of relevance, some other privacy-preserving localization schemes were presented in Song et al. (2020), Wang et al. (2018b), and Das et al. (2017). Nevertheless, these terrestrial environment-suited localization solutions cannot be directly applied to the localization of USNs. Compared with the radio wave channels, the asynchronous clock and stratification effect make the privacy-preserving localization of USNs much more difficult.

In this chapter, we first design a privacy-preserving asynchronous transmission protocol to collect timestamp measurements. For such transmission protocol, the malicious attacks can be detected and the straight-line localization bias can be compensated in this chapter, which are not available in Chap. 5. Based on this, a least squares estimator is developed to estimate the position of target. Main contributions of this chapter are shown as follows.

(1) **Privacy-preserving asynchronous transmission protocol.** A privacy-preserving asynchronous transmission protocol is developed. Compared with the existing protocols (Shu et al. 2015; Shi and Wu 2018; Wang et al. 2018a; Song et al. 2020; Wang et al. 2018b; Das et al. 2017), the clocks of target and anchors in this chapter are not required to be well synchronized. Besides, it can detect the malicious anchor nodes and hide the position information of anchor nodes, which is not available in the transmission protocols (Mortazavi et al. 2017; Yan et al. 2018, 2019, 2020).

(2) **Privacy-preserving estimator with ray compensation.** In order to remove the localization bias from assuming the straight-line transmission, a privacy-preserving localization estimator with ray compensation is designed to localize the underwater target. Per knowledge of the authors, this chapter is the first work that jointly adopts asynchronous clock and stratification effect into the privacy-preserving asynchronous localization of USNs.

6.2 Network Model and Problem Formulation

6.2.1 Network Architecture

To enable privacy-preserving localization, a network architecture including three different types of nodes is presented.

- **Surface Buoys**. With the assistance of electromagnetic channels, surface buoys employ GPS technology to acquire their accurate time references and positions, as shown in Fig. 6.1. Their purpose is to supply self-localization and clock synchronization services for anchor nodes.
- **Anchor Nodes**. Anchor nodes communicate directly surface buoys. Specifically, we consider the scenario where surface buoys and anchor nodes trust with each other. Based on the above, it is assumed that the clocks of anchor nodes are synchronized and the positions are pre-known via some existing technology, e.g., Mortazavi et al. (2017), Yan et al. (2018), Yan et al. (2019), and Yan et al. (2020). In particular, the function of anchors is to provide localization references for target.
- **Target**. A static target (e.g., oil pipeline or platform) is considered, and it is provide with edge computing system (e.g., distributed sensing, communication and control unit) to make local decision. In particular, its depth information can be measured by depth sensor, while its horizontal position is demanded to be estimated. In addition, the clock of target is not well synchronized with the real time. For the security of USNs, target and anchor nodes are both considered on position privacy. Thereby, it is essential to hide the position information for both target and anchor nodes.

6.2.2 Clock and Stratification Models

It is assumed that $m \geq 4$ anchor nodes are arranged in target's neighboring range, and the index set of anchor nodes is represented as $\mathcal{I} = \{1, ..., m\}$. Unlike the simplified clock model as proposed in Carroll et al. (2014) and Yan et al. (2018), a clock model including both clock skew and clock offset is considered, i.e.,

$$T = \alpha t + \beta, \tag{6.1}$$

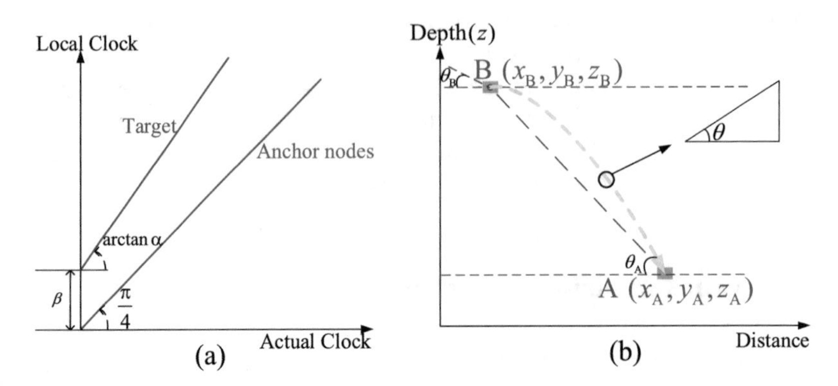

Fig. 6.2 (**a**) Description of the asynchronous clock model. (**b**) Stratification effect between receiving node A and the sending node B

where T is the local clock of target, and t represents the actual clock. Meanwhile, α and β are the unknown clock skew and clock offset, respectively. Notely, clock skew lies in $[1 - 2\mathrm{e}^{-4},\ 1 + 2\mathrm{e}^{-4}]$, which is less than 200 ppm (Gong et al. 2018). A description of the asynchronous clock model is given in Fig. 6.2a.

We presume that the sound speed is linear to the depth. This assumption has been widely applied in many existing works, e.g., Liu et al. (2016) and Yan et al. (2021). In this condition, the sound speed at depth z can be represented as $\vartheta(z) = b + az$, where b denotes the speed of sound on the surface of the water, a represents a scalar determined by the environment, and z denotes the depth of sensor node. Based above, the propagation delay between the receiving node A and the sending node B can be described as

$$\tau_{\mathrm{A,B}} = -\frac{1}{a}\left(\ln\frac{1+\sin\theta_{\mathrm{B}}}{\cos\theta_{\mathrm{B}}} - \ln\frac{1+\sin\theta_{\mathrm{A}}}{\cos\theta_{\mathrm{A}}}\right), \tag{6.2}$$

where $\theta_{\mathrm{A}} = \phi_0 + \varphi_0$ and $\theta_{\mathrm{B}} = \phi_0 - \varphi_0$ denote the acoustic ray angles at the receiving node A and the sending node B, respectively. To be specific, $\phi_0 = \arctan(\frac{z_{\mathrm{B}}-z_{\mathrm{A}}}{\sqrt{(x_{\mathrm{B}}-x_{\mathrm{A}})^2+(y_{\mathrm{B}}-y_{\mathrm{A}})^2}})$ represents the angle of the straight line between the node A and the node B, with respect to the horizontal axis. In the same way, $\varphi_0 = \arctan(\frac{0.5a*\sqrt{(x_{\mathrm{B}}-x_{\mathrm{A}})^2+(y_{\mathrm{B}}-y_{\mathrm{A}})^2}}{b+0.5a(z_{\mathrm{B}}+z_{\mathrm{A}})})$ is the angle of the ray trajectory deviating from the straight line. Besides, θ represents the angle of the point along the travel path of the acoustic ray, as depicted in Fig. 6.2b. It is worth recalling that, $\mathbf{p}_{\mathrm{A}} = (x_{\mathrm{A}}, y_{\mathrm{A}}, z_{\mathrm{A}})$ and $\mathbf{p}_{\mathrm{B}} = (x_{\mathrm{B}}, y_{\mathrm{B}}, z_{\mathrm{B}})$ represent the position vectors of nodes A and B, respectively.

6.2.3 Attack and Privacy Models

The openness of USNs makes underwater localization system easily vulnerable to forging attack from malicious anchor nodes. Therefore, the signal characteristics such as timestamps can be easily falsified by increasing malicious noises (Ferrag et al. 2017). Based on this, we propose a forging attack model for USNs. Therein, the timestamps during transmission procedure are falsified. Let node C indicates the sending node, while node B denotes the receiving node. Under the proposed communication protocol (see Sect. 6.3.1), node C needs to send its local timestamp t_C to node B. In a non-adversarial environment, the received timestamp on node B is t_C. In an adversarial environment, we deem that node C is captured by forging attack, as depicted by Fig. 6.3a. In view of the above description, the received timestamp on node B from node C in adversarial environment can be written as

$$\mathbf{t}_C = t_C + \gamma_C, \tag{6.3}$$

where γ_C represents the malicious noise. Different from Liu et al. (2019), γ_C does not require to satisfy Gaussian distribution.

As mentioned above, the position information demands to be protected. As the reason referred above, we consider a scenario where the position difference $\mathbf{p}_B - \mathbf{p}_A$ requires to be calculated, as expressed by Fig. 6.3b. In order to protect the position information (i.e., $\mathbf{p}_B$ and $\mathbf{p}_A$), a privacy-preserving difference model is proposed in this chapter. In particular, node C sends out a message that includes a pre-known vector $\mathbf{V}_C$ to node B. When the message from node C has been received, node B sends out its message that includes the value of $\mathbf{p}_B + \mathbf{V}_C$ to node A. Subsequently, node A calculates the value of $\mathbf{p}_B + \mathbf{V}_C - \mathbf{p}_A$, and sends it to node C. After that, node C calculates the following position difference

$$\mathbf{p}_B + \mathbf{V}_C - \mathbf{p}_A - \mathbf{V}_C = \mathbf{p}_B - \mathbf{p}_A. \tag{6.4}$$

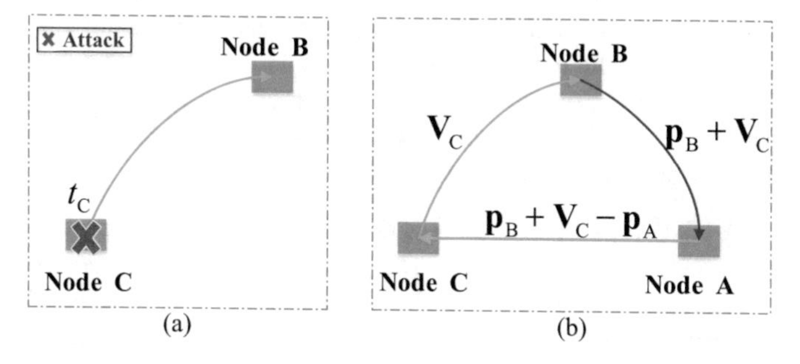

Fig. 6.3 Description of the attack and privacy models

Obviously, the above privacy-preserving difference model can protect the position information (i.e., $\mathbf{p}_{\mathrm{B}}$ and $\mathbf{p}_{\mathrm{A}}$), because $\mathbf{p}_{\mathrm{B}}$ or $\mathbf{p}_{\mathrm{A}}$ is not revealed to the other nodes except itself. It is noted that node A contains about the privacy position information of node B, i.e., the value of $\mathbf{p}_{\mathrm{B}}$ is coinhere in the value of $\mathbf{p}_{\mathrm{B}} + \mathbf{V}_{\mathrm{C}}$. Intuitively, the higher the randomness of $\mathbf{V}_{\mathrm{C}}$, the better privacy protection of $\mathbf{p}_{\mathrm{B}}$. To quantitatively describe this relationship, we define the mutual information between $\mathbf{p}_{\mathrm{B}}$ and $\mathbf{p}_{\mathrm{B}} + \mathbf{V}_{\mathrm{C}}$. In particular, mutual information is a measure of the amount of information (Thomas and Joy 2006), in which a random variable comprises about another one. Based on above, mutual information can be used as the privacy notion, as pointed out by the related works in Wang et al. (2016). With the above analyses, the mutual information between node A and node B in terms of the position $\mathbf{p}_{\mathrm{B}}$ can be represented as $I_{\mathbf{p}_{\mathrm{B}}}(\mathrm{A}, \mathrm{B})$,

$$I_{\mathbf{p}_{\mathrm{B}}}(\mathrm{A}, \mathrm{B}) = H(\mathbf{p}_{\mathrm{B}} + \mathbf{V}_{\mathrm{C}}) - H(\mathbf{p}_{\mathrm{B}} + \mathbf{V}_{\mathrm{C}} \,|\mathbf{p}_{\mathrm{B}}), \tag{6.5}$$

where $H(\mathbf{p}_{\mathrm{B}} + \mathbf{V}_{\mathrm{C}})$ denotes the entropy for random vector $\mathbf{p}_{\mathrm{B}} + \mathbf{V}_{\mathrm{C}}$, and $H(\mathbf{p}_{\mathrm{B}} + \mathbf{V}_{\mathrm{C}} \,|\mathbf{p}_{\mathrm{B}})$ denotes the condition entropy for $\mathbf{p}_{\mathrm{B}} + \mathbf{V}_{\mathrm{C}}$. Besides, $\mathbf{V}_{\mathrm{C}}$ represents a Gaussian random vector with zero mean and variance N_0.

Notely, $I_{\mathbf{p}_{\mathrm{B}}}(\mathrm{A}, \mathrm{B}) \geq 0$ with equality if and only if the values of $\mathbf{p}_{\mathrm{B}} + \mathbf{V}_{\mathrm{C}}$ and $\mathbf{V}_{\mathrm{C}}$ are independent. In the below, a special case is discussed, i.e., $\mathbf{V}_{\mathrm{C}}$ is a Gaussian random vector and $\mathbf{p}_{\mathrm{B}}$ is a constant (i.e., $\mathbf{p}_{\mathrm{B}}$ known by node B). Therefore, mutual information $I_{\mathbf{p}_{\mathrm{B}}}(\mathrm{A}, \mathrm{B})$ defined in (6.5) is calculated as

$$\begin{aligned} I_{\mathbf{p}_{\mathrm{B}}}(\mathrm{A}, \mathrm{B}) &= H(\mathbf{V}_{\mathrm{C}}) - H(\mathbf{p}_{\mathrm{B}} + \mathbf{V}_{\mathrm{C}} \,|\mathbf{p}_{\mathrm{B}}) \\ &= H(\mathbf{V}_{\mathrm{C}}) - H(\mathbf{V}_{\mathrm{C}}) \\ &= 0. \end{aligned} \tag{6.6}$$

Furthermore, readers can refer to Thomas and Joy (2006) and Wang et al. (2016) for mutual information.

Correspondingly, we give the definition of privacy preservation. Before proceeding, let $\mathbf{p}_i = (x_i, y_i, z_i)$ represents the position vector of anchor node $i \in \mathcal{I}$, where x_i, y_i and z_i denote the positions on X, Y and Z axes, respectively. As such, $\mathbf{p}_{\text{target}} = (x_{\mathrm{g}}, y_{\mathrm{g}}, z_{\mathrm{g}})$ denotes the position vector of target, where x_{g}, y_{g} and z_{g} are the positions on X, Y and Z axes, respectively. Subsequently, the definition of privacy preservation is given as follows.

Definition 6.1 (Privacy Preservation) Through the transmission protocol between anchors and target, the $\mathbf{p}_i$ for $\forall i \in \mathcal{I}$ (or $\mathbf{p}_{\text{target}}$) cannot be exposed to the other nodes except itself, while each node cannot calculate an estimate of the other node's position. The privacy between any two nodes in terms of one node's position is defined by mutual information (6.5). Referring to Thomas and Joy (2006) and Wang et al. (2016), one can further get that the mutual information is highly correlated with the privacy strength, i.e., the more mutual information between the two nodes, the less privacy, and vice versa.

6.2.4 Problem Formulation

The goal of this chapter is to design a privacy-preserving asynchronous transmission protocol, through which a privacy-preserving localization estimator with ray compensation is sought to localize the target in USNs. Therefore, the following two preformation metrics are performed:

1. ***Correctness.*** With consideration of the forging attack, privacy preservation, clock skew, clock offset and stratification effect, the target position calculated in this chapter is equal to the ones with the privacy-lacking localization schemes, such as, Liu et al. (2016), Mortazavi et al. (2017), and Yan et al. (2020).
2. ***Privacy Preservation.*** We consider the privacy preserving localization for USNs, that is, a node's coordinate is never disclosed to other nodes. For this reason, the position vector $\mathbf{p}_i$ for each anchor node $i \in \mathcal{I}$ cannot be revealed to the target. At the same time, the position vector $\mathbf{p}_{\text{target}}$ for target cannot be revealed to each anchor node.

6.3 Privacy-Preserving Localization for USNs

6.3.1 Privacy-Preserving Asynchronous Transmission Protocol

Without losing generality, it is assuming that the IDs of neighboring anchor nodes are pre-known to target. The above supposing has been proposed in many existing works, which can be realized by handshake protocol (Diamant et al. 2016; Chen et al. 2018).

Initially, target sends out an initiator message to anchor nodes. Upon receiving it, anchor nodes record and reply the target. Especially, the message transmission process is shown in Fig. 6.4. It is noted that the timestamps may be falsified in

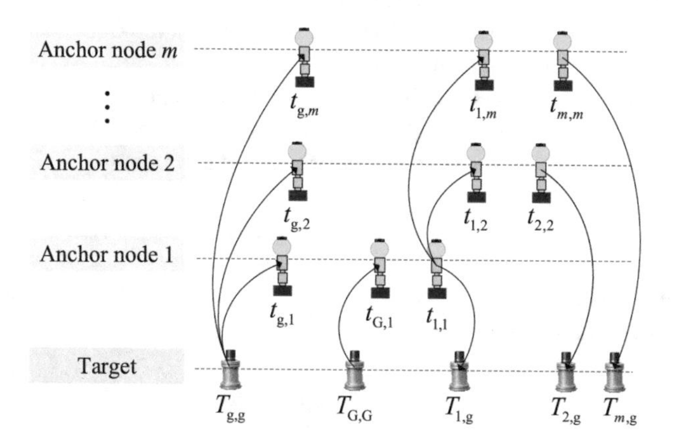

Fig. 6.4 Description of the asynchronous transmission protocol

insecure communication channel, however the RSS is not easy to be falsified (Luo et al. 2016). In view of this, a RSS-based strategy is incorporated into the transmission protocol, such that impact of malicious anchor can be removed before carrying out the localization estimation. Therefore, the details of the transmission protocol are described below.

(a) At timestamp $T_{g,g}$, target sends out an initiator message to anchor nodes, including the sending order of anchor nodes, i.e., $i = 1, \ldots, m$. At timestamp $t_{g,i}$, anchor node i receives the initiator message for $i \in \mathcal{I}$. After the initialization is achieved, target sends out a random matrix $\mathbf{E} \in \mathcal{R}^{(m-1)\times 3}$ to anchor node 1 at timestamp $T_{G,G}$. After then, target switches into the listening mode and waits for the replies from anchor nodes.
(b) At timestamp $t_{G,1}$, anchor node 1 receives the random matrix from target. After then, at timestamp $t_{1,1}$, anchor node 1 sends out a message to the other anchor nodes. This message includes the *pseudo* position matrix $\widetilde{\mathbf{P}} \in \mathcal{R}^{(m-1)\times 3}$, the position square sum S_1, the received signal strength $\mathcal{P}_1$, and the local clock $t_{g,1}$. In particular, $\widetilde{\mathbf{P}}$ and S_1 are expressed as

$$\begin{aligned} \widetilde{\mathbf{P}} &= [\mathbf{p}_1 + \mathbf{E}_1; \ \ldots; \ \mathbf{p}_1 + \mathbf{E}_{m-1}], \\ S_1 &= x_1^2 + y_1^2 + z_1^2, \end{aligned} \tag{6.7}$$

where $\mathbf{E}_j$ represents the j-th row of $\mathbf{E}$ for $j \in \mathcal{I}/\{m\}$.
(c) At timestamp $t_{1,l}$, anchor $l \in \mathcal{I}/\{1\}$ receives the message from anchor node 1. When the message from anchor node 1 has been received, anchor node l sends out its message at time $t_{l,l}$. This message includes the position square sum S_l, the local clock $t_{g,l}$, the *pseudo* position difference $\Delta\mathbf{p}_l$, the received signal strength $\mathcal{P}_l$ and the *pseudo* Z-position sum Δz_l. In particular, $S_l = x_l^2 + y_l^2 + z_l^2$. Besides, $\Delta\mathbf{p}_l$ and Δz_l are given as

$$\Delta\mathbf{p}_l = \widetilde{\mathbf{P}}_{l-1} - \mathbf{p}_l, \ \Delta z_l = \widetilde{\mathbf{P}}_{l-1,3} + \mathbf{p}_{l,3}, \tag{6.8}$$

where $\widetilde{\mathbf{P}}_{l-1}$ denotes the element in the $(l-1)$-th row of $\widetilde{\mathbf{P}}$. Similarly, $\widetilde{\mathbf{P}}_{l-1,n}$ denotes the element in the $(l-1)$-th row and n-th column of $\widetilde{\mathbf{P}}$. In addition, $\mathbf{p}_{l,n}$ denotes the n-th column of $\mathbf{p}_l$ for $n = \{1, 2, 3\}$.
(d) At timestamp $T_{i,g}$, target receives the reply from anchor node i, where $i \in \mathcal{I}$. After receiving replies from all anchor nodes, target ends the listening mode. With the above process, target collects the following messages

$$\{\Delta\mathbf{p}_l, \Delta z_l\}_{l=2}^m, \{S_i, t_{g,i}, T_{i,g}, \mathcal{P}_i\}_{i=1}^m, T_{g,g}, T_{G,G}. \tag{6.9}$$

(e) With the collected messages in (6.9), target calculates the following time difference, i.e., for $l \in \mathcal{I}/\{1\}$

$$\Delta t_{l,1} = t_{g,l} - t_{g,1}, \tag{6.10}$$

where $t_{g,l}$ and $t_{g,1}$ are timestamps as defined in *Step a*. For the time difference $\Delta t_{l,1}$, the target can employ the RSS technique to calculate its ray length difference $\Delta d_{l,1}$, and hence one can further have

$$\Delta d_{l,1} = \breve{d}_{g,l} - \breve{d}_{g,1}, \tag{6.11}$$

where $\breve{d}_{g,l}$ denotes the ray length between target and anchor node $l \in \mathcal{I}/\{1\}$. In addition, $\breve{d}_{g,1}$ is the ray length between target and anchor node 1. Notely, $\breve{d}_{g,l}$ and $\breve{d}_{g,1}$ can be calculated by Newton iterative method, i.e.,

$$\breve{d}^{\rho+1}_{g,l} = \breve{d}^{\rho}_{g,l} - \frac{\breve{d}^{\rho}_{g,l} 10^{\breve{d}^{\rho}_{g,l}} - \mathcal{P}_s/10^{\epsilon \times 10^{-4}} \mathcal{P}_l}{10^{\breve{d}^{\rho}_{g,l}} + \breve{d}^{\rho}_{g,l} 10^{\breve{d}^{\rho}_{g,l}} \ln 10}, \tag{6.12}$$

$$\breve{d}^{\rho+1}_{g,1} = \breve{d}^{\rho}_{g,1} - \frac{\breve{d}^{\rho}_{g,1} 10^{\breve{d}^{\rho}_{g,1}} - \mathcal{P}_s/10^{\epsilon \times 10^{-4}} \mathcal{P}_1}{10^{\breve{d}^{\rho}_{g,1}} + \breve{d}^{\rho}_{g,1} 10^{\breve{d}^{\rho}_{g,1}} \ln 10}, \tag{6.13}$$

where $\breve{d}^{\rho}_{g,l}$ and $\breve{d}^{\rho}_{g,1}$ represent the ρ-th iterations of $\breve{d}_{g,l}$ and $\breve{d}_{g,1}$, respectively. In addition, $\mathcal{P}_s$ represents the transmitted signal strength of target, and ϵ denotes absorption coefficient as defined in Dong et al. (2019).

(f) Based on (6.10) and (6.11), the estimated average propagation velocity can be calculated as

$$\bar{\vartheta}_{l,1} = (\breve{d}_{g,l} - \breve{d}_{g,1})/(t_{g,l} - t_{g,1}). \tag{6.14}$$

It is worth mentioning that the actual average propagation velocity is 1500 m/s (Carroll et al. 2014). In this context, the following rules can be executed to judge and remove the malicious anchor nodes, i.e.,

$$\begin{cases} \bar{\vartheta}_{l,1} \in \left[1500 - \vartheta_g,\ 1500 + \vartheta_g\right], & \text{declare } \mathcal{H}_0 \\ \bar{\vartheta}_{l,1} \notin \left[1500 - \vartheta_g,\ 1500 + \vartheta_g\right], & \text{declare } \mathcal{H}_1 \end{cases} \tag{6.15}$$

ϑ_g is the tolerance threshold. Hypothesis $\mathcal{H}_0$ denotes that anchor node is not captured by malicious node, and $\mathcal{H}_1$ denotes that anchor is malicious.

(g) After the malicious anchor nodes are deleted, a privacy-preserving estimator with ray compensation can be implemented to localize the target, as shown in Sect. 6.3.2.

In the above protocol, we adopt a privacy-preserving difference policy to hide private position information. More specifically, anchor node $i \in \mathcal{I}$ has a concern on its privacy position $\mathbf{p}_i$, while target requires to acquire the difference matrix $\Delta\mathbf{P} = [\mathbf{p}_1 - \mathbf{p}_2; \ldots; \mathbf{p}_1 - \mathbf{p}_m]$ and the sum vector $\Delta\mathbf{Z} = [z_1 + z_2; \ldots; z_1 + z_m]$. With regard to this, target generates a random matrix $\mathbf{E}$ and sends it to anchor node 1, as illustrated by *Step a* in the transmission protocol. Therefore, anchor node 1 sends $\widetilde{\mathbf{P}} = [\mathbf{p}_1 + \mathbf{E}_1; \ \ldots; \ \mathbf{p}_1 + \mathbf{E}_{m-1}]$ to anchor node l (see *Step b*), and then anchor node l constructs $\Delta\mathbf{p}_l$ by deducting $\mathbf{p}_l$ (see *Step c*). And then, target receives the reply messages from all anchor nodes, and these messages include $\{\Delta\mathbf{p}_l, \ \Delta z_l\}_{l=2}^{m}$ and $\{S_i, t_{g,i}\}_{i=1}^{m}$ (see *Step d*). Therefore, target computes the following results

$$\begin{aligned}\Delta\mathbf{P} &= [\widetilde{\mathbf{P}}_1 - \mathbf{p}_2 - \mathbf{E}_1; \ldots; \widetilde{\mathbf{P}}_{m-1} - \mathbf{p}_m - \mathbf{E}_{m-1}] \\ &= \begin{bmatrix} \mathbf{p}_1 + \mathbf{E}_1 - \mathbf{p}_2 - \mathbf{E}_1 \\ \vdots \\ \mathbf{p}_1 + \mathbf{E}_{m-1} - \mathbf{p}_m - \mathbf{E}_{m-1} \end{bmatrix} \\ &= [\mathbf{p}_1 - \mathbf{p}_2; \ldots; \mathbf{p}_1 - \mathbf{p}_m],\end{aligned} \tag{6.16}$$

$$\begin{aligned}\Delta\mathbf{Z} &= \begin{bmatrix} \widetilde{\mathbf{P}}_{1,3} + \mathbf{p}_{2,3} - \mathbf{E}_{1,3} \\ \vdots \\ \widetilde{\mathbf{P}}_{m-1,3} + \mathbf{p}_{m,3} - \mathbf{E}_{m-1,3} \end{bmatrix} \\ &= \begin{bmatrix} \mathbf{p}_{1,3} + \mathbf{E}_{1,3} + \mathbf{p}_{2,3} - \mathbf{E}_{1,3} \\ \vdots \\ \mathbf{p}_{1,3} + \mathbf{E}_{m-1,3} + \mathbf{p}_{m,3} - \mathbf{E}_{m-1,3} \end{bmatrix} \\ &= [\mathbf{p}_{1,3} + \mathbf{p}_{2,3}; \ldots; \mathbf{p}_{1,3} + \mathbf{p}_{m,3}] \\ &= [z_1 + z_2; \ldots; z_1 + z_m],\end{aligned} \tag{6.17}$$

where $\mathbf{E}_{j,n}$ denotes the element in the j-th row and n-th column of $\mathbf{E}$ for $j \in \mathcal{I}/\{m\}$, $n = \{1, 2, 3\}$.

Thus, from (6.16) and (6.17), one can obtain $\Delta\mathbf{P}$ and $\Delta\mathbf{Z}$ without knowing any anchor node's privacy position, i.e., $\mathbf{p}_i$ for $\forall i \in \mathcal{I}$. To clear description, an example of the privacy-preserving difference strategy is shown in Fig. 6.5.

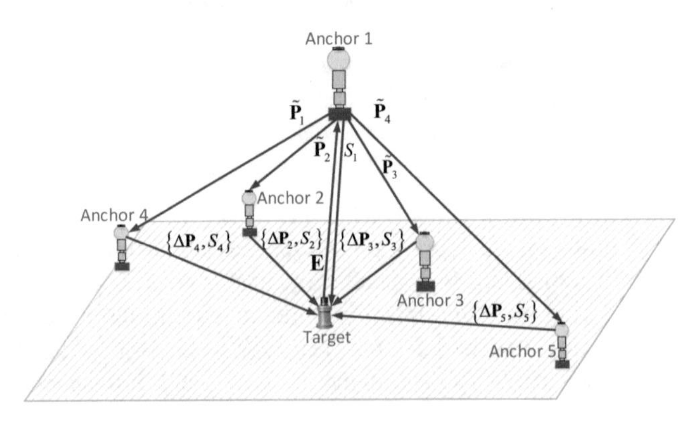

Fig. 6.5 Example of the privacy-preserving difference strategy

6.3.2 Privacy-Preserving Estimator with Ray Compensation

It is assuming that all the nodes have the same measurement quality. Specially, the measurement noise of each measurement is a random variable with zero mean and variance ϱ_{mea}^2. By adopting (6.10), the connection between time differences and propagation delays is constructed as

$$\begin{aligned}\Delta t_{l,1} &= (t_{\text{g,g}} + \tau_{\text{g},l} + \varpi_{\text{g},l}) - (t_{\text{g,g}} + \tau_{\text{g},1} + \varpi_{\text{g},1}) \\ &= \tau_{\text{g},l} - \tau_{\text{g},1} + \varpi_{l,1},\end{aligned} \tag{6.18}$$

$\tau_{\text{g},l}$ is the one-way propagation delay between target and anchor node l. Beyond that, $\tau_{\text{g},1}$ denotes the one-way propagation delay between target and anchor node 1, and $t_{\text{g,g}}$ represents the real time when initiator message is sent to the networks. $\varpi_{\text{g},l}$ and $\varpi_{\text{g},1}$ are the measurement noises, which conform to the distributions of $\varpi_{\text{g},l} \sim \mathcal{N}(0, \varrho_{\text{mea}}^2)$ and $\varpi_{\text{g},1} \sim \mathcal{N}(0, \varrho_{\text{mea}}^2)$. Thus, the reconstructed noise $\varpi_{l,1}$ satisfies the distribution of $\varpi_{l,1} \sim \mathcal{N}(0, 2\varrho_{\text{mea}}^2)$.

By substituting (6.2) into (6.18), we can get

$$\begin{aligned}\Delta t_{l,1} &= -\frac{1}{a}\left(\ln\frac{1+\sin\theta_{\text{g},l}}{\cos\theta_{\text{g},l}} - \ln\frac{1+\sin\theta_l}{\cos\theta_l}\right) \\ &\quad + \frac{1}{a}\left(\ln\frac{1+\sin\theta_{\text{g},1}}{\cos\theta_{\text{g},1}} - \ln\frac{1+\sin\theta_1}{\cos\theta_1}\right) + \varpi_{l,1},\end{aligned} \tag{6.19}$$

where $\theta_{\text{g},i} = \phi_i - \varphi_i$ and $\theta_i = \phi_i + \varphi_i$ for $\forall i \in \mathcal{I}$. $\phi_i = \arctan(\frac{z_{\text{g}}-z_i}{\sqrt{(x_{\text{g}}-x_i)^2+(y_{\text{g}}-y_i)^2}})$ and $\varphi_i = \arctan(\frac{a\sqrt{(x_{\text{g}}-x_i)^2+(y_{\text{g}}-y_i)^2}}{2b+a(z_{\text{g}}+z_i)})$, as described in Fig. 6.6.

It is seen from (6.19) that

$$\begin{aligned}&a(\Delta t_{l,1} - \varpi_{l,1}) \\ &= \ln\frac{1+\sin\theta_{\text{g},1}}{\cos\theta_{\text{g},1}} + \ln\frac{1+\sin\theta_l}{\cos\theta_l} - \ln\frac{1+\sin\theta_1}{\cos\theta_1} - \ln\frac{1+\sin\theta_{\text{g},l}}{\cos\theta_{\text{g},l}} \\ &= \ln\left[\left(\frac{1+\sin\theta_{\text{g},1}}{\cos\theta_{\text{g},1}}\frac{1+\sin\theta_l}{\cos\theta_l}\right) \Big/ \left(\frac{1+\sin\theta_{\text{g},l}}{\cos\theta_{\text{g},l}}\frac{1+\sin\theta_1}{\cos\theta_1}\right)\right],\end{aligned} \tag{6.20}$$

which implies that for $l \in \mathcal{I}/\{1\}$,

$$\begin{aligned}&e^{a(\Delta t_{l,1} - \varpi_{l,1})} \\ &= \left(\frac{1+\sin\theta_{\text{g},1}}{\cos\theta_{\text{g},1}}\frac{1+\sin\theta_l}{\cos\theta_l}\right) \Big/ \left(\frac{1+\sin\theta_{\text{g},l}}{\cos\theta_{\text{g},l}}\frac{1+\sin\theta_1}{\cos\theta_1}\right).\end{aligned} \tag{6.21}$$

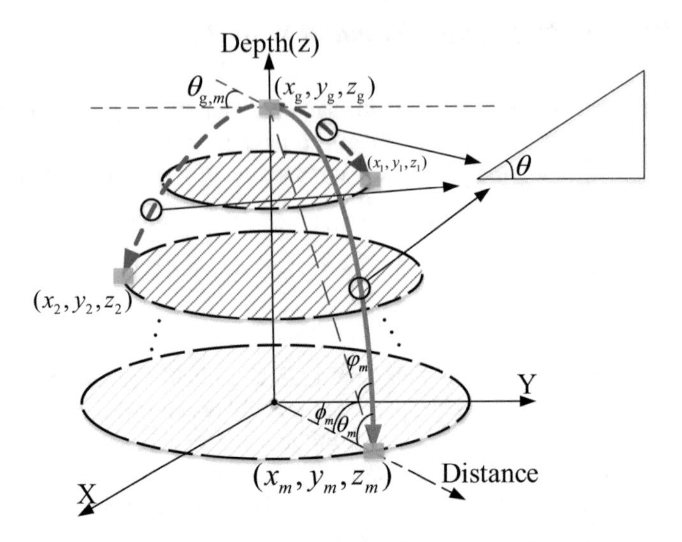

Fig. 6.6 Schematic diagram of the multi-ray collaboration

By employing factorization, $\mathrm{e}^{a(\Delta t_{l,1}-\varpi_{l,1})}$ is

$$\begin{aligned}
&\mathrm{e}^{a(\Delta t_{l,1}-\varpi_{l,1})}\\
&=\left(\frac{1}{\cos\theta_{g,1}}+\tan\theta_{g,1}\right)\left(\frac{1}{\cos\theta_{l}}+\tan\theta_{l}\right)/\Bigg[\left(\frac{1}{\cos\theta_{g,l}}\right.\\
&\left.+\tan\theta_{g,l}\right)\left(\frac{1}{\cos\theta_{1}}+\tan\theta_{1}\right)\Bigg]\\
&=\left(\frac{1}{\cos\left(\arctan\left(\lambda_{g,1}\right)\right)}+\lambda_{g,1}\right)\left(\frac{1}{\cos\left(\arctan\left(\lambda_{l}\right)\right)}+\lambda_{l}\right)\\
&/\left[\left(\frac{1}{\cos\left(\arctan\left(\lambda_{g,l}\right)\right)}+\lambda_{g,l}\right)\left(\frac{1}{\cos\left(\arctan\left(\lambda_{1}\right)\right)}+\lambda_{1}\right)\right],
\end{aligned} \tag{6.22}$$

where $\lambda_{g,l}=\frac{(z_g-z_l)(2b+a(z_g+z_l))-aR_l^2}{2R_l\vartheta(z_g)}$ and $\lambda_l=\frac{(z_g-z_l)(2b+a(z_g+z_l))+aR_l^2}{2R_l\vartheta(z_l)}$ for $l\in\mathcal{I}/\{1\}$. Similarly, $\lambda_{g,1}=\frac{(z_g-z_1)(2b+a(z_g+z_1))-aR_1^2}{2R_1\vartheta(z_g)}$ and $\lambda_1=\frac{(z_g-z_1)(2b+a(z_g+z_1))+aR_1^2}{2R_1\vartheta(z_1)}$. For $i\in\mathcal{I}$, we define $R_i=\sqrt{(x_g-x_i)^2+(y_g-y_i)^2}$ as the horizontal relative distance between target and anchor node i. In the meantime, $\vartheta(z_g)=b+az_g$, and $\vartheta(z_i)=b+az_i$ for $i\in\mathcal{I}$. Beyond that, the depths of anchor nodes are given by $z_1=\frac{1}{2}(\Delta\mathbf{p}_{l,3}+\Delta z_l-2\mathbf{E}_{l-1,3})$ and $z_l=\frac{1}{2}(\Delta z_l-\Delta\mathbf{p}_{l,3})$ for $l\in\mathcal{I}/\{1\}$, as deduced by Corollary 6.1.

It is noticed that the unknown terms in (6.22) include R_1 and R_l. Subsequently, by employing the methods of variable separation, we can get $R_l = f(R_1)$, where $f(R_1)$ is a nonlinear function with respect to R_1. Due to the high complexity of (6.22), the edge computing system in target can adopt MATLAB to separate the variables. To be specific, the methods of variable separation is to use R_1 to represent R_l. Subsequently, we obtain

$$\left(\frac{1}{\cos(\arctan(\lambda_l))} + \lambda_l\right) / \left(\frac{1}{\cos(\arctan(\lambda_{g,l}))} + \lambda_{g,l}\right) = U_1, \tag{6.23}$$

where

$$\begin{aligned} \lambda_l &= (\lambda_{g,l}\vartheta(z_g) + aR_l)/\vartheta(z_l), \\ U_1 &= e^{a(\Delta t_{l,1} - \varpi_{l,1})} \frac{\dfrac{1}{\cos(\arctan(\lambda_1))} + \lambda_1}{\dfrac{1}{\cos(\arctan(\lambda_{g,1}))} + \lambda_{g,1}}. \end{aligned} \tag{6.24}$$

By further simplification, formula (6.23) can be written as

$$\frac{\lambda_l + \sqrt{\lambda_l^2 + 1}}{\lambda_{g,l} + \sqrt{\lambda_{g,l}^2 + 1}} = U_1. \tag{6.25}$$

Following we define

$$\lambda_l + \sqrt{\lambda_l^2 + 1} = U_1 \left(\lambda_{g,l} + \sqrt{\lambda_{g,l}^2 + 1}\right) = U_2, \tag{6.26}$$

therefore, we obtained

$$\lambda_l = \begin{cases} \sqrt{\dfrac{U_2^4 - 2U_2^2 + 1}{4U_2^2}} \text{if} \sqrt{\dfrac{U_2^4 - 2U_2^2 + 1}{4U_2^2}} + 1 + \sqrt{\dfrac{U_2^4 - 2U_2^2 + 1}{4U_2^2}} = U_2 \\ \varnothing \qquad \text{otherwise} \end{cases} \tag{6.27}$$

Finally, we can get R_l by calculation.

Based above, the distance between target and anchor node i is given as

$$d_i^2 = R_i^2 + (z_g - z_i)^2, \; i \in \mathcal{I}, \tag{6.28}$$

which implies

$$d_1^2 = (x_g - x_1)^2 + (y_g - y_1)^2 + (z_g - z_1)^2, \tag{6.29}$$

and

$$d_l^2 = (x_{\rm g} - x_l)^2 + (y_{\rm g} - y_l)^2 + (z_{\rm g} - z_l)^2,\ l \in \mathcal{I}/\{1\}. \tag{6.30}$$

Subtracting (6.29) from (6.30), we have

$$\begin{aligned} & d_l^2 - d_1^2 \\ &= S_l - S_1 - 2(x_{\rm g}x_l + y_{\rm g}y_l + z_{\rm g}z_l) + 2(x_{\rm g}x_1 + y_{\rm g}y_1 + z_{\rm g}z_1), \end{aligned} \tag{6.31}$$

subsequently, by employing the asynchronous transmission protocol as proposed in Sect. 6.3.1, one has

$$d_l^2 - d_1^2 = S_l - S_1 + 2x_{\rm g}\Delta x_{1,l} + 2y_{\rm g}\Delta y_{1,l} + 2z_{\rm g}\Delta z_{1,l}, \tag{6.32}$$

where $\Delta x_{1,l} = \widetilde{\mathbf{P}}_{1,1} - \mathbf{p}_{2,1} - \mathbf{E}_{1,1}$, $\Delta y_{1,l} = \widetilde{\mathbf{P}}_{1,2} - \mathbf{p}_{2,2} - \mathbf{E}_{1,2}$, and $\Delta z_{1,l} = \widetilde{\mathbf{P}}_{1,3} - \mathbf{p}_{2,3} - \mathbf{E}_{1,3}$.

In addition, as illustrated in (6.28), we know

$$\begin{aligned} & d_l^2 - d_1^2 \\ &= R_l^2 - R_1^2 + 2z_{\rm g}z_1 - 2z_{\rm g}z_l + z_l^2 - z_1^2 \\ &= R_l^2 - R_1^2 + (2z_{\rm g} - \Delta\mathbf{Z}_{l-1})\Delta z_{1,l} \\ &= f^2(R_1) - R_1^2 + (2z_{\rm g} - \Delta\mathbf{Z}_{l-1})\Delta z_{1,l}. \end{aligned} \tag{6.33}$$

In order to estimate $\mathbf{p}_{\rm target}$, we represent the equilibrium point of $f^2(R_1) - R_1^2 + (2z_{\rm g} - \Delta\mathbf{Z}_{l-1})\Delta z_{1,l} = 0$ as $R_1 = R_{1,\rm e}$. According to, we define $C = \left.\frac{\partial(f^2(R_1) - R_1^2 + (2z_{\rm g} - \Delta\mathbf{Z}_{l-1})\Delta z_{1,l})}{\partial R_1}\right|_{R_1 = R_{1,\rm e}}$, and $h(\Delta t_{l,1}) = f^2(R_1)\big|_{R_1=0}$. Therefore, Eq. (6.33) can be linearized as

$$\begin{aligned} & d_l^2 - d_1^2 \\ &= CR_1 + h(\Delta t_{l,1}) + (2z_{\rm g} - \Delta\mathbf{Z}_{l-1})\Delta z_{1,l}, \end{aligned} \tag{6.34}$$

where $\Delta\mathbf{Z}_{l-1}$ denotes the element in the $(l-1)$-th row of $\Delta\mathbf{Z}$.

Substituting (6.34) into (6.32), one can gain

$$\begin{aligned} & h(\Delta t_{l,1}) + S_1 - S_l - \Delta\mathbf{Z}_{l-1}\Delta z_{1,l} \\ &= \begin{bmatrix} 2\Delta x_{1,l} & 2\Delta y_{1,l} & -C \end{bmatrix} \begin{bmatrix} x_{\rm g} \\ y_{\rm g} \\ R_1 \end{bmatrix}. \end{aligned} \tag{6.35}$$

Algorithm 3: Design process of the least-squares estimator

Input: Message collection as proposed in (6.9), and the iteration number is set as **k**
Output: Estimated position of the target
1 **for** $k = 1 : \mathbf{k}$ **do**
2 Update noise $\varpi_{l,1}$ and separate variable R_l
3 With MATLAB, the k-th update is given as R_l^k
4 $R_l \leftarrow \frac{1}{\mathbf{k}} \sum_1^{\mathbf{k}} R_l^k$
5 Under $f^2(R_1) - R_1^2 + (2z_g - \Delta \mathbf{Z}_{l-1})\Delta z_{1,l} = 0$, one calculates $R_{1,e}$, and constant term $h(\Delta t_{l,1})$
6 Calculate C, then $\mathbf{A}$ and $\mathbf{B}$ are constructed
7 Acquire the estimated position of target through (6.36)
8 **return** $\left[\hat{x}_g,\ \hat{y}_g,\ \hat{R}_1\right]^T$

For the target, we note that the terms of $h(\Delta t_{l,1}) + S_1 - S_l - \Delta \mathbf{Z}_{l-1} \Delta z_{1,l}$ and C for $l \in \mathcal{I}/\{1\}$ can be calculated by (6.32) and (6.34). Besides, the terms of $x_1 - x_l$ and $y_1 - y_l$ can be collected by the transmission protocol, as presented in Sect. 6.3.1. Based on this, the least-squares method is employed by target to estimate x_g, y_g and R_1. More specially, the least-squares estimator can be designed as

$$\left[\hat{x}_g,\ \hat{y}_g,\ \hat{R}_1\right]^{\mathrm{T}} = (\mathbf{A}^{\mathrm{T}}\mathbf{A})^{-1}\mathbf{A}^{\mathrm{T}}\mathbf{B}, \tag{6.36}$$

with

$$\begin{aligned} \mathbf{A} &= \begin{bmatrix} 2(x_1 - x_2) & 2(y_1 - y_2) & -C \\ \vdots & \vdots & \vdots \\ 2(x_1 - x_m) & 2(y_1 - y_m) & -C \end{bmatrix}, \\ \mathbf{B} &= \begin{bmatrix} h(\Delta t_{2,1}) + S_1 - S_2 - (z_1 + z_2)(z_1 - z_2) \\ \vdots \\ h(\Delta t_{m,1}) + S_1 - S_m - (z_1 + z_m)(z_1 - z_m) \end{bmatrix}, \end{aligned} \tag{6.37}$$

where $\hat{x}_g$, $\hat{y}_g$ and $\hat{R}_1$ are the estimations of x_g, y_g and R_1. In particular, main design process of the least-squares estimator is shown in Algorithm 3.

6.4 Performance Analyses

6.4.1 Equivalence with the Privacy-Lacking Estimation

We first verify the equivalence of the designed estimator (6.36). Before processing, we consider a traditional scenario, that is, the privacy preservation is ignored during the target localization procedure. Thereby, by adopting (6.31) and referring to Liu et al. (2016) and Mortazavi et al. (2017), one gets the following estimator

$$\left[\hat{x}_{g\#},\ \hat{y}_{g\#},\ \hat{R}_{1\#}\right]^{\mathrm{T}} = (\mathbf{A}_{\mathrm{c}}^{\mathrm{T}}\mathbf{A}_{\mathrm{c}})^{-1}\mathbf{A}_{\mathrm{c}}^{\mathrm{T}}\mathbf{B}_{\mathrm{c}}, \tag{6.38}$$

with

$$\begin{aligned} \mathbf{A}_{\mathrm{c}} &= \begin{bmatrix} 2x_1 - 2x_2 & 2y_1 - 2y_2 & -C \\ \vdots & \vdots & \vdots \\ 2x_1 - 2x_m & 2y_1 - 2y_m & -C \end{bmatrix}, \\ \mathbf{B}_{\mathrm{c}} &= \begin{bmatrix} h(\Delta t_{2,1}) + S_1 - S_2 - z_1^2 + z_2^2 \\ \vdots \\ h(\Delta t_{m,1}) + S_1 - S_m - z_1^2 + z_m^2 \end{bmatrix}, \end{aligned} \tag{6.39}$$

$\hat{x}_{g\#}$, $\hat{y}_{g\#}$ and $\hat{R}_{1\#}$ are the privacy-ignoring estimations of x_g, y_g and R_1.

In (6.38), anchors send their position information (i.e., x_i, y_i and z_i for $i \in \mathcal{I}$) to target directly, through which matrices $\mathbf{A}_{\mathrm{c}}$ and $\mathbf{B}_{\mathrm{c}}$ are structured. Obviously, the private position information of anchors is leaked. In contrast to this characteristic, the proposed estimator (6.36) in this chapter can hide private position information, because the pseudo position information is sent to target, as shown in Sect. 6.3.1. Thus, the following theorem is given.

Theorem 6.1 *With the privacy-preserving asynchronous transmission protocol in Sect. 6.3.1, the privacy-preserving estimation* $[\hat{x}_g,\ \hat{y}_g,\ \hat{R}_1]$ *in (6.36) is equivalent to the privacy-lacking estimation* $[\hat{x}_{g\#},\ \hat{y}_{g\#},\ \hat{R}_{1\#}]$ *in (6.38), and more significantly, the private positions of anchors in (6.36) can be hidden.*

Proof To beginning with, we define $\overrightarrow{\mathbf{A}} = \mathbf{A}^{\mathrm{T}}\mathbf{A}$ and $\overrightarrow{\mathbf{B}} = \mathbf{A}^{\mathrm{T}}\mathbf{B}$. Therefore, the estimator in (6.36) can be rearranged as

$$\begin{aligned} &[\hat{x}_g,\ \hat{y}_g,\ \hat{R}_1]^{\mathrm{T}} \\ &= \begin{bmatrix} \mathbf{A}_{1,1} & \mathbf{A}_{1,2} & \mathbf{A}_{1,3} \\ \mathbf{A}_{2,1} & \mathbf{A}_{2,2} & \mathbf{A}_{2,3} \\ \mathbf{A}_{3,1} & \mathbf{A}_{3,2} & \mathbf{A}_{3,3} \end{bmatrix}^{-1} \begin{bmatrix} \mathbf{B}_{1,1} \\ \mathbf{B}_{2,1} \\ \mathbf{B}_{3,1} \end{bmatrix}, \end{aligned} \tag{6.40}$$

where $\mathbf{A}_{q,r}$ denotes the element in the q-th row and r-th column of $\overrightarrow{\mathbf{A}}$ for $q \in \{1, 2, 3\}$ and $r \in \{1, 2, 3\}$. Similarly, $\mathbf{B}_{q,1}$ represents the element in the q-th row and the first column of $\overrightarrow{\mathbf{B}}$ for $q \in \{1, 2, 3\}$.

For $l \in \mathcal{I}/\{1\}$, we note that the value of $\widetilde{\mathbf{P}}_{l-1} - \mathbf{p}_l - \mathbf{E}_{l-1}$ can be directly acquired by target, as described in *Step c* of the transmission protocol in Sect. 6.3.1. As a result, from (6.16), we can easily acquire

$$\begin{aligned} & 4\sum_{l=2}^{m}(\widetilde{\mathbf{P}}_{l-1,1} - \mathbf{p}_{l,1} - \mathbf{E}_{l-1,1})^2 \\ &= 4\sum_{l=2}^{m}(x_1 + \mathbf{E}_{l-1,1} - x_l - \mathbf{E}_{l-1,1})^2 \\ &= 4\sum_{l=2}^{m}(x_1 - x_l)(x_1 - x_l) \\ &= \mathbf{A}_{1,1}. \end{aligned} \tag{6.41}$$

It is seen from (6.41) that $\mathbf{A}_{1,1}$ can be indirectly obtained without knowing the privacy position information of anchor nodes. On the other hand, when the privacy preservation is ignored, i.e., x_i, y_i and z_i are directly sent to the networks, target can easily construct $x_1 - x_l$. Therefore, one has

$$4\sum_{l=2}^{m}(x_1 - x_l)(x_1 - x_l) = \mathbf{A}^{c}_{1,1}, \tag{6.42}$$

where $\mathbf{A}^{c}_{q,r}$ denotes the element in the q-th row and r-th column of $\mathbf{A}_c^{T}\mathbf{A}_c$ for q, $r \in \{1, 2, 3\}$.

From (6.41) and (6.42), one has the conclusions as below: (1) the element $\mathbf{A}_{1,1}$ in (6.36) can hide private position information, while the element $\mathbf{A}^{c}_{1,1}$ in (6.38) cannot do this; (2) the value of $\mathbf{A}_{1,1}$ is equivalent to the one of $\mathbf{A}^{c}_{1,1}$.

Subsequently, we consider the other elements of $\overrightarrow{\mathbf{A}}$, then one has

$$\begin{aligned} & 4\sum_{l=2}^{m}(\widetilde{\mathbf{P}}_{l-1,1} - \mathbf{p}_{l,1} - \mathbf{E}_{l-1,1})(\widetilde{\mathbf{P}}_{l-1,2} - \mathbf{p}_{l,2} - \mathbf{E}_{l-1,2}) \\ &= 4\sum_{l=2}^{m}(x_1 - x_l)(y_1 - y_l) \\ &= \mathbf{A}_{1,2} = \mathbf{A}^{c}_{1,2}, \\ & 4\sum_{l=2}^{m}(\widetilde{\mathbf{P}}_{l-1,2} - \mathbf{p}_{l,2} - \mathbf{E}_{l-1,2})^2 \\ &= 4\sum_{l=2}^{m}(y_1 - y_l)(y_1 - y_l) \\ &= \mathbf{A}_{2,2} = \mathbf{A}^{c}_{2,2}, \\ & 2\sum_{l=2}^{m}(\widetilde{\mathbf{P}}_{l-1,1} - \mathbf{p}_{l,1} - \mathbf{E}_{l-1,1})C \\ &= 2\sum_{l=2}^{m}(x_1 - x_l)\,C \\ &= \mathbf{A}_{1,3} = \mathbf{A}^{c}_{1,3}, \\ & 2\sum_{l=2}^{m}(\widetilde{\mathbf{P}}_{l-1,2} - \mathbf{p}_{l,2} - \mathbf{E}_{l-1,2})C \\ &= 2\sum_{l=2}^{m}(y_1 - y_l)\,C \\ &= \mathbf{A}_{2,3} = \mathbf{A}^{c}_{2,3}, \end{aligned} \tag{6.43}$$

Similar to (6.43), the equivalence between $\overrightarrow{\mathbf{B}}$ and $\mathbf{A}_{\mathrm{c}}^{\mathrm{T}}\mathbf{B}_{\mathrm{c}}$ can also be verified. Thereby, we can have the conclusions as below: (1) the calculation of $(\mathbf{A}^{\mathrm{T}}\mathbf{A})^{-1}\mathbf{A}^{\mathrm{T}}\mathbf{B}$ in (6.36) can hide private position information during the localization procedure; (2) the value of $(\mathbf{A}^{\mathrm{T}}\mathbf{A})^{-1}\mathbf{A}^{\mathrm{T}}\mathbf{B}$ is equivalent to the one of $(\mathbf{A}_{\mathrm{c}}^{\mathrm{T}}\mathbf{A}_{\mathrm{c}})^{-1}\mathbf{A}_{\mathrm{c}}^{\mathrm{T}}\mathbf{B}_{\mathrm{c}}$. That completes the proof. □

6.4.2 Influencing Factors of Localization Errors

In general, the localization error can be divided into three categories, i.e., the error related with timestamp measurement, the error related with anchor's position, and the error related with sound speed measurement. Referring to Liu et al. (2018) and Thomson et al. (2018), the above three categories can be detailed as follows.

The error connected with timestamp measurement is defined as

$$e_{\mathrm{mea}}^2 = \sum\nolimits_{l=2}^{m} \left(1 \Big/ \frac{\partial \Delta t_{l,1}}{\partial \mathbf{p}_{\mathrm{target}}}\right)^2 \varrho_{\mathrm{t}}^2 \tag{6.44}$$

with

$$\begin{aligned}\frac{\partial \Delta t_{l,1}}{\partial \mathbf{p}_{\mathrm{target}}} = \frac{1}{a}\Bigg(&\frac{1}{\cos\theta_{\mathrm{g},1}}\frac{\partial \theta_{\mathrm{g},1}}{\partial \mathbf{p}_{\mathrm{target}}} + \frac{1}{\cos\theta_l}\frac{\partial \theta_l}{\partial \mathbf{p}_{\mathrm{target}}}\\ &-\frac{1}{\cos\theta_1}\frac{\partial \theta_1}{\partial \mathbf{p}_{\mathrm{target}}} - \frac{1}{\cos\theta_{\mathrm{g},l}}\frac{\partial \theta_{\mathrm{g},l}}{\partial \mathbf{p}_{\mathrm{target}}}\Bigg),\end{aligned} \tag{6.45}$$

where $\frac{\partial \theta_{\mathrm{g},i}}{\partial \mathbf{p}_{\mathrm{target}}} = \frac{1}{1+\mu_{\mathrm{g},i}^2}\frac{\partial \mu_{\mathrm{g},i}}{\partial \mathbf{p}_{\mathrm{target}}} - \frac{1}{1+\eta_{\mathrm{g},i}^2}\frac{\partial \eta_{\mathrm{g},i}}{\partial \mathbf{p}_{\mathrm{target}}}$, and $\frac{\partial \theta_i}{\partial \mathbf{p}_{\mathrm{target}}} = \frac{1}{1+\mu_{\mathrm{g},i}^2}\frac{\partial \mu_{\mathrm{g},i}}{\partial \mathbf{p}_{\mathrm{target}}} + \frac{1}{1+\eta_{\mathrm{g},i}^2}\frac{\partial \eta_{\mathrm{g},i}}{\partial \mathbf{p}_{\mathrm{target}}}$ for $i \in \mathcal{I}$ and $l \in \mathcal{I}/\{1\}$. Notely, ϱ_{t}^2 is the variance of time difference measurement error at each anchor node, i.e., $\varrho_{\mathrm{t}}^2 = 2\varrho_{\mathrm{mea}}^2$.

The error connected with anchor's position is defined as

$$e_{\mathrm{anchor}}^2 = \left(\frac{\partial \mathbf{p}_{\mathrm{target}}}{\partial \mathbf{p}_1}\right)^2 \varrho_{\mathrm{p}}^2 + \sum\nolimits_{l=2}^{m}\left(\frac{\partial \mathbf{p}_{\mathrm{target}}}{\partial \mathbf{p}_l}\right)^2 \varrho_{\mathrm{p}}^2, \tag{6.46}$$

with

$$\frac{\partial \mathbf{p}_{\mathrm{target}}}{\partial \mathbf{p}_1} = \frac{\partial \mathbf{p}_{\mathrm{target}}}{\partial \Delta t_{l,1}}\frac{\partial \Delta t_{l,1}}{\partial \mathbf{p}_1}, \quad \frac{\partial \mathbf{p}_{\mathrm{target}}}{\partial \mathbf{p}_l} = \frac{\partial \mathbf{p}_{\mathrm{target}}}{\partial \Delta t_{l,1}}\frac{\partial \Delta t_{l,1}}{\partial \mathbf{p}_l} \tag{6.47}$$

where $\frac{\partial \Delta t_{l,1}}{\partial \mathbf{p}_1} = \frac{1}{a}\left(\frac{1}{\cos\theta_{\mathrm{g},1}}\frac{\partial \theta_{\mathrm{g},1}}{\partial \mathbf{p}_1} - \frac{1}{\cos\theta_1}\frac{\partial \theta_1}{\partial \mathbf{p}_1}\right)$, $\frac{\partial \Delta t_{l,1}}{\partial \mathbf{p}_l} = -\frac{1}{a}\left(\frac{1}{\cos\theta_{\mathrm{g},l}}\frac{\partial \theta_{\mathrm{g},l}}{\partial \mathbf{p}_l} - \frac{1}{\cos\theta_l}\frac{\partial \theta_l}{\partial \mathbf{p}_l}\right)$, $\frac{\partial \theta_{\mathrm{g},i}}{\partial \mathbf{p}_i} = \frac{1}{1+\mu_{\mathrm{g},i}^2}\frac{\partial \mu_{\mathrm{g},i}}{\partial \mathbf{p}_i} - \frac{1}{1+\eta_{\mathrm{g},i}^2}\frac{\partial \eta_{\mathrm{g},i}}{\partial \mathbf{p}_i}$, $\frac{\partial \theta_i}{\partial \mathbf{p}_i} = \frac{1}{1+\mu_{\mathrm{g},i}^2}\frac{\partial \mu_{\mathrm{g},i}}{\partial \mathbf{p}_i} + \frac{1}{1+\eta_{\mathrm{g},i}^2}\frac{\partial \eta_{\mathrm{g},i}}{\partial \mathbf{p}_i}$, $\mu_{\mathrm{g},i} = \tan\phi_i$,

$\eta_{g,i} = \tan\varphi_i$ for $i \in \mathcal{I}$ and $l \in \mathcal{I}/\{1\}$. In addition, ϱ_p^2 denotes the variance of each anchor's location estimation error.

In addition, the error connected with sound speed measurement is

$$e_{\text{sound}}^2 = \left(\frac{\partial \mathbf{p}_{\text{target}}}{\partial \Delta t_{l,1}} \frac{\partial \Delta t_{l,1}}{\partial b}\right)^2 \varrho_b^2 + \left(\frac{\partial \mathbf{p}_{\text{target}}}{\partial \Delta t_{l,1}} \frac{\partial \Delta t_{l,1}}{\partial a} z\right)^2 \varrho_a^2, \tag{6.48}$$

with

$$\begin{aligned}\frac{\partial \Delta t_{l,1}}{\partial a} &= -\frac{1}{a^2}\left(\ln\frac{1+\sin\theta_{g,1}}{\cos\theta_{g,1}} + \ln\frac{1+\sin\theta_l}{\cos\theta_l}\right.\\ &\left.\quad - \ln\frac{1+\sin\theta_1}{\cos\theta_1} - \ln\frac{1+\sin\theta_{g,l}}{\cos\theta_{g,l}}\right)\\ &\quad + \frac{1}{a}\left(\frac{1}{\cos\theta_{g,1}}\frac{\partial\theta_{g,1}}{\partial a} + \frac{1}{\cos\theta_l}\frac{\partial\theta_l}{\partial a}\right.\\ &\left.\quad - \frac{1}{\cos\theta_1}\frac{\partial\theta_1}{\partial a} - \frac{1}{\cos\theta_{g,l}}\frac{\partial\theta_{g,l}}{\partial a}\right),\end{aligned} \tag{6.49}$$

$$\begin{aligned}\frac{\partial \Delta t_{l,1}}{\partial b} &= \frac{1}{a}\left(\frac{1}{\cos\theta_{g,1}}\frac{\partial\theta_{g,1}}{\partial b} + \frac{1}{\cos\theta_l}\frac{\partial\theta_l}{\partial b}\right.\\ &\left.\quad - \frac{1}{\cos\theta_1}\frac{\partial\theta_1}{\partial b} - \frac{1}{\cos\theta_{g,l}}\frac{\partial\theta_{g,l}}{\partial b}\right),\end{aligned} \tag{6.50}$$

where $\frac{\partial\theta_{g,i}}{\partial a} = -\frac{1}{1+\eta_{g,i}^2}\frac{\partial\eta_{g,i}}{\partial a} = -\frac{1}{1+\eta_{g,i}^2}\frac{2bR_i}{(2b+a(z_g+z_i))^2}$, $\frac{\partial\theta_i}{\partial a} = \frac{1}{1+\eta_{g,i}^2}\frac{\partial\eta_{g,i}}{\partial a} = \frac{1}{1+\eta_{g,i}^2}\frac{2bR_i}{(2b+a(z_g+z_i))^2}$, $\frac{\partial\theta_{g,i}}{\partial b} = -\frac{1}{1+\eta_{g,i}^2}\frac{\partial\eta_{g,i}}{\partial b} = \frac{-1}{1+\eta_{g,i}^2}\frac{-2aR_i}{(2b+a(z_g+z_i))^2}$, and $\frac{\partial\theta_i}{\partial b} = \frac{1}{1+\eta_{g,i}^2}\frac{\partial\eta_{g,i}}{\partial b} = \frac{1}{1+\eta_{g,i}^2}\frac{-2aR_i}{(2b+a(z_g+z_i))^2}$. Notely, ϱ_a^2 and ϱ_b^2 denote the error variances of parameters a and b, respectively.

By adopting (6.44), (6.46) and (6.48), we can obtain the total localization error e_{total}, which is constructed as

$$e_{\text{total}} = \sqrt{e_{\text{mea}}^2 + e_{\text{anchor}}^2 + e_{\text{sound}}^2}. \tag{6.51}$$

6.4.3 *Privacy-Preserving Property*

As demonstrated in Theorem 6.1, the pseudo position information of anchor nodes is sent to target. Next, we consider a general case, such that the target is able to calculate an estimate of the other node's position based on the collected messages. For such case, we have the following corollary.

Corollary 6.1 *When the target has calculation ability, the positions of anchor nodes on Z-axis can be calculated.*

Proof From *Step c* of the asynchronous transmission protocol, target receives the messages of $\Delta \mathbf{p}_l = \widetilde{\mathbf{P}}_{l-1} - \mathbf{p}_l$ and $\Delta z_l = \widetilde{\mathbf{P}}_{l-1,3} + z_l$. Therefore, the element in the third column of $\Delta \mathbf{p}_l$ can be written as

$$\Delta \mathbf{p}_{l,3} = \widetilde{\mathbf{P}}_{l-1,3} - z_l. \tag{6.52}$$

Accordingly, it is noticed that the random matrix $\mathbf{E}$ is pre-known to the target. Therefore, the target subtracts $\mathbf{E}_{l-1,3}$ from $\Delta \mathbf{p}_{l,3}$ and Δz_l. Correspondingly, the target can obtain

$$\begin{aligned} \Delta \mathbf{p}_{l,3} - \mathbf{E}_{l-1,3} &= \widetilde{\mathbf{P}}_{l-1,3} - z_l - \mathbf{E}_{l-1,3} \\ &= z_1 - z_l, \end{aligned} \tag{6.53}$$

$$\begin{aligned} \Delta z_l - \mathbf{E}_{l-1,3} &= \widetilde{\mathbf{P}}_{l-1,3} + z_l - \mathbf{E}_{l-1,3} \\ &= z_1 + z_l. \end{aligned} \tag{6.54}$$

Adding (6.53) to (6.54), one has

$$z_1 = \frac{1}{2}(\Delta \mathbf{p}_{l,3} + \Delta z_l - 2\mathbf{E}_{l-1,3}). \tag{6.55}$$

Similarly, subtracting (6.53) from (6.54), one has

$$z_l = \frac{1}{2}(\Delta z_l - \Delta \mathbf{p}_{l,3}), \ l \in \mathcal{I}/\{1\}. \tag{6.56}$$

Obviously, the positions of anchor nodes on Z-axis can be computed when the target has calculation ability. □

In the following, we consider that the anchor nodes have calculation ability, and hence the following corollary is developed.

Corollary 6.2 *For anchor node $i \in \mathcal{I}$, it has calculation ability, and meanwhile it does not share its extra information to the target and other anchors beyond that already received by the transmission protocol. In this context, anchor node i cannot compute $\mathbf{p}_{target}$ and $\mathbf{p}_{\tilde{l}}$ for $\tilde{l} \in \mathcal{I}/\{i\}$.*

Proof For anchor node 1, the received message from target is $\mathbf{E}$, as presented in *Step a* of the asynchronous transmission protocol. In order to calculate the position of target, anchor node 1 needs to reconstruct matrices $\mathbf{A}$ and $\mathbf{B}$, such that $\mathbf{p}_{\text{target}}$ can be estimated through

$$[\hat{x}_g, \hat{y}_g, \hat{R}_1]^{\mathrm{T}} = (\mathbf{A}^{\mathrm{T}}\mathbf{A})^{-1}\mathbf{A}^{\mathrm{T}}\mathbf{B}, \tag{6.57}$$

where $\mathbf{A}$ and $\mathbf{B}$ are defined in (6.36).

Nevertheless, anchor node 1 cannot obtain some elements of **A** and **B**, i.e.,

$$\{x_1 - x_l, y_1 - y_l\}_{l=2}^{m}, \quad (6.58)$$

$$\{\Delta t_{l,1}, S_l, z_1 + z_l, z_1 - z_l\}_{l=2}^{m}. \quad (6.59)$$

From (6.58) and (6.59), one knows the estimation in (6.57) cannot be achieved, i.e., anchor node 1 cannot compute $\mathbf{p}_{\text{target}}$. For the other anchor nodes, they do not receive message from target besides that the initiator message, and thereby, anchor node $l \in \mathcal{I}/\{1\}$ cannot reconstruct $\mathbf{p}_{\text{target}}$.

On the other hand, anchor node 1 send its *pseudo* position matrix $\widetilde{\mathbf{P}}$ and the position square sum S_1 to anchor node $l \in \mathcal{I}/\{1\}$, as proposed in *Step b* of the asynchronous transmission protocol. In order to calculate the position of anchor node 1, anchor node l requires to know the random matrix **E**, such that $\mathbf{p}_1$ can be estimated through

$$[\mathbf{p}_1; \cdots; \mathbf{p}_1] = \widetilde{\mathbf{P}} - \mathbf{E}. \quad (6.60)$$

Noticed that anchor node l does not receive random matrix **E,** and hence it cannot reconstruct $\mathbf{p}_1$. Meanwhile, it does not share message to anchor node $\breve{l} \in \mathcal{I}/\{l\}$. Therefore, anchor $l \in \mathcal{I}/\{1\}$ cannot construct $\mathbf{p}_{l_*}$ for $l_* \in \mathcal{I}/\{l, 1\}$, and anchor 1 cannot construct $\mathbf{p}_l$ for $l \in \mathcal{I}/\{1\}$ because the communication between anchor nodes l and l_* is lacked. Consequently, anchor node i cannot compute $\mathbf{p}_{\text{target}}$ and $\mathbf{p}_{\tilde{l}}$ for $\tilde{l} \in \mathcal{I}/\{i\}$. □

Remark 6.1 From Corollary 6.1, one notes that the positions of anchor nodes on Z-axis can be computed by the target. At the same time, the positions on X and Y axes cannot be inferred since the coordinate summations (i.e., $x_1 + x_l$ and $y_1 + y_l$) are not transmitted in the asynchronous transmission protocol. It is worth mentioning that, the volume of underwater area is very vast. More importantly, it is sufficient to obfuscate just two coordinates rather than all three coordinates to achieve reasonable location privacy preservation, as pointed out by Song et al. (2019). In view of above, one can acquire that the leakage of Z position does not affect the whole safety of the USNs.

In the following, we consider a distinct scenario, i.e., part of anchor nodes involve in collusion. For such scenario, each colluding anchor node exchanges its extra information to the other colluding anchor nodes beyond that already received through the transmission protocol. With these considerations in mind, the following theorem is developed.

Theorem 6.2 *When anchor node* 1 *does not involve in collusion, anchor node* $l \in \mathcal{I}/\{1\}$ *cannot reconstruct* $\mathbf{p}_{target}$ *even though anchor nodes* $2, \ldots, m$ *involve in collusion. When anchor node* 1 *involves in collusion with three or more colluding anchor nodes,* $\mathbf{p}_{target}$ *can be reconstructed.*

Proof To estimate $\mathbf{p}_{\text{target}}$, anchor node $l \in \mathcal{I}/\{1\}$ needs to construct matrices $\mathbf{A}$ and $\mathbf{B}$, such that a least-squares localization estimator (6.36) can be presented. With respect to this, the following two cases are considered.

(a) Anchor Node 1 *Does not Involve in Collusion*
In this case, anchor node $l \in \mathcal{I}/\{1\}$ receives the pseudo position difference $\Delta\mathbf{p}_l$ from anchor node 1. Furthermore, the extra information (e.g., S_l and $\Delta t_{l,1}$) can be exchanged by colluding with the other anchor nodes. However, the above messages cannot support the construction of $\mathbf{A}$, since $\mathbf{E}_{l-1}$ is lacked during the constructions of $x_1 - x_l$ and $y_1 - y_l$ from $\Delta\mathbf{p}_l - \mathbf{E}_{l-1}$. Thereby, matrix $\mathbf{B}$ cannot be constructed, since S_1 is lacked for all anchor nodes. In conclusion, one knows that anchor node $l \in \mathcal{I}/\{1\}$ cannot reconstruct $\mathbf{p}_{\text{target}}$ even though anchor nodes $2, \ldots, m$ involve in collusion.

(b) Anchor Node 1 *Involves in Collusion*
In this case, anchor node $l \in \mathcal{I}/\{1\}$ receives $\Delta\mathbf{p}_l$ from anchor node 1. Meanwhile, the extra information, e.g., S_1, $\mathbf{E}_{l-1}$, S_l and $\Delta t_{l,1}$, can be exchanged through collusion. In this context, the collected messages can support the constructions of $\mathbf{A}$ and $\mathbf{B}$, since $x_1 - x_l$, $y_1 - y_l$, S_1, S_l and $\Delta t_{l,1}$ can be directly acquired. Particularly, anchor node $i \in \mathcal{I}$ requires to solve the following equation, i.e.,

$$\mathbf{A}[\hat{x}_{\text{g}},\ \hat{y}_{\text{g}},\ \hat{R}_1]^{\text{T}} = \mathbf{B}. \tag{6.61}$$

Based on above, the following matrix can be defined to determine whether (6.61) is solvable, i.e.,

$$\mathbf{F} = \begin{bmatrix} 2(x_1 - x_2) & 2(y_1 - y_2) & -C & \mathbf{B}_1 \\ \vdots & \vdots & \vdots & \vdots \\ 2(x_1 - x_m) & 2(y_1 - y_m) & -C & \mathbf{B}_{m-1} \end{bmatrix}, \tag{6.62}$$

where $\mathbf{B}_j$ represents the j-th row of $\mathbf{B}$ for $j \in \mathcal{I}/\{m\}$.

Consequently, the solution to (6.61) can be given as follows

$$\begin{cases} \text{No solution}, & m \leq 3 \\ K\mathbf{x}_{\text{s}} + [\hat{x}_{\text{g}}^*,\ \hat{y}_{\text{g}}^*,\ \hat{R}_1^*]^{\text{T}}, & m > 3 \end{cases} \tag{6.63}$$

where $[\hat{x}_{\text{g}}^*,\ \hat{y}_{\text{g}}^*,\ \hat{R}_1^*]^{\text{T}}$ is a special solution. Besides, $\mathbf{x}_{\text{s}}$ is the homogeneous solution to $\mathbf{A}[\hat{x}_{\text{g}},\ \hat{y}_{\text{g}},\ \hat{R}_1]^{\text{T}} = 0$, where K is the constant. Hence, one knows $\mathbf{p}_{\text{target}}$ can be reconstructed when anchor node 1 involves in collusion with three or more colluding anchor nodes. That completes the proof. □

6.4.4 Tradeoff Between Privacy and Transmission Cost

It is noticed that the privacy-preserving model (6.4) introduces Gaussian random vector $\mathbf{V}_{\mathrm{C}}$. Intuitively, the higher the randomness of $\mathbf{V}_{\mathrm{C}}$, the better privacy protection of $\mathbf{p}_{\mathrm{B}}$. However, the introduction of high randomness can increase the transmission load. Obviously, there is a tradeoff between privacy and transmission cost. In order to quantify the above tradeoff, we propose the mutual information between nodes A and B. Recalling (6.5) and referring to Prelov and Verdu (2004), we assume that the probability distribution of $\mathbf{p}_{\mathrm{B}}$ satisfies the condition

$$P\left\{\|\mathbf{p}_{\mathrm{B}}\| > \delta\right\} \leq \exp\left\{-\delta^{\upsilon}\right\}, \text{ for all } \delta > \delta_0 \tag{6.64}$$

where $\delta_0 > 0$ and $\upsilon > 0$ denote positive constants.

Referring to Prelov and Verdu (2004), one defines $\boldsymbol{\Delta}(\mathbf{p}_{\mathrm{B}}; \mathbf{p}_{\mathrm{B}}) = -\|\mathrm{cov}(\mathbf{p}_{\mathrm{B}})\|^2$. As $N_0 \to \infty$, $I_{\mathbf{p}_{\mathrm{B}}}(\mathrm{A}, \mathrm{B})$ can be expressed by

$$\begin{aligned} I_{\mathbf{p}_{\mathrm{B}}}(\mathrm{A,B}) &= \frac{\log \mathrm{e}}{N_0} \mathrm{var}\left[\mathbf{p}_{\mathrm{B}}\right] + \frac{\log \mathrm{e}}{2N_0^2} \boldsymbol{\Delta}(\mathbf{p}_{\mathrm{B}}; \mathbf{p}_{\mathrm{B}}) + o\left(N_0^{-2}\right) \\ &= \frac{\log \mathrm{e}}{N_0} \mathrm{var}\left[\mathbf{p}_{\mathrm{B}}\right] - \frac{\log \mathrm{e}}{2N_0^2} \|\mathrm{cov}(\mathbf{p}_{\mathrm{B}})\|^2 + o\left(N_0^{-2}\right) \\ &= \frac{\log \mathrm{e}}{N_0} \mathrm{var}\left[\mathbf{p}_{\mathrm{B}}\right] - \frac{\log \mathrm{e}}{2N_0^2} \mathrm{var}^2\left[\mathbf{p}_{\mathrm{B}}\right] + o\left(N_0^{-2}\right), \end{aligned} \tag{6.65}$$

where $\mathrm{cov}(\mathbf{p}_{\mathrm{B}})$ and $\mathrm{var}[\mathbf{p}_{\mathrm{B}}]$ denote the covariance matrix and variance of $\mathbf{p}_{\mathrm{B}}$, respectively. Notely, the above equation holds for any proper $\mathbf{p}_{\mathrm{B}}$ and noise $\mathbf{V}_{\mathrm{C}}$ such that $\mathrm{E}[\|\mathbf{p}_{\mathrm{B}}\|^{4+\varepsilon}] < \infty$, where ε represent a positive constant. Besides, $\mathbf{p}_{\mathrm{B}}$ can be regarded as a random vector, which follows a normal distribution with an equal number of measurements. N_0 is assumed to be greater than $\mathrm{var}[\mathbf{p}_{\mathrm{B}}]$ and close to ∞.

When the value of $\mathrm{var}[\mathbf{p}_{\mathrm{B}}]$ is fixed, one can know that the higher value of N_0 in the interval $[\mathrm{var}[\mathbf{p}_{\mathrm{B}}], \infty)$, the smaller value of $I_{\mathbf{p}_{\mathrm{B}}}(\mathrm{A}, \mathrm{B})$. Therefore, one can conclude that the high value of N_0 can enhance the privacy strength when $N_0 \in [\mathrm{var}[\mathbf{p}_{\mathrm{B}}], \infty)$.

In the following, we propose the transmission cost for the added noise vector $\mathbf{V}_{\mathrm{C}}$, where the confidence level of $\mathbf{V}_{\mathrm{C}}$ is assumed to be 95%. On this context, we refer to the result in Wan et al. (2019), and thereby, the transmission cost with confidence interval in $\left[-2\sqrt{N_0}, 2\sqrt{N_0}\right]$ can be expressed as

$$\mathcal{W}\left(\mathbf{V}_{\mathrm{C}}\right) = \left\lceil \log_2\left(\frac{4\sqrt{N_0}}{\zeta} + 1\right)\right\rceil, \tag{6.66}$$

where $\lceil \cdot \rceil$ represents the ceiling operator, and ζ is the resolution.

From (6.66), one notes that the larger value of N_0, the larger value of $\mathcal{W}\left(\mathbf{V}_{\mathrm{C}}\right)$. That means that the high value of N_0 can increase the transmission load. Recalling the conclusion in (6.65), one knows that the tradeoff between privacy and transmission cost is mainly depended on the value of N_0. Noteworthy, the selection criteria of N_0 is out of the scope of this chapter, and it will be investigated in our future work.

6.5 Simulation and Experiment Results

6.5.1 Simulation Studies

In this section, we assume that the nodes can always efficiently communicate with the other ones. Particularly, five anchor nodes are deployed in an area of 600 m×600 m×50 m to localize the target. The positions of anchor nodes are set as $\mathbf{p}_1 = [550, 504, -40]$, $\mathbf{p}_2 = [520, 470, -35]$, $\mathbf{p}_3 = [500, 400, -30]$, $\mathbf{p}_4 = [480, 430, -20]$, and $\mathbf{p}_5 = [456, 360, -10]$. Meanwhile, target is located at ten different points in proper sequence. Besides that, some parameters used in the simulation are given in Table 6.1.

(1) Robust of the Localization Solution with Forging Attack
It is noticed that part of the anchor nodes may be captured as malicious nodes. Inspired by this, we set anchor node 5 as the malicious node, where its timestamps are falsified. In this context, the proposed localization protocol in this chapter is conducted to localize the target, where a RSS-based detection strategy is incorporated to remove the malicious anchor node. Meanwhile, the privacy-preservation protocol in Zhao et al. (2020) is also employed, in which the attack detection strategy was not considered. Based on this, the actual and localized positions of target are shown in Fig. 6.7a. For clear description, we denote $\hat{\mathbf{p}}_{\text{target}}$ as the estimation of $\mathbf{p}_{\text{target}}$, and then its localization error is defined by $err_1 = \left\| \mathbf{p}_{\text{target}} - \hat{\mathbf{p}}_{\text{target}} \right\|$, as shown in Fig. 6.7b.

From Fig. 6.7a, b, one knows the localization accuracy in this chapter can be guaranteed, while the localization accuracy in Zhao et al. (2020) is significantly degraded. The reason associated with this phenomenon is that the RSS-based detection strategy in this chapter can remove the influence of malicious anchor before carrying out the localization estimation. Particularly, the estimated average propagation velocities with the proposed localization protocol in this chapter are shown by Fig. 6.7c, and it indicates that anchor node 5 is a malicious node. Therefore, it is necessary to incorporate the attack detection strategy into the

Table 6.1 Some parameters used in the simulation

Parameter	Value	Parameter	Value
α	1.01	β	0.2 s
b	1473 m/s	a	0.017 s^{-1}
m	5	k	6
ϱ_t	3	ϱ_p^2	60
ϑ_g	220 m/s	ϱ_a	3

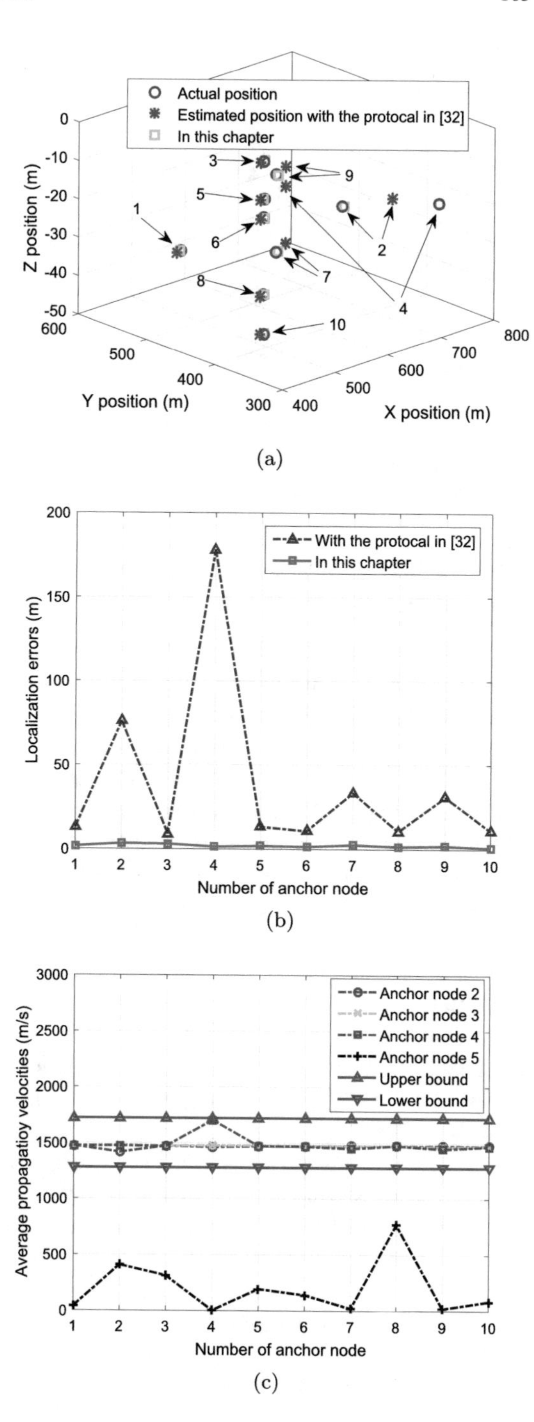

Fig. 6.7 Simulation results for robust with forging attack. (**a**) Estimated positions with forging attack. (**b**) Localization errors with forging attack. (**c**) Estimated average propagation velocities

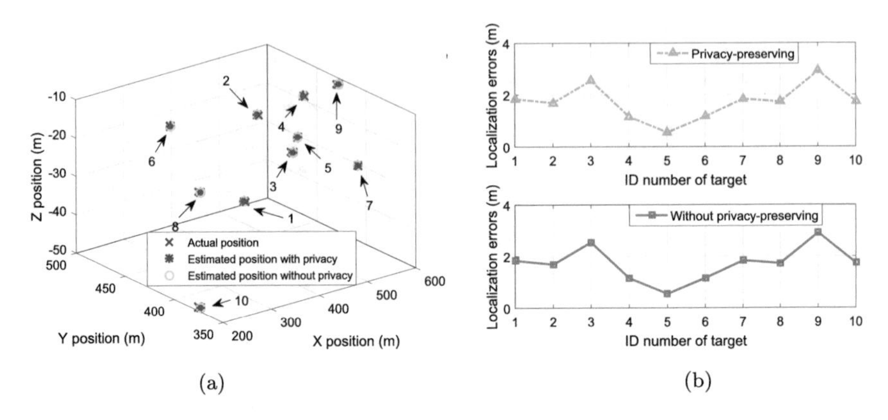

Fig. 6.8 Simulation results for the equivalence. (**a**) Localized positions with two methods. (**b**) Localization errors with two methods

privacy-preservation transmission protocol, such that the robust of the localization solution can be enhanced.

(2) Equivalence Between Our Method and Traditional Method

In this chapter, a privacy-preserving localization solution involving asynchronous clock, stratification effect and forging attack is given to localize the underwater target. In order to verify the effectiveness of the proposed localization solution, the actual and localized positions of underwater target are shown in Fig. 6.8a. Meanwhile, the traditional least squares estimator, e.g., Liu et al. (2016), is also applied to localize the target, as shown in Fig. 6.8a. It is noted that the privacy preservation of anchor nodes is not considered in Liu et al. (2016). Correspondingly, the localization errors are shown in Fig. 6.8b. From Fig. 6.8a, b, we can see the localization accuracies obtained by our solution and traditional least squares estimator are completely the same. The above results verify the effectiveness of Theorem 6.1.

(3) Comparison with Synchronous Localization Algorithm

In Shi and Wu (2018) and Wang et al. (2018a), the privacy-preserving summation was incorporated into the target localization procedure. It should be emphasized that the clocks between anchor nodes and target were synchronized, i.e., $\alpha = 1$ and $\beta = 0$. The above scheme can effectively achieve localization task when the clocks are synchronous, however it cannot estimate accurate location information when the clocks are asynchronous. In order to verify this judgement, we consider the following two scenarios: (1) $\alpha = 1$ and $\beta = 0$; (2) $\alpha \neq 1$ and $\beta = 0$. Accordingly, the actual and localized positions of target in Scenario 1 are shown in Fig. 6.9a, whose localization errors are presented in Fig. 6.9b. Clearly, the localization task in Scenario 1 can be achieved, since location errors are confined into a desired bound. For Scenario 2, the positions and errors are shown in Fig. 6.9c, d, respectively. It is known that, the asynchronous clocks seriously affect the localization accuracy and

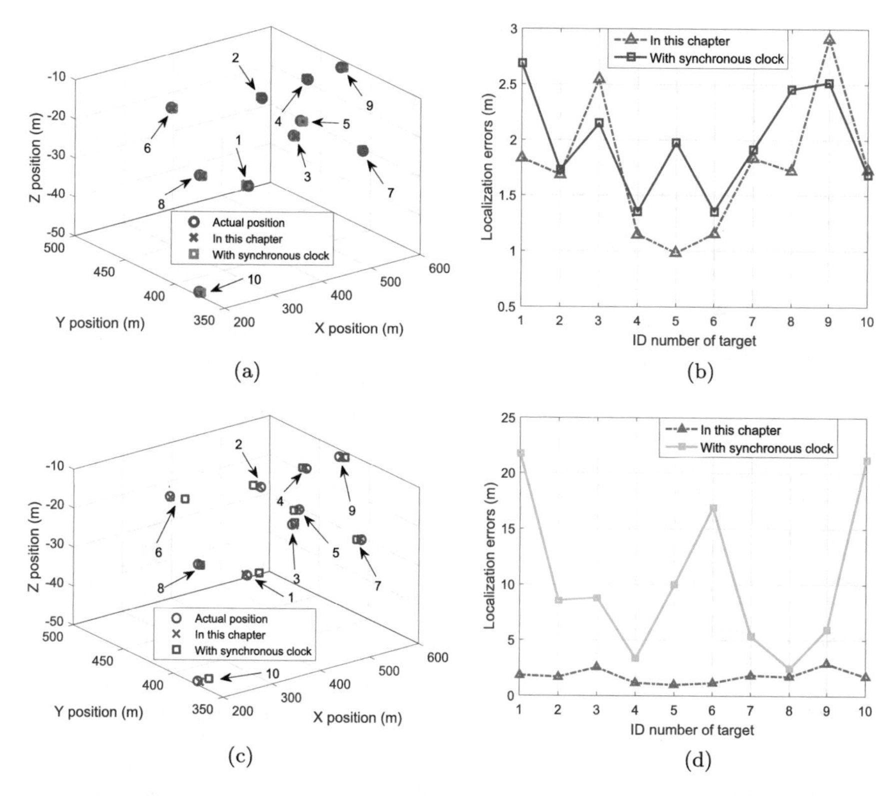

Fig. 6.9 Simulation results for the comparison with synchronous localization algorithm. (**a**) Actual and located positions in scenario 1. (**b**) Localization errors in scenario 1. (**c**) Actual and located positions in scenario 2. (**d**) Localization errors in scenario 2

the localization task is not achieved. Through this comparison, one obtains that the consideration of asynchronous clocks in USNs is necessary.

(4) Comparison with Asynchronous Localization Algorithm

In order to eliminate the influence of asynchronous clocks, two asynchronous localization algorithms were developed in Carroll et al. (2014) and Yan et al. (2018), where the clock skew was ignored by the localization procedure, i.e., $\alpha = 1$. With the above assumption, the positions and localization errors of target are presented in Fig. 6.10a, b, respectively. It is noted that the localization task can be achieved. Next, we consider a general case, i.e., $\alpha \neq 1$ and $\beta \neq 0$. Particularly, the clock skew is set as $\alpha = 1.01$, while the clock offset is set as $\beta = 0.2$s. Accordingly, the localized positions of target are shown in Fig. 6.10c, and the localization errors are given in Fig. 6.10d. From Fig. 6.10c, d, one knows the clock skew reduces the localization accuracy. Alternatively, the asynchronous localization method in this chapter can eliminate the effect of clock offset and clock skew. Thereby, the localization task in this chapter can be achieved. Through comparisons, we obtain that the consideration

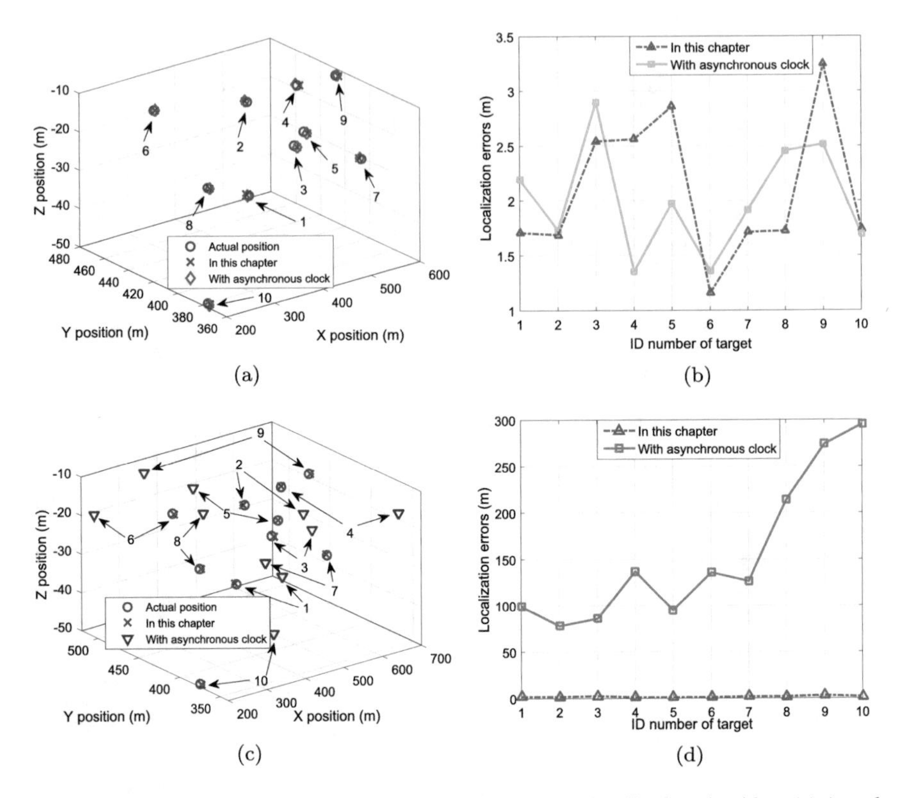

Fig. 6.10 Simulation results for comparison with asynchronous localization algorithm. (**a**) Actual and localized positions of target. (**b**) Localization errors without clock skew. (**c**) Actual and localized positions of target. (**d**) Localization errors with clock skew

of clock skew is necessary, and the proposed localization method in this chapter can effectively eliminate the impact of the clock skew.

(5) Factors Related to the Localization Accuracy

As it is mentioned above, the localization error of target is related with anchor's position estimation, timestamp measurement and sound speed measurement. In order to verity this conclusion, we first study the influence associated with anchor's position estimation. In view of this, the following scenario is considered, i.e., the position estimations of anchor nodes are not accurate, and then stochastic noises are added to the position estimations of anchor nodes. Regarding to the noises, the noises can be set as the following two cases, i.e., $\varrho_{\mathrm{p}}^2 = 1$ for case I and $\varrho_{\mathrm{p}}^2 = 60$ for case II. Accordingly, the localized positions of target are shown in Fig. 6.11a, whose localization errors are presented in Fig. 6.11b. It is clear that, the localization accuracy is significantly reduced when the position estimation of anchor nodes is not accurate. Similarly, Fig. 6.12a, b are presented to show the impacts of timestamp measurement and sound speed measurement, respectively. Clearly, the inaccurate

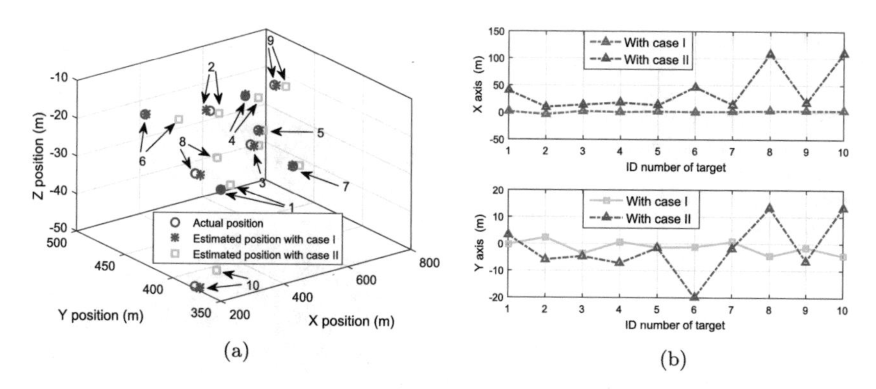

Fig. 6.11 Simulation results with the positions of anchors. (**a**) Actual and localized positions of target. (**b**) Localization errors with anchors' positions

timestamp measurement and inaccurate sound speed measurement can lead to the decrease of target localization accuracy. Combining the above factors together, the total localization errors of target can be shown in Fig. 6.12c. This result verifies the effectiveness of the error analysis as presented in Sect. 6.3.2.

(6) Performance for the Stratification Effect
It is worth mentioning that the water medium is inhomogeneous. To be specific, the propagation delay of underwater acoustic wave varies in different depth levels. Nevertheless, the stratification effect in Carroll et al. (2014) and Yan et al. (2018) was ignored, which can make the localization errors increase with the change of node depth. In order to verify this judgement, we consider the following two scenarios: (1) the sound speed is fixed, i.e., $\vartheta(z) = 1473$m/s; (2) the sound speed is linear to the depth, i.e., $\vartheta(z) = b + az$. Accordingly, the actual and localized positions of target in Scenario 1 are shown in Fig. 6.13a, whose localization errors are presented in Fig. 6.13b. It is clear that, the localization task in Scenario 1 can be achieved, because location errors are confined into a desired bound. For Scenario 2, the positions and errors of target are shown in Fig. 6.13c, d, respectively. It is known that, the stratification effect seriously affect the localization accuracy and the localization task is not achieved. Through this comparison, one can obtain that the consideration of stratification effect in USNs is necessary and meaningful.

6.5.2 Experimental Studies

In this section, the experimental studies are conducted. Specially, one target is deployed on the bottom of pool, i.e., $z_g = -1$m, which means the number of anchor nodes can be reduced from m to $m-1$. Based on this, four anchor nodes are deployed to provide localization reference for target, as depicted in Fig. 6.14a. Of note, anchor nodes employ the ultra-wideband (UWB) technology (Wang et al. 2018c) to acquire

Fig. 6.12 Simulation results for the factors related to the localization accuracy. (**a**) Localization errors with timestamps. (**b**) Localization errors with speed measurements. (**c**) Total localization errors

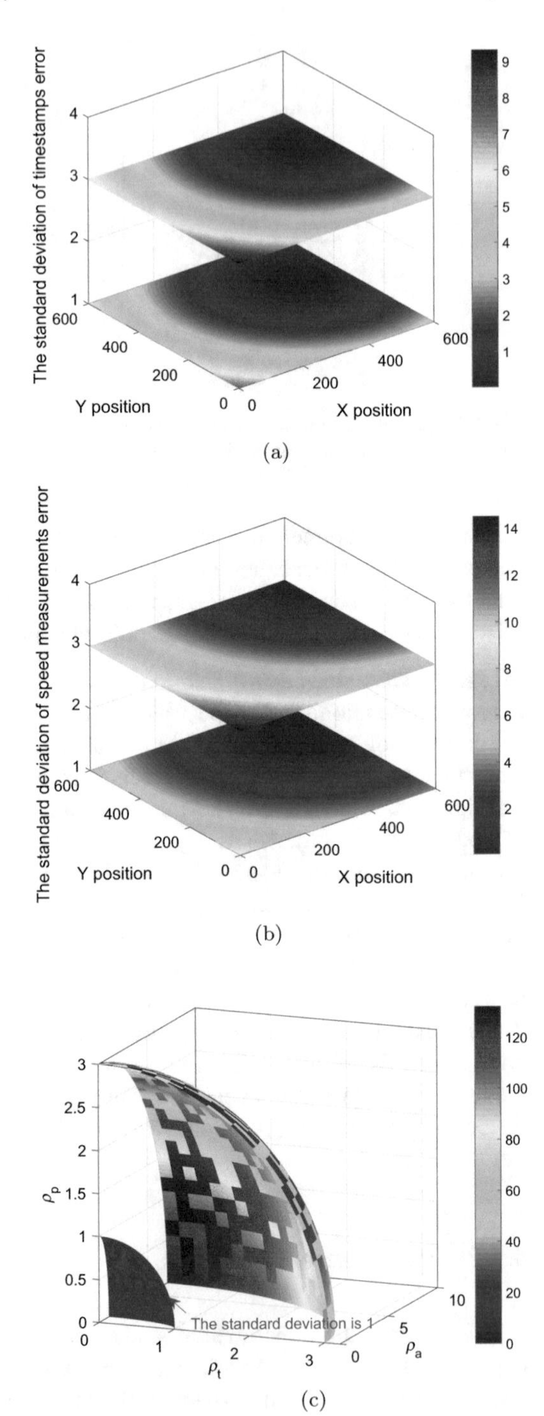

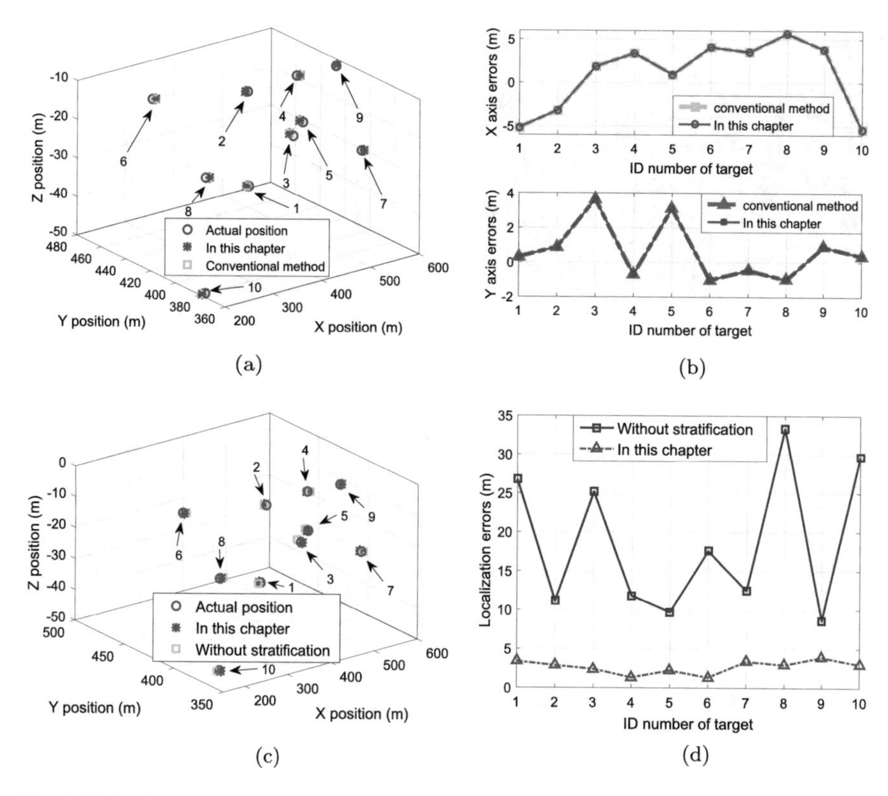

Fig. 6.13 Performance for the stratification effect. (**a**) Estimated positions in scenario 1. (**b**) Localization errors in scenario 1. (**c**) Estimated positions in scenario 2. (**d**) Localization errors in scenario 2

their positions. It is designed that four Readers are connected to one buoy. Upon receiving self-localization request, a localization tag installed on each buoy node sends out UWB pulses to the Readers. Then, Readers receive these UWB pulses and down-convert them into baseband pulses. Subsequently, the clock module triggers the UWB pulse generator, and then the pulses are amplified via a 3.1 to 5 GHz bandpass filter. The schematic for the above transmission procedure is depicted in Fig. 6.14b. Correspondingly, the estimated positions of anchor nodes are shown in Fig. 6.14c, whose estimation errors are given in Fig. 6.14d. From Fig. 6.14c, d, the self-localization of anchor nodes can be achieved.

Upon achieving self-localization, anchor nodes provide localization reference for underwater target. In order to guarantee reliable acoustic communication, a modem is required to provide communication ability for the target-to-anchor wireless links. Based on this, an acoustic modem that includes transmitter, receiver, and STM32 processor is developed in our lab. With the edge computing system, we employ the privacy-preserving localization approach proposed in this chapter to localize the target. Since the pool depth is shallow ($\approx$ 1.5 m), the acoustic speed in the pool can

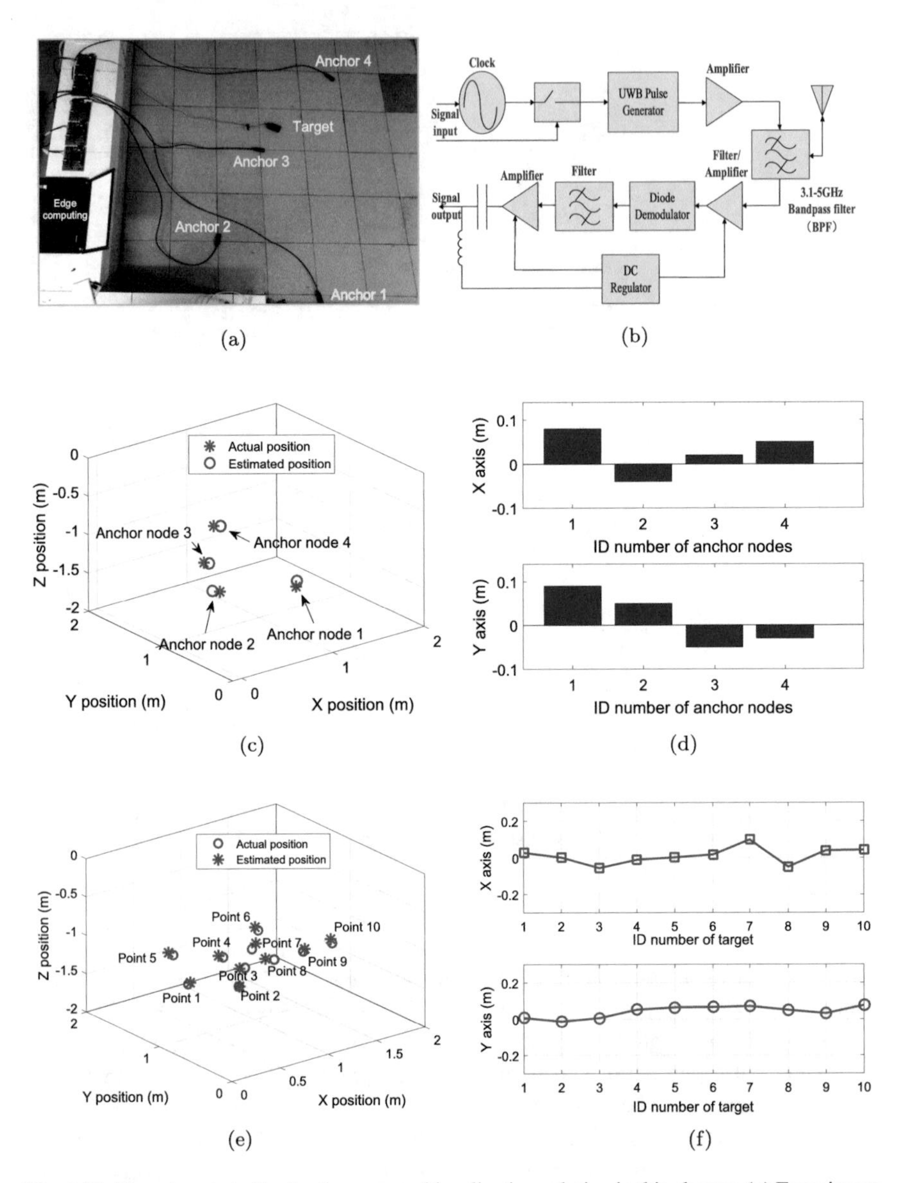

Fig. 6.14 Experiment studies for the proposed localization solution in this chapter. (**a**) Experiment setup in a pool. (**b**) Schematic for transmission procedure with UWB. (**c**) Actual and estimated positions of anchor nodes. (**d**) Localization errors of anchor nodes. (**e**) Actual and estimated positions of the target. (**f**) Localization errors of the target

Table 6.2 Time-difference measurements between two nodes

Distance (m)	Time difference (ms)	Estimated velocity (m/s)
0.3	0.206	1456.311
0.6	0.409	1466.993
0.9	0.616	1461.039
1.2	0.819	1465.201
1.5	1.026	1461.988
1.8	1.230	1463.415
2.1	1.439	1459.347

be considered as fixed value. In this case, we conduct some validation experiments, and the measured data is presented in Table 6.2. As a result, the acoustic speed in the pool is estimated as 1462 m/s. Through the above design, the actual and localized positions of target are shown in Fig. 6.14e. Correspondingly, the localization errors are depicted in Fig. 6.14f. It is clear that, the localization task for target can be achieved since the localization errors are very close to zero. These results reflect that the privacy-preserving asynchronous localization approach developed in Sect. 6.3 is effective and meaningful.

6.6 Conclusion

This chapter investigates a privacy-preserving localization issue for USNs. With consideration of clock skew, clock offset, stratification effect and forging attack, a privacy-preserving asynchronous transmission protocol is presented. Based on this, we design a privacy-preserving estimator to localize the target, where the performance analyses are also presented. Finally, simulation and experimental results are both conducted to verify the effectiveness of the proposed localization approach in this chapter.

References

Ardagna C, Cremonini M, Vimercati S, Samarati P (2011) An obfuscation-based approach for protecting location privacy. IEEE Trans Depend Secure Comput 8(1):13–27

Buccafurri F, Lax G (2009) An efficient k-anonymous localization technique for assistive environments. In: International Conference on PErvasive Technologies Related to Assistive Environments ACM, pp 1–8

Carroll P, Mahmood K, Zhou S, Zhou H, Xu X, Cui J (2014) On-demand asynchronous localization for underwater sensor networks. IEEE Trans Signal Process 62(13):3337–3348

Chen Y, Ji F, Guan Q, Wang Y, Chen F, Yu H (2018) Adaptive RTO for handshaking-based MAC protocols in underwater acoustic networks. Future Gener Comput Syst 86(1):1185–1192

Das A, Pathak P, Chuah C, Mohapatra P (2017) Privacy-aware contextual localization using network traffic analysis. Comput Netw 118(1):24–36

Diamant R, Casari P, Campagnaro F, Zorzi M (2016) A handshake-based protocol exploiting the near-far effect in underwater acoustic networks. IEEE Wireless Commun Lett 5(3):308–311

Dong L, Liu J, Su Y, Fu X (2019) Underwater multi-target position method based on received signal strength. In: OCEANS, France, pp 1–4

Ferrag M, Maglaras L, Ahmim A (2017) Privacy-preserving schemes for Ad Hoc social networks: a survey. IEEE Commun Surv Tutorials 19(4):3015–3045

Gong Z, Li C, Jiang F (2018) AUV-aided joint localization and time synchronization for underwater acoustic sensor networks. IEEE Signal Process Lett 25(4):477–481

Liu J, Wang Z, Cui J, Zhou S, Yang B (2016) A joint time synchronization and localization design for mobile underwater sensor networks. IEEE Trans Mob Comput 15(3):530–543

Liu Y, Fang G, Chen H, Xie L, Fan R, Su X (2018) Error analysis of a distributed node positioning algorithm in underwater acoustic sensor networks. In: International Conference on wireless Communications, Signal Processing (WCSP), Hangzhou

Liu X, Su S, Han F, Liu Y, Pan Z (2019) A range-based secure localization algorithm for wireless sensor networks. IEEE Sensors J 19(2):785–796

Luo Y, Pu L, Peng Z, Shi Z (2016) RSS-based secret key generation in underwater acoustic networks: advantages, challenges, and performance improvements. IEEE Commun Mag 54(2):32–38

Mortazavi E, Javidan R, Dehghani M, Kavoosi V (2017) A robust method for underwater wireless sensor joint localization and synchronization. Ocean Eng 137(1):276–286

Prelov V, Verdu S (2004) Second-order asymptotics of mutual information. IEEE Trans. Inf. Theory 50(8):1567–1580

Samarati P, Sweeney L (1998) Generalizing data to provide anonymity when disclosing information. In: Proceedings of 17th ACM sigart-sigmod-sigart symposium. Principles database System (PODS), pp 1–13

Shi X, Wu J (2018) To hide private position information in localization using time difference of arrival. IEEE Trans Signal Process 66(18):4946–4956

Shu T, Chen Y, Yang J (2015) Protecting multi-lateral localization privacy in pervasive environments. IEEE/ACM Trans Netw 23(5):1688–1701

Song Y, Wang C, Tay W (2019) Compressive privacy for a linear dynamical system. IEEE Trans Inf Forensics Secur 15(1):895–910

Song Y, Wang C, Tay W (2020) Compressive privacy for a linear dynamical system. IEEE Trans Inf Forensics Secur 15(1):895–910

Sweeney L (2002) *k*-anonymity: a model for protecting privacy. Int. J. Uncertain Fuzz Knowl Based Syst 10(5):557–570

Thomas M, Joy A (2006) Element of information theory. Hoboken, Wiley

Thomson D, Dosso S, Barclay D (2018) Modeling AUV localization error in a long baseline acoustic positioning system. IEEE J Ocean Eng 43(4):955–968

Wan Y, Yan J, Lin Z, Sheth V, Das SK (2019) On the structural perspective of computational effectiveness for quantized consensus in layered UAV networks. IEEE Trans Control Netw Syst 6(1):276–288

Wang W, Ying L, Zhang J (2016) On the relation between identifiability, differential privacy, and mutual-information privacy. IEEE Trans Inf Theory 62(9):5018–5029

Wang G, He J, Shi X, Pan J, Shen S (2018a) Analyzing and evaluating efficient privacy-preserving localization for pervasive computing. IEEE Internet Things J 5(4):2993–3007

Wang Y, Huang M, Jin Q, Ma J (2018b) DP3: a differential privacy-based privacy-preserving indoor localization mechanism. IEEE Commun Lett 22(12):2547–2550

Wang W, Wang D, Zhang B, Li T, Jiang S (2018c) Through-wall multistatus target identification in smart and autonomous systems with UWB radar. IEEE Internet Things J. 5(5):3278–3288

Yan J, Zhang X, Luo X, Wang Y, Chen C, Guan X (2018) Asynchronous localization with mobility prediction for underwater acoustic sensor networks. IEEE Trans Veh Technol 67(3):2543–2556

Yan J, Ban H, Luo X, Zhao H, Guan X (2019) Joint localization and tracking design for AUV with asynchronous clocks and state disturbances. IEEE Trans Veh Technol 68(5):4707–4720

Yan J, Zhao H, Luo X, Wang Y, Chen C, Guan X (2020) Asynchronous localization of underwater target using consensus-based unscented Kalman filtering. IEEE J Ocean Eng 45(4):1466–1481

Yan J, Meng Y, Yang X, Luo X, Guan X (2021) Privacy-preserving localization for underwater sensor networks via deep reinforcement learning. IEEE Trans Inf Forensics Secur 16(1):1880–1895

Zhao H, Yan J, Luo X, Guan X (2020) Privacy preserving solution for the asynchronous localization of underwater sensor networks. IEEE/CAA J Autom Sin 7(6):1511–1527

Chapter 7
Deep Reinforcement Learning Based Privacy Preserving Localization of USNs

Abstract This chapter is concerned with a privacy-preserving localization solution for USNs in inhomogeneous underwater medium. Besides, an honest-but-curious model is proposed to develop a privacy-preserving localization protocol. Based on the above, a localization problem is developed for sensor nodes to minimize the sum of all measurement errors, where a ray compensation strategy is designed to remove the localization bias. In order to make the above problem tractable, we represent the unsupervised, supervised and semisupervised scenarios, through which deep reinforcement learning (DRL) based localization estimators are used to estimate the positions of sensor nodes. Of note, the proposed localization issue in this chapter can hide the private position information of USNs, and besides that, it is robust to local optimum for nonconvex and nonsmooth localization solution in inhomogeneous underwater medium. Finally, simulations are given to show the position privacy can be protected, while the localization accuracy can be enhanced as compared with the other existing works.

Keywords Deep reinforcement learning · Inhomogeneous · Localization · Privacy preserving · Underwater sensor networks (USNs)

7.1 Introduction

In order to perform accurate localization for USNs (Luo et al. 2018; Xiao et al. 2019; Han et al. 2015), a large number of underwater localization schemes have been developed, e.g., Gong et al. (2020), Carroll et al. (2014), and Saeed et al. (2020). Typically, these schemes include the following three steps, i.e., *information collection*, *distance measurement* and *position estimation*. In the first step, sensor nodes collect position-related information such as the positions of anchor nodes, TDOA, TOF, TOA and RSS. With the collected message, the relative distances between sensor nodes and anchor nodes are measured in the second step, through which localization estimators are designed in the third step to calculate the positions of sensor nodes. For the above three steps, the position information leakage is unavoidable, since the relative distances are revealed to the networks while the

J. Yan et al., *Localization in Underwater Sensor Networks*, Wireless Networks,
https://doi.org/10.1007/978-981-16-4831-1_7

positions of anchor nodes are fed into the localization estimators. It is worth mentioning that USNs are usually arranged in harsh environment, and the neglecting of privacy preservation can make localization system easily vulnerable to many attacks. For instance, the enemies can easily attack the anchor nodes and wreck the entire localization system when the position information of anchor nodes has been obtained. Another example is that the enemies can infer the positions of sensor nodes by correlating the distance measurements with the monitoring area. Consequently, it is necessary to protect the private positions of both anchor and sensor nodes during the localization procedure.

Since considering the privacy-preserving localization, we note that many privacy-preserving localization protocols have been designed for terrestrial networks. For example, a homomorphic encryption based localization protocol that can propose different levels of privacy preservation was designed in Shu et al. (2014). In Yang and Javinen (2018), the paillier encryption technology was used to solve the privacy preservation of WiFi fingerprint localization. Despite the encryption technique can provide strong privacy level, its communication and computing overheads are high (Li et al. 2015), which is not suitable for USNs due to the limited bandwidth and battery power of underwater devices. In contrast, Shi and Wu (2018) and Wang et al. (2018) employed the information-hiding technique, e.g., PPS and PPDP, to hide the private position information of anchor nodes and target. Compared with encryption technique, the information-hiding technique not only has the advantage of computation efficiency, but also can avoid additional localization errors. However, the localization protocols in Shi and Wu (2018) and Wang et al. (2018) depend on the assumption of straight-line data transmission. This assumption is realistic in terrestrial environment, but, it does not exist in underwater environment since the water medium is inhomogeneous and the propagation paths of sound waves are curved (Jiang 2019). Ignore the inhomogeneity characteristic of underwater medium can increase ranging errors, which in turn reduce the localization accuracy. Parallel to this, the consideration of inhomogeneity characteristic can lead to the failure of information-hiding technique, because the additional noises cannot be eliminated ultimately. With regard to the inhomogeneity characteristic, how to utilize the information-hiding technology to design a privacy-preserving localization protocol for USNs is largely unsolved.

At present, many localization estimators have been designed in the existing literatures. Generally, these estimators can be classified into the following two categories, i.e., *least squares-based estimators* (Ramezani et al. 2013; Liu et al. 2016; Mortazavi et al. 2017) and *convex optimization-based estimators* (Zheng et al. 2018; Soares et al. 2017; Jia et al. 2019). The least squares-based estimators have the merits of theoretical simplicity and operation convenience, nevertheless, they are easy to trap in local optimum. Otherwise, the convex optimization-based estimators can overcome the local optimum by using convex relaxation. Despite all this, it is not an easy task to relax nonconvex problem into convex optimization problem, in especial, when the inhomogeneity characteristic of underwater medium is considered. In order to handle this issue, we note that the deep reinforcement learning (DRL) can provide us with a simple and effective solution. In particular,

DRL is an advanced branch of the artificial intelligence algorithm, in which neural networks are trained to assist the agent to make global optimal decision in a real-time manner (Luong et al. 2019). The merits of DRL are that the convex relaxation is inessential and the global optimal solution can be acquired. Accordingly, it is still an openness issue to adopt DRL to solve the privacy-preserving localization for USNs.

Notice that Chaps. 5 and 6 employ the least squares-based estimators to acquire the position information. Neverthless, the least squares-based localization estimators can easily fall into local minimum. To solve the above isseu, this chapter proposes a DRL-based privacy-preserving localization solution for USNs. An honest-but-curious model is considered for USNs, such that a ray compensation strategy is developed to remove the effect of inhomogeneous water medium. Based on the above, DRL-based localization estimators are designd to seek the positions of sensor nodes. Main contributions of this chapter can be summarized as below.

1. **Privacy-Preserving Localization Protocol.** Without using any encryption technique, a privacy-preserving localization protocol is designed in this chapter to protect the private position information of anchor and sensor nodes. Compared with the existing works (Shi and Wu 2018; Wang et al. 2018), the assumption of straight-line data transmission can be relaxed in this chapter. Besides, the private position information of anchor and sensor nodes can be preserved in this chapter, which is different from the ones in Ramezani et al. (2013) and Ameer and Jacob (2010).
2. **DRL-Based Localization Estimators**. DRL-based localization estimators are developed in this chapter, wherein the unsupervised, supervised and semisupervised scenarios are considered, respectively. Compared with the least squares-based estimators in Ramezani et al. (2013), Liu et al. (2016), and Mortazavi et al. (2017), the DRL-based localization estimators in this chapter can conquer the local optimum and acquire the global optimal solution. In addition, it has much faster learning speed than the reinforcement learning (RL) based localization estimator, e.g., Yan et al. (2020b).

7.2 Network Architecture and Problem Formulation

7.2.1 Network Architecture

In order to carry out the privacy-preserving localization, a network that consists of three different types of nodes is considered.

- **Surface Buoys.** Surface buoys can make communication with satellite directly, and they employ the global positioning system (GPS) to obtain their accurate position information. It is assumed that the position information of surface buoys does not need to be protected, and their role is to provide self-localization service for anchor nodes.

- **Anchor Nodes**. Anchor nodes can make communication with surface buoys directly, through which they are able to self-localization by using some existing localization schemes, such as Zheng et al. (2018), Soares et al. (2017), and Jia et al. (2019). The role of anchor nodes is to supply localization reference for sensor nodes.
- **Sensor Nodes**. Sensor nodes are static nodes, whose position information requires to be estimated by interacting with anchor nodes. It is worth mentioning that sensor and anchor nodes are both concerned on their position privacy. Accordingly, it is necessary to hide the position privacy for both sensor and anchor nodes during the localization procedure.

Generality, we assume that the depth information of sensor nodes can be measured by depth units accurately, and thereby, at least three non-collinear anchor nodes are needed to estimate the positions of sensor nodes on X and Y axes. Based on the above, we consider that $M \geq 3$ anchor nodes and $N \geq 2$ sensor nodes are developed in the underwater monitoring area. In particular, the index sets of anchor nodes and sensor nodes are denoted by $\mathcal{I}_{\rm A} = \{1, \cdots, M\}$ and $\mathcal{I}_{\rm S} = \{1, \cdots, N\}$, respectively. A popular isogradient model (Ramezani et al. 2013; Ameer and Jacob 2010) is adopted to describe the sound speed profile (SSP), i.e., the sound speed is linear to the depth. Thus, the sound speed can be modeled as

$$\vartheta(z) = b + az, \tag{7.1}$$

where b is the sound speed on water surface, a indicates a scalar related to the environment, and z represents the depth.

In view of (7.1), a ray tracing approach is utilized to model the acoustic propagation between any two nodes, as shown in Fig. 7.1. Let $\mathbf{P}_{\rm A} = [x_{\rm A}, y_{\rm A}, z_{\rm A}]^{\rm T}$ and $\mathbf{P}_{\rm B} = [x_{\rm B}, y_{\rm B}, z_{\rm B}]^{\rm T}$ be the position vectors of sender node A and receiver node B, respectively. By adopting Snell's law (Liu et al. 2018), one acquires

$$\frac{\cos\theta}{\vartheta(z)} = \frac{\cos\theta_{\rm A}}{\vartheta(z_{\rm A})} = \frac{\cos\theta_{\rm B}}{\vartheta(z_{\rm B})}, \tag{7.2}$$

where $\theta_{\rm A} \in [-\frac{\pi}{2}, \frac{\pi}{2}]$ and $\theta_{\rm B} \in [-\frac{\pi}{2}, \frac{\pi}{2}]$ indicate ray angles at the positions of sender node A and receiver node B, respectively.

From (7.2), we can further obtain $\mathrm{d}l = \frac{\mathrm{d}z}{\sin\theta}$ and $\frac{\mathrm{d}l}{\mathrm{d}t} = \vartheta(z)$, where l is the path length between nodes A and B. Consequently, t denotes the travel time of l. Thus, the travel time between sender node A and receiver node B is calculated as Ramezani et al. (2013)

$$t = -\frac{1}{a}\left(\ln\frac{1+\sin\theta_{\rm B}}{\cos\theta_{\rm B}} - \ln\frac{1+\sin\theta_{\rm A}}{\cos\theta_{\rm A}}\right), \tag{7.3}$$

where $\theta_{\rm A} = \beta_0 + \alpha_0$ and $\theta_{\rm B} = \beta_0 - \alpha_0$ are the acoustic ray angles at the sender node A and the receiver node B, respectively. In specifical, $\beta_0 =$

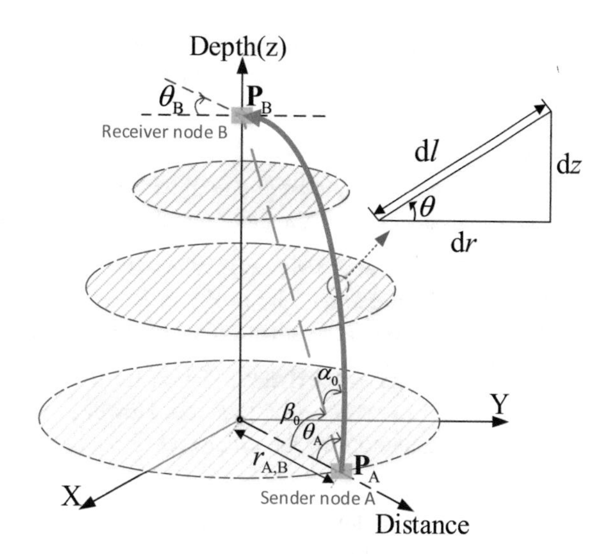

Fig. 7.1 Description of the non-straight acoustic propagation

$\arctan(\frac{z_B - z_A}{\sqrt{(x_B - x_A)^2 + (y_B - y_A)^2}})$ represents the angle of the straight line between the sender node A and the receiver node B, with respect to the distance axis. Accordingly, $\alpha_0 = \arctan(\frac{a\sqrt{(x_B - x_A)^2 + (y_B - y_A)^2}}{2b + a(z_B + z_A)})$ represents the angle of the ray that deviates from the straight line.

7.2.2 *Adversary and Privacy Models*

The openness of USNs makes localization system easily vulnerable to attack from adversary nodes. In this case, an honest-but-curious model (Shu et al. 2015; Im et al. 2020) is considered for USNs, where each node honestly follows the specified localization protocol, however, it is curious about other nodes' position information. Furthermore, we assume the communications between sensor and anchor nodes are secure, which means the private information cannot be eavesdropped through communication links. The above hypothesis has been employed in some existing works, e.g., Shi and Wu (2018), which can be realized by covert transmission. Therefore, the adversary model is presented as follows.

Definition 7.1 (Adversary Model) A node can be regarded as adversary if it colludes with some nodes trying to obtain some other nodes' position information. For such situation, colluded nodes are capable of exchanging their illegitimate information to calculate the positions of others.

Based on the above mentioned model, two scenarios are considered in the privacy analysis, such as independent nodes and colluding nodes. For the first scenario, a node can compute other nodes' positions only by the legal information it receives.

For the second scenario, colluded nodes can establish a side channel to reveal more information about the others. In order to cope with this, we define the following three levels of position privacy.

Definition 7.2 (Level-1 Privacy) When the localization algorithm ending, sensor node $j \in \mathcal{I}_S$ can estimate its position. Nevertheless, sensor node $j \in \mathcal{I}_S$ or anchor node $i \in \mathcal{I}_A$ cannot know the other nodes' position information by itself.

Definition 7.3 (Level-2 Privacy) When the localization algorithm ends, sensor node $j \in \mathcal{I}_S$ can estimate its position. Besides, sensor node $j \in \mathcal{I}_S$ or anchor node $i \in \mathcal{I}_A$ can know the other nodes' positions by colluding with other nodes.

Definition 7.4 (Level-3 Privacy) When the localization algorithm ends, sensor node $j \in \mathcal{I}_S$ can estimate its position. However, sensor node $j \in \mathcal{I}_S$ or anchor node $i \in \mathcal{I}_A$ cannot know the other nodes' positions even if it colludes with other nodes.

It makes sense to define the above privacy levels. Especially, Level-1 and Level-2 privacies are applicable to cases when all nodes are unlikely to collude. It is noticed that Level-1 privacy is the basic requirement for all privacy-preserving localization systems. Furthermore, Level-3 privacy is the most demanding one, and it is suitable for case when each node needs high location privacy even in the presence of collusion attacks.

7.2.3 Scenario Description

In this chapter, we consider a cooperative localization network, where sensor nodes send localization requests to anchor nodes. Anchor nodes reply position-related information back to sensor nodes, through which localization estimation is conducted by sensor nodes. It is noted that some knowledge of the underwater environment can be pre-known to sensor nodes at times. For instance, an AUV is patrolling environment to collect the data at different positions, and then it periodically conveys the collected data to sensor nodes. In such a network, the collected data is named as 'position label', and we believe that the localization performance can benefit a lot from position labels. Based on this, we consider the following three scenarios:

S1 All data is unlabelled, i.e., sensor nodes do not have any information on position labels, as described in Fig. 7.2a.

S2 The labelled data occupies the majority of environment information, such as sensor nodes have most of the information on the position labels, as describe in Fig. 7.2b.

S3 The unlabelled data occupies the majority of environment information, i.e., sensor nodes have few information on the position labels, as depicted in Fig. 7.2c.

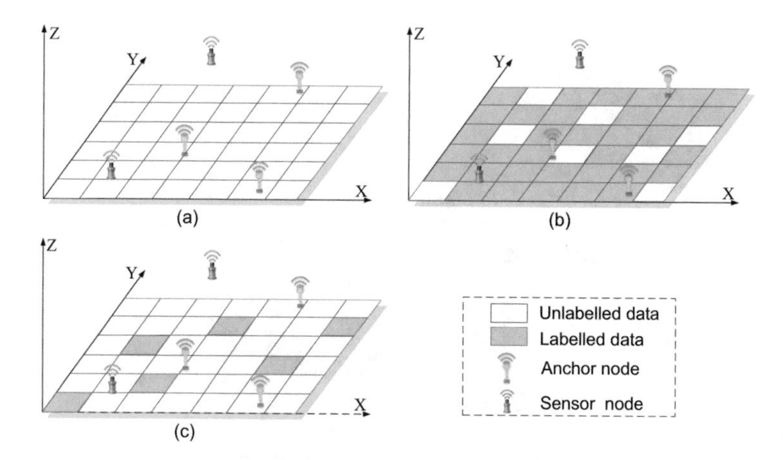

Fig. 7.2 Overview of the three localization scenarios. (**a**) S1: All the data is unlabelled. (**b**) S2: Labelled data occupies the majority. (**c**) S3: Unlabelled data occupies the majority

7.2.4 *Problem of Interest*

The purpose is to design a privacy-preserving localization protocol to hide the private position information of anchor and sensor nodes, through which DRL-based localization estimators are used to localize sensor nodes. Accordingly, the following three preformation metrics are given:

(I) **Privacy Preservation.** In general, let $\mathbf{P}_{\mathrm{A},i} = [x_{\mathrm{A},i}, y_{\mathrm{A},i}, z_{\mathrm{A},i}]^{\mathrm{T}}$ and $\mathbf{P}_{\mathrm{S},j} = [x_{\mathrm{S},j}, y_{\mathrm{S},j}, z_{\mathrm{S},j}]^{\mathrm{T}}$ be the position vectors of anchor node $i \in \mathcal{I}_{\mathrm{A}}$ and sensor node $j \in \mathcal{I}_{\mathrm{S}}$, respectively. For any anchor node i, its position vector $\mathbf{P}_{\mathrm{A},i}$ cannot be learned by sensor nodes. Furthermore, the position vector $\mathbf{P}_{\mathrm{S},j}$ for sensor node j cannot be learned by anchor nodes.

(II) **Correctness**. With privacy preserving in inhomogeneous water medium, the sensor positions calculated in this chapter are equivalent to the ones by using privacy-lacking localization solutions, e.g., Li et al. (2020), Zhang et al. (2014), and Jondhale et al. (2020).

(III) **Global Optimum**. By considering the privacy preservation in inhomogeneous water medium, a global optimal position solution can be obtained.

7.3 Privacy-Preserving Localization Protocol

To beginning with, sensor nodes send localization requests to anchor nodes. After receiving the requests, anchor nodes record and reply *pseudo* position information back to sensor nodes. In special, the transmission process between sensor node $j \in \mathcal{I}_{\mathrm{S}}$ and anchor nodes is shown in Fig. 7.3. Furthermore, the privacy-preserving

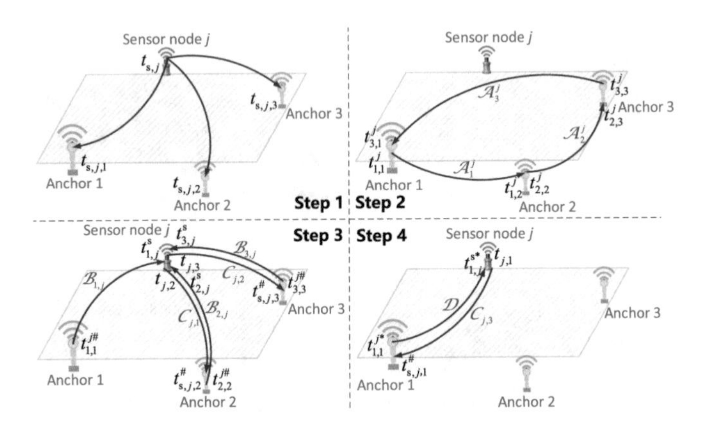

Fig. 7.3 Description of the privacy-preserving transmission process

localization protocol between sensor node $j \in \mathcal{I}_\text{S}$ and anchor nodes can be detailed as below.

0. Suppose the monitoring region is a large cuboid $\mathbf{D}$, whose volume is $S \times W \times C$. In particular, $\mathbf{D}$ is partitioned into n subcuboids, and the volume of each subcuboid is $s \times w \times c$. For subcuboid $\check{i} \in \mathcal{I}_\text{P} = \{1, \cdots, \check{n}\}$, its center position is $\mathbf{P}_{\text{P},\check{i}}$, where $\check{n} \in [1, \cdots, \frac{SWC}{swc}]$ denotes the number of available position labels. When a mobile node (e.g., AUV) is staying in subcuboid $\check{i}$, the propagation delay between AUV and anchor node $i \in \mathcal{I}_\text{A}$ is $\tau_{\text{P},\check{i},i}$. In this case, we define $\mathbf{P}_{\text{P},\check{i}}$ and $\tau_{\text{P},\check{i},i}$ as the 'position label' for subcuboid $\check{i}$. Notely, the position labels are not prerequisites for the localization system.
1. At time $t_{\text{s},j}$, sensor node j broadcasts an initiator packet to the networks, which is the delivery order of anchors, i.e, $i \in \{1, \cdots, M\}$. Then, sensor node j switches into listen mode, and waits for the replies from anchors. In special, anchor node $i \in \mathcal{I}_\text{A}$ receives the initiator packet at time $t_{\text{s},j,i}$.
2. Anchor node $h \in \mathcal{I}_\text{A}/\{M\}$ sends message $\mathcal{A}_h^j$ to anchor node $h+1$ at time $t_{h,h}^j$, through which anchor node $h+1$ receives this message at time $t_{h,h+1}^j$. Meanwhile, anchor node M sends message $\mathcal{A}_M^j$ to anchor node 1 at time $t_{M,M}^j$, where anchor node 1 receives message $\mathcal{A}_M^j$ at time $t_{M,1}^j$. Noteworthy, $\mathcal{A}_i^j$ is a random vector $\boldsymbol{\gamma}_{j,i} \in \mathcal{R}^{3\times 1}$ for $i \in \mathcal{I}_\text{A}$.
3. After the above procedure is completed, anchor node $h \in \mathcal{I}_\text{A}/\{M\}$ replies message $\mathcal{B}_{h,j}$ to sensor node j at time $t_{h,h}^{j\#}$. At time $t_{h,j}^\text{s}$, sensor node j receives message $\mathcal{B}_{h,j}$ from anchor node $h \in \mathcal{I}_\text{A}/\{M\}$. Next, sensor node j transmits message $\mathcal{C}_{j,h}$ to anchor node $h+1$ at time $t_{j,h+1}$, and anchor node $h+1$ receives it at time $t_{\text{s},j,h+1}^\#$. After receiving the message $\mathcal{C}_{j,M-1}$ from sensor node j at time

$t^{\#}_{\mathrm{s},j,M}$, anchor node M replies message $\mathcal{B}_{M,j}$ to sensor node j at time $t^{j\#}_{M,M}$. In particular, the messages $\mathcal{B}_{i,j}$ and $\mathcal{C}_{j,h}$ can be defined

$$\begin{aligned}\mathcal{B}_{1,j} &= \{t_{\mathrm{s},j,1},\ t^{j\#}_{1,1},\ \mathrm{L}_{j,1},\},\\ \mathcal{B}_{i,j} &= \{t_{\mathrm{s},j,i},\ t^{j\#}_{i,i},\ \mathrm{L}_{j,i},\ \Delta\mathrm{P}_{j,1,i-1},\ \Delta x_{j,1,i-1},\\ &\qquad \Delta y_{j,1,i-1},\ \Delta x'_{j,1,i-1},\ \Delta y'_{j,1,i-1}\}_{i\geq 2},\\ \mathcal{C}_{j,h} &= \{\hat{\mathbf{P}}_{\mathrm{S},j,1} - \mathrm{L}_{j,h},\ s,\ w\},\end{aligned} \tag{7.4}$$

with $\forall i \in \mathcal{I}_{\mathrm{A}}$

$$\begin{aligned}\mathrm{L}_{j,i} &= \mathbf{P}_{\mathrm{A},i} - \boldsymbol{\gamma}_{j,i},\\ \Delta\mathrm{P}_{j,1,i} &= \{\Delta R_{j,1,i},\ z_{\mathrm{S},j} - z_{\mathrm{A},i}\},\\ \Delta R_{j,1,i} &= \sqrt{(\hat{x}_{\mathrm{S},j,1} - x_{\mathrm{A},i})^2 + (\hat{y}_{\mathrm{S},j,1} - y_{\mathrm{A},i})^2},\\ \Delta x_{j,1,i} &= (\hat{x}_{\mathrm{S},j,1} - x_{\mathrm{A},i})^2 + \varepsilon_{i,\mathrm{x}},\\ \Delta y_{j,1,i} &= (\hat{y}_{\mathrm{S},j,1} - y_{\mathrm{A},i})^2 + \varepsilon_{i,\mathrm{y}},\\ \Delta x'_{j,1,i} &= 2s(\hat{x}_{\mathrm{S},j,1} - x_{\mathrm{A},i}) + s^2 - (\varepsilon_{i,\mathrm{x}} + \varepsilon_{i,\mathrm{y}}),\\ \Delta y'_{j,1,i} &= 2w(\hat{y}_{\mathrm{S},j,1} - y_{\mathrm{A},i}) + w^2 - (\varepsilon_{i,\mathrm{x}} + \varepsilon_{i,\mathrm{y}}),\end{aligned} \tag{7.5}$$

where $\hat{\mathbf{P}}_{\mathrm{S},j,1} = [\hat{x}_{\mathrm{S},j,1},\ \hat{y}_{\mathrm{S},j,1},\ z_{\mathrm{S},j}]^{\mathrm{T}}$. Especially, $\hat{x}_{\mathrm{S},j,1}$ and $\hat{y}_{\mathrm{S},j,1}$ denote the initial estimated positions on X and Y axes, respectively. $\varepsilon_{i,\mathrm{x}}$ and $\varepsilon_{i,\mathrm{y}}$ are random constants.

4. At time $t_{j,1}$, sensor node j sends end message $\mathcal{C}_{j,M}$ to anchor node 1. Anchor node 1 replies message $\mathcal{D}$ to sensor node j at time $t^{j*}_{1,1}$, and then the transmission process for sensor node $j \in \mathcal{I}_{\mathrm{S}}$ is ended. $\mathcal{C}_{j,M}$ and $\mathcal{D}$ are defined

$$\begin{aligned}\mathcal{C}_{j,M} &= \{\hat{\mathbf{P}}_{\mathrm{S},j,1} - \mathrm{L}_{j,M},\ s,\ w\},\\ \mathcal{D} &= \{t^{\#}_{\mathrm{s},j,1},\ \Delta\mathrm{P}_{j,1,M},\ \Delta x_{j,1,M},\ \Delta y_{j,1,M},\\ &\qquad \Delta x'_{j,1,M},\ \Delta y'_{j,1,M}\}.\end{aligned} \tag{7.6}$$

5. Meanwhile, sensor node j collects the following messages

$$\begin{aligned}&\begin{cases}\varnothing, & \text{if labels are unavailable}\\ \{\mathbf{P}_{\mathrm{P},\check{i}},\ \tau_{\mathrm{P},\check{i},i}\}_{\check{i}=1}^{\check{n}}, & \text{if labels are available}\end{cases}\\ &t_{\mathrm{s},j},\ \mathcal{D},\ \{t^{\mathrm{s}}_{i,j},\ t_{j,i},\ \mathcal{B}_{i,j}\}_{i=1}^{M}.\end{aligned} \tag{7.7}$$

With the collected messages in (7.7), the time difference between sensor $j \in \mathcal{I}_{\mathrm{S}}$ and anchor $i \in \mathcal{I}_{\mathrm{A}}$ is presented as

$$\Delta t_{j,i} = [(t_{\mathrm{s},j,i} - t_{\mathrm{s},j}) + (t_{i,j}^{\mathrm{s}} - t_{i,i}^{j\#})]/2. \tag{7.8}$$

It is assumed that the measurement noise of each local measurement is a random variable with zero mean and variance σ_{mea}^2. By adopting (7.8), the relationship between time difference and travel time is constructed as

$$\Delta t_{j,i} = \tau_{j,i} + \omega_{j,i}, \tag{7.9}$$

where $\tau_{j,i}$ represents the travel time between sensor node j and anchor node i, as calculated in (7.3). In addition, $\omega_{j,i}$ denotes the measurement noise, satisfying the distribution of $\omega_{j,i} \sim \mathcal{N}(0, \sigma_{\mathrm{mea}}^2)$.

Let $\hat{x}_{\mathrm{S},j}$ and $\hat{y}_{\mathrm{S},j}$ be the estimations of $x_{\mathrm{S},j}$ and $y_{\mathrm{S},j}$, respectively. Consequently, the likelihood function of $x_{\mathrm{S},j}$ and $y_{\mathrm{S},j}$ for sensor node $j \in \mathcal{I}_{\mathrm{S}}$ can be formulated as

$$\Lambda_{\mathrm{S},j} = \frac{1}{\sqrt{2\pi}\sigma_{\mathrm{mea}}} \exp\left\{\frac{1}{\sigma_{\mathrm{mea}}^2} \sum_{i=1}^{M} [\Delta t_{j,i} - \tau_{j,i}]^2\right\}. \tag{7.10}$$

From (7.10), the localization optimization problem for sensor node $j \in \mathcal{I}_{\mathrm{S}}$ can be developed as

$$\begin{aligned}(x_{\mathrm{S},j}^*, y_{\mathrm{S},j}^*) &= \operatorname{argmin}_{(\hat{x}_{\mathrm{S},j}, \hat{y}_{\mathrm{S},j})} \mathcal{H}(\hat{x}_{\mathrm{S},j,k}, \hat{y}_{\mathrm{S},j,k}) \\ &= \operatorname{argmin}_{(\hat{x}_{\mathrm{S},j}, \hat{y}_{\mathrm{S},j})} \frac{1}{\sigma_{\mathrm{mea}}^2} \sum_{i=1}^{M} [\Delta t_{j,i} - \tau_{j,i}]^2.\end{aligned} \tag{7.11}$$

7.4 DRL-Based Localization Estimator

In this section, we adopt DRL to seek the solution for the optimization problem in (7.11). For this reason, three DRL-based localization estimators are developed to estimate $x_{\mathrm{S},j}$ and $y_{\mathrm{S},j}$.

7.4.1 Localization when All Data Is Unlabelled

In this scenario, sensor node $j \in \mathcal{I}_{\mathrm{S}}$ does not have any information on position labels, and it can be regarded as unsupervised scenario. The deep Q-network (DQN), which is an important branch of DRL (Mnih et al. 2015; Meng et al. 2020), can be employed to estimate $x_{\mathrm{S},j}$ and $y_{\mathrm{S},j}$. We considered the localization estimation as a Markov decision process, and thereby, the following dynamical system is given to describe the transition procedure for the localization estimator of sensor node $j \in \mathcal{I}_S$

$$\hat{\bar{\mathbf{P}}}_{S,j,k+1} = \hat{\bar{\mathbf{P}}}_{S,j,k} + \bar{\mathbf{u}}_{S,j,k}, \tag{7.12}$$

where $\hat{\bar{\mathbf{P}}}_{S,j,k} = [\hat{x}_{S,j,k},\ \hat{y}_{S,j,k}]^{\mathrm{T}}$ is the estimated horizontal position vector of sensor node j at iteration step k. In addition, $\bar{\mathbf{u}}_{S,j,k} \in \mathcal{U}$ is the action for the localization estimator of sensor node j at iteration step k. In particular, $\mathcal{U}$ denotes the action space, which includes staying at the same subcuboid and moving toward east, west, north and south for a subcuboid, i.e., $\mathcal{U} = \{[0, 0]^{\mathrm{T}},\ [s,\ 0]^{\mathrm{T}},\ [-s, 0]^{\mathrm{T}},\ [0, w]^{\mathrm{T}}, [0, -w]^{\mathrm{T}}\}$.

In order to assess the action performance, the reward function learned from $\hat{\bar{\mathbf{P}}}_{S,j,k}$ to $\hat{\bar{\mathbf{P}}}_{S,j,k+1}$ is defined as

$$r_k = \begin{cases} \dfrac{1}{\mathcal{H}(\hat{x}_{S,j,k}, \hat{y}_{S,j,k})}, & \text{if } \mathcal{H}_{\Delta,k} > 0 \\ -\mathcal{H}(\hat{x}_{S,j,k}, \hat{y}_{S,j,k}), & \text{else} \end{cases} \tag{7.13}$$

where $\mathcal{H}_{\Delta,k} = \mathcal{H}(\hat{x}_{S,j,k}, \hat{y}_{S,j,k}) - \mathcal{H}(\hat{x}_{S,j,k+1}, \hat{y}_{S,j,k+1})$, and $\mathcal{H}(\hat{x}_{S,j,k}, \hat{y}_{S,j,k})$ indicates the quadratic form at step k.

The calculation of $\mathcal{H}(\hat{x}_{S,j,k}, \hat{y}_{S,j,k})$ demands the horizontal distance $\Delta R_{j,k,i} = \sqrt{(\hat{x}_{S,j,k} - x_{A,i})^2 + (\hat{y}_{S,j,k} - y_{A,i})^2}$, which can leak position privacy, since $x_{A,i}$ and $y_{A,i}$ are required to be known. In order to solve this issue, the following four cases are given to indirectly calculate $\Delta R_{j,k,i}$.

Case 1. Sensor node j moves to east, i.e., $\bar{\mathbf{u}}_{S,j,k} = [s,\ 0]^{\mathrm{T}}$. Express $k_1 = (\hat{x}_{S,j,k-1} - \hat{x}_{S,j,1})/s$, and then one obtain

$$\begin{aligned} &(\hat{x}_{S,j,k} - x_{A,i})^2 + (\hat{y}_{S,j,k} - y_{A,i})^2, \\ &= (\hat{x}_{S,j,k-1} + s - x_{A,i})^2 + (\hat{y}_{S,j,k-1} - y_{A,i})^2, \\ &= \Delta R^2_{j,k-1,i} + \Delta x'_{j,1,i} + 2k_1 s^2 \\ &\quad + \Delta x_{j,1,i} + \Delta y_{j,1,i} - \Delta R^2_{j,1,i}. \end{aligned} \tag{7.14}$$

Case 2. Sensor node j moves to west for a subcuboid, i.e., $\bar{\mathbf{u}}_{S,j,k} = [-s,\ 0]^{\mathrm{T}}$, and thereby, one has

$$\begin{aligned} &(\hat{x}_{S,j,k} - x_{A,i})^2 + (\hat{y}_{S,j,k} - y_{A,i})^2, \\ &= (\hat{x}_{S,j,k-1} - s - x_{A,i})^2 + (\hat{y}_{S,j,k-1} - y_{A,i})^2, \\ &= \Delta R^2_{j,k-1,i} - \Delta x'_{j,1,i} + 2(-k_1 + 1)s^2 \\ &\quad - \Delta x_{j,1,i} - \Delta y_{j,1,i} + \Delta R^2_{j,1,i}. \end{aligned} \tag{7.15}$$

Case 3. Sensor node j moves to north, i.e., $\bar{\mathbf{u}}_{\mathrm{S},j,k} = [0,\ w]^{\mathrm{T}}$. Indicate $k_2 = (\hat{y}_{\mathrm{S},j,k-1} - \hat{y}_{\mathrm{S},j,1})/w$, and then one acquire

$$\begin{aligned}
&\left(\hat{x}_{\mathrm{S},j,k} - x_{\mathrm{A},i}\right)^2 + \left(\hat{y}_{\mathrm{S},j,k} - y_{\mathrm{A},i}\right)^2, \\
&= \left(\hat{x}_{\mathrm{S},j,k-1} - x_{\mathrm{A},i}\right)^2 + \left(\hat{y}_{\mathrm{S},j,k-1} + w - y_{\mathrm{A},i}\right)^2, \\
&= \Delta R^2_{j,k-1,i} + \Delta y'_{j,1,i} + 2k_2 w^2 \\
&\quad + \Delta x_{j,1,i} + \Delta y_{j,1,i} - \Delta R^2_{j,1,i}.
\end{aligned} \tag{7.16}$$

Case 4. Sensor node j moves to south for a subcuboid, i.e., $\bar{\mathbf{u}}_{\mathrm{S},j,k} = [0, -w]^{\mathrm{T}}$, and therefore, one has

$$\begin{aligned}
&(\hat{x}_{\mathrm{S},j,k} - x_{\mathrm{A},i})^2 + (\hat{y}_{\mathrm{S},j,k} - y_{\mathrm{A},i})^2, \\
&= (\hat{x}_{\mathrm{S},j,k-1} - x_{\mathrm{A},i})^2 + (\hat{y}_{\mathrm{S},j,k-1} - w - y_{\mathrm{A},i})^2, \\
&= \Delta R^2_{j,k-1,i} - 2w(\hat{x}_{\mathrm{S},j,k-1} - x_{\mathrm{A},i}) + w^2, \\
&= \Delta R^2_{j,k-1,i} - \Delta y'_{j,1,i} + 2(-k_2 + 1)w^2 \\
&\quad - \Delta x_{j,1,i} - \Delta y_{j,1,i} + \Delta R^2_{j,1,i}.
\end{aligned} \tag{7.17}$$

By adopting (7.13), the Q function is constructed as

$$\begin{aligned}
&Q(\hat{\mathbf{P}}_{\mathrm{S},j,k}, \mathbf{u}_{\mathrm{S},j,k}; \eta) \\
&= \mathbf{E}\left\{r_k + \ell Q(\hat{\mathbf{P}}_{\mathrm{S},j,k+1}, \mathbf{u}_{\mathrm{S},j,k+1}; \eta)\right\},
\end{aligned} \tag{7.18}$$

where $\hat{\mathbf{P}}_{\mathrm{S},j,k} = [\hat{\bar{\mathbf{P}}}^{\mathrm{T}}_{\mathrm{S},j,k}, z_{\mathrm{S},j}]^{\mathrm{T}}$ and $\mathbf{u}_{\mathrm{S},j,k} = [\bar{\mathbf{u}}^{\mathrm{T}}_{\mathrm{S},j,k}, 0]^{\mathrm{T}}$. In addition, $\ell \in (0, 1)$ represents the discount factor. η denotes the weight parameter set of the Q-network, required to be learned.

From (7.18), the Bellman optimality equation can defined as

$$\begin{aligned}
&Q^*(\hat{\mathbf{P}}_{\mathrm{S},j,k}, \mathbf{u}_{\mathrm{S},j,k}; \eta) \\
&= \mathbf{E}\left\{r_k + \ell \max_{\mathbf{u}_{\mathrm{S},j,k+1}} Q^*(\hat{\mathbf{P}}_{\mathrm{S},j,k+1}, \mathbf{u}_{\mathrm{S},j,k+1}; \eta)\right\},
\end{aligned} \tag{7.19}$$

which yields the optimal policy

$$\pi^*(\hat{\mathbf{P}}_{\mathrm{S},j,k}) = \mathrm{argmax}_{\mathbf{u}_{\mathrm{S},j,k} \in \mathcal{U}}\, Q^*(\hat{\mathbf{P}}_{\mathrm{S},j,k}, \mathbf{u}_{\mathrm{S},j,k}; \eta). \tag{7.20}$$

Algorithm 4: Localization estimator when all data is unlabelled

Input: The messages in (7.7), the capacity of memory pool N_{m}, total exploration step N_{e} and probability ϵ

Output: Estimated position of sensor node j

1 Sensor node j selects a random position vector $\hat{\mathbf{P}}_{\mathrm{S},j,1}$, and then it explores the environment
2 Put $\{\hat{\mathbf{P}}_{\mathrm{S},j,k^*}, \mathbf{u}_{\mathrm{S},j,k^*}, r_{k^*}, \hat{\mathbf{P}}_{\mathrm{S},j,k^*+1}\}_{k^*=1}^{N_{\mathrm{e}}}$ in memory pool
3 Train the DQN by combining random samples into memory pool and target $\hat{Q}$-network
4 **while** $\mathcal{H}(\hat{x}_{S,j,k+1}, \hat{y}_{S,j,k+1}) \geq \Delta\mathcal{H}$ *or* $k+1 > K$ **do**
5   **if** N_p *steps are explored* **then**
6   Target $\hat{Q}$-network replicates the DQN, and updates the parameters of DQN
7   **if** *within probability* $1-\epsilon$ **then**
8   Select the optimal policy $\pi^*(\hat{\mathbf{P}}_{\mathrm{S},j,k})$
9   **else**
10   Randomly select action in action space
11   Update $\hat{\mathbf{P}}_{\mathrm{S},j,k+1}$ and $\mathcal{H}(\hat{x}_{\mathrm{S},j,k+1}, \hat{y}_{\mathrm{S},j,k+1})$
12 **return** *Position vector of sensor node* j*, i.e.,* $\hat{\mathbf{P}}_{S,j,k+1}$

Using recursive mechanism, Q-function is replaced as

$$
\begin{aligned}
& Q(\hat{\mathbf{P}}_{\mathrm{S},j,k}, \mathbf{u}_{\mathrm{S},j,k}; \eta) \\
& = (1-\alpha_k) Q(\hat{\mathbf{P}}_{\mathrm{S},j,k}, \mathbf{u}_{\mathrm{S},j,k}; \eta) \\
& + \alpha_k (r_k + \ell \max Q(\hat{\mathbf{P}}_{\mathrm{S},j,k+1}, \mathbf{u}_{\mathrm{S},j,k+1}; \eta)),
\end{aligned} \tag{7.21}
$$

where α_k indicates learning rate, $\alpha_k \in (0, 1)$.

Remark 7.1 If the number of available actions and positions is small, we can employ RL (Yan et al. 2020b) to obtain the optimal policy $\pi^*(\hat{\mathbf{P}}_{\mathrm{S},j,k})$. Nevertheless, in some specified underwater scenarios, the number of available actions and positions is very large. This promotes us to adopt the DQN to seek the optimal policy, since DQN combines the RL with DNN where Q function can be approximated by network parameter η.

With regard to (7.19) and (7.20), a DQN-based localization estimator is proposed to seek the optimal solution, as presented in Algorithm 4. As a result, its architecture is described in Fig. 7.4, and the main process can be described in detail as follows.

Step 1. Exploring the Environment

Initially, sensor node $j \in \mathcal{I}_{\mathrm{S}}$ selects a random position vector $\hat{\hat{\mathbf{P}}}_{\mathrm{S},j,1}$, through which an unexplored random action $\bar{\mathbf{u}}_{\mathrm{S},j,1}$ is obtained to update (7.12). In the following, by employing (7.13), the reward r_1 is calculated. At iteration step $k^* > 1$, sensor node $j \in \mathcal{I}_{\mathrm{S}}$ chooses another unexplored random action $\bar{\mathbf{u}}_{\mathrm{S},j,k^*}$, and updates its reward r_{k^*}. Repeating the above process until $k^* = N_{\mathrm{e}}$, one can have the following tuples, i.e., $\{\hat{\mathbf{P}}_{\mathrm{S},j,k^*}, \mathbf{u}_{\mathrm{S},j,k^*}, r_{k^*}, \hat{\mathbf{P}}_{\mathrm{S},j,k^*+1}\}_{k^*=1}^{N_{\mathrm{e}}}$, and thereby, these tuples are stored in the memory pool, where $N_{\mathrm{e}} \in [1, \cdots, N_{\mathrm{m}}]$ is the total exploration step and N_{m} represents the capacity of memory pool.

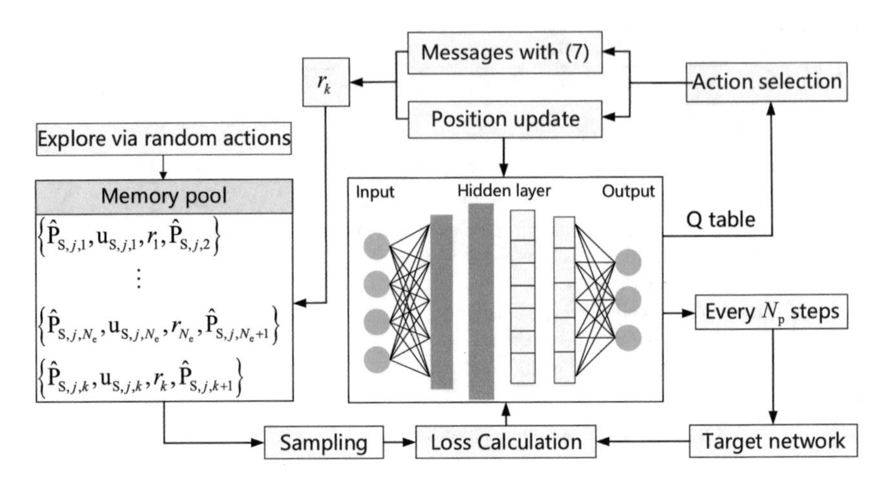

Fig. 7.4 Architecture of the DQN-based localization estimator

Step 2. Training the Entire Network
After the environment is explored, the network parameter η needs to be optimized, such that the mean square error (MSE) between $y_{j,k}$ and $Q(\hat{\mathbf{P}}_{S,j,k}, \mathbf{u}_{S,j,k}; \eta)$ can be minimized. Note that, $y_{j,k}$ is the target value, and $Q(\hat{\mathbf{P}}_{S,j,k}, \mathbf{u}_{S,j,k}; \eta)$ is the Q-value calculated from the previous training. Thus, the loss function at the k-th iteration is given as

$$L(\eta) = \mathbf{E}[(y_{j,k} - Q(\hat{\mathbf{P}}_{S,j,k}, \mathbf{u}_{S,j,k}; \eta))^2], \tag{7.22}$$

where $y_{j,k} = r_k + \ell \max_{\mathbf{u}'_{S,j,k+1}} \hat{Q}(\hat{\mathbf{P}}_{S,j,k+1}, \mathbf{u}'_{S,j,k+1}; \hat{\eta})$. In particular, $\hat{Q}(\hat{\mathbf{P}}_{S,j,k+1}, \mathbf{u}'_{S,j,k+1}; \hat{\eta})$ is the Q-value of target $\hat{Q}$-network. Moreover, $\hat{\eta}$ represents the network parameter of target $\hat{Q}$-network. It is worth mentioning that the target $\hat{Q}$-network is copied from the DQN at every N_p steps.

Consequently, the stochastic gradient descent method is adopted to train the entire network, and thereby, η can be optimized while the loss function in (7.22) can be minimized.

Step 3. Position Estimation
With the optimized η, we take into account $\hat{\mathbf{P}}_{S,j,k}$ the input of DQN, and then update the Q table. Based on the above, the optimal strategy $\pi^*(\hat{\mathbf{P}}_{S,j,k})$ can be acquired. In order to avoid local maxima, the ϵ-greedy strategy is employed to select action, i.e.,

$$\mathbf{u}_{S,j,k} = \begin{cases} \pi^*(\hat{\mathbf{P}}_{S,j,k}), & \text{with probability } 1-\epsilon \\ \text{random}, & \text{otherwise} \end{cases} \tag{7.23}$$

where $\epsilon > 0$ denotes a very small probability.

By employing (7.23), the position vector $\hat{\mathbf{P}}_{S,j,k+1}$ is obtained, and then $\mathcal{H}(\hat{x}_{S,j,k+1}, \hat{y}_{S,j,k+1})$ is calculated. Next, we define $\Delta\mathcal{H}$ as the error tolerance threshold. If $\mathcal{H}(\hat{x}_{S,j,k+1}, \hat{y}_{S,j,k+1})$ is smaller than $\Delta\mathcal{H}$, we regard $\hat{\mathbf{P}}_{S,j,k+1}$ as the final position vector of sensor node $j \in \mathcal{I}_S$. Else, repeat the above procedure until $\mathcal{H}(\hat{x}_{S,j,k+1}, \hat{y}_{S,j,k+1}) < \Delta\mathcal{H}$ or $k+1 > K$. Note that, K is the total number of iteration that is connected with the required precision.

7.4.2 Localization when Labelled Data Occupies the Majority

For Scenario S2, the labelled data occupies the majority, and sensor node $j \in \mathcal{I}_S$ has most of the information on position labels, i.e., $\mathbf{n} \to n$. Such situation can be regarded as supervised scenario. Obviously, the DQN-based estimator is not suitable for the localization system, since the environment information is pre-known and the implementation of environment exploring is not necessary. With regard to the above, the deep neural networks, which is a core technology part of DQN, can be adopted to estimate $x_{S,j}$ and $y_{S,j}$. Specifically, we consider the time difference measurements as the input of DNN, while the estimations $x_{S,j}$ and $y_{S,j}$ are regarded as the output. As shown in Fig. 7.5, we employ the labelled data $\{\mathbf{P}_{P,\mathbf{i}}, \tau_{P,\mathbf{i},i}\}_{\mathbf{i}=1}^{\mathbf{n}}$ to train the DNN, whose aim is to approximate the nonlinear mapping relationship between time difference and position estimation. Hence, the nonlinear mapping relationship between time difference and position estimation is rearranged as

$$\hat{\mathbf{P}}_{S,j}^{\zeta} = \sum_{\mu=1}^{\mu_{\max}} \Gamma_{\mu} g_{\mu,\zeta} + \varphi_{\zeta}, \tag{7.24}$$

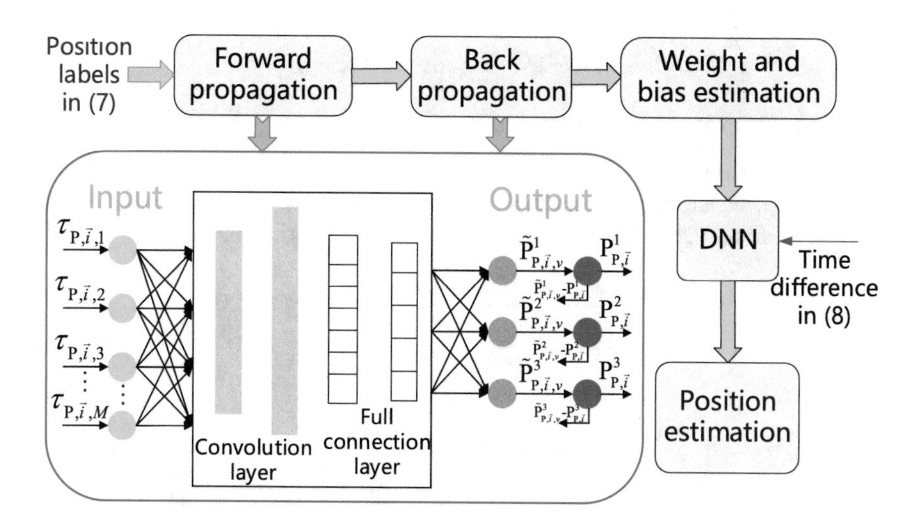

Fig. 7.5 Structure of the DNN-based localization estimator

where $\Gamma_\mu = \frac{1}{1+\exp(-(\sum_{i=1}^{M} \Delta t_{j,i} d_{i,\mu} + \phi_\mu))}$. In special, $\hat{\mathbf{P}}_{\mathrm{S},j}^{\zeta}$ is the ζ-th element of $\hat{\mathbf{P}}_{\mathrm{S},j}$ for $\zeta \in \{1, 2, 3\}$, $\mu_{\max}$ is the number of hidden units, and $g_{\mu,\zeta}$ is the weight from hidden unit Γ_μ to output unit $\hat{\mathbf{P}}_{\mathrm{S},j}^{\zeta}$. Besides, φ_ζ is the trained bias of output unit $\hat{\mathbf{P}}_{\mathrm{S},j}^{\zeta}$, and $d_{i,\mu}$ denotes the weight from the input unit $\Delta t_{j,i}$ to the hidden unit Γ_μ. Besides, ϕ_μ is the trained bias of hidden unit Γ_μ.

From Fig. 7.5, one notices that the above DNN model is divided into input, hidden and output layers. In order to achieve ubiquitous localization, the localization problem of sensor node $j \in \mathcal{I}_{\mathrm{S}}$ can be descend to estimate parameters $d_{i,\mu}$, $g_{\mu,\zeta}$, ϕ_μ and φ_ζ. Therefore, the main process of DNN-based localization estimator is depicted as follows.

Step 1. Forward Propagation
To estimate parameters $d_{i,\mu}$, $g_{\mu,\zeta}$, ϕ_μ and φ_ζ, we employ the labelled data $\{\mathbf{P}_{\mathrm{P},\mathbf{i}}, \tau_{\mathrm{P},\mathbf{i},i}\}_{\mathbf{i}=1}^{\mathbf{n}}$ to train DNN. Therefore, the label $\mathbf{P}_{\mathrm{P},\mathbf{i}}^{\zeta}$ can be estimated as

$$\tilde{\mathbf{P}}_{\mathrm{P},\mathbf{i},v}^{\zeta} = \sum_{\mu=1}^{\mu_{\max}} H_{\mu,v} g_{\mu,\zeta,v} + \varphi_{\zeta,v}, \tag{7.25}$$

where $H_{\mu,v} = \frac{1}{1+\exp(-(\sum_{i=1}^{M} \tau_{\mathrm{P},\mathbf{i},i} d_{i,\mu,v} + \phi_{\mu,v}))}$. In particular, $d_{i,\mu,v}$, $g_{\mu,\zeta,v}$, $\phi_{\mu,v}$ and $\varphi_{\zeta,v}$ are the estimations of $d_{i,\mu}$, $g_{\mu,\zeta}$, ϕ_μ and φ_ζ at training step v.

From (7.25), the estimation error can be defined as

$$P_{\mathrm{err},v} = \frac{1}{2} \sum_{\zeta=1}^{3} (\mathbf{P}_{\mathrm{P},\mathbf{i}}^{\zeta} - \tilde{\mathbf{P}}_{\mathrm{P},\mathbf{i},v}^{\zeta})^2, \tag{7.26}$$

and hence, the parameters $d_{i,\mu}$, $g_{\mu,\zeta}$, ϕ_μ and φ_ζ can be obtained by minimizing the following problem

$$[d_{i,\mu}, g_{\mu,\zeta}, \phi_\mu, \varphi_\zeta] = \operatorname{argmin}_{[d_{i,\mu,v}, g_{\mu,\zeta,v}, \phi_{\mu,v}, \varphi_{\zeta,v}]} P_{\mathrm{err},v}, \tag{7.27}$$

where $\mathbf{P}_{\mathrm{P},\mathbf{i}}^{\zeta}$ denotes the ζ-th element of $\mathbf{P}_{\mathrm{P},\mathbf{i}}$.

Step 2. Back Propagation
From (7.27), we adopt the stochastic gradient descent method to update the weights and biases. In particular, the derivation process of weights can be given as

$$\begin{aligned} \frac{\partial P_{\mathrm{err},v}}{\partial d_{i,\mu,v}} &= \frac{\partial P_{\mathrm{err},v}}{\partial H_{\mu,v}} \frac{\partial H_{\mu,v}}{\partial d_{i,\mu,v}}, \\ \frac{\partial P_{\mathrm{err},v}}{\partial g_{\mu,\zeta,v}} &= (\mathbf{P}_{\mathrm{P},\mathbf{i}}^{\zeta} - \tilde{\mathbf{P}}_{\mathrm{P},\mathbf{i},v}^{\zeta})(-H_{\mu,v}), \end{aligned} \tag{7.28}$$

where $\frac{\partial P_{\mathrm{err},v}}{\partial H_{\mu,v}} = -\sum_{\zeta=1}^{3} (\mathbf{P}_{\mathrm{P},\mathbf{i}}^{\zeta} - \tilde{\mathbf{P}}_{\mathrm{P},\mathbf{i},v}^{\zeta}) g_{\mu,\zeta,v}$ and $\frac{\partial H_{\mu,v}}{\partial d_{i,\mu,v}} = H_{\mu,v}(1 - H_{\mu,v})\tau_{\mathrm{P},\mathbf{i},i}$. Similarly, the derivation process of biases are given as $\frac{\partial P_{\mathrm{err},v}}{\partial \phi_{\mu,v}} = \frac{\partial P_{\mathrm{err},v}}{\partial H_{\mu,v}} \frac{\partial H_{\mu,v}}{\partial \phi_{\mu,v}}$ and $\frac{\partial P_{\mathrm{err},v}}{\partial \varphi_{\zeta,v}} = -(\mathbf{P}_{\mathrm{P},\mathbf{i}}^{\zeta} - \tilde{\mathbf{P}}_{\mathrm{P},\mathbf{i},v}^{\zeta})$, where $\frac{\partial H_{\mu,v}}{\partial \phi_{\mu,v}} = H_{\mu,v}(1 - H_{\mu,v})$.

Consequently, the weights and biases are updated as

$$\begin{aligned} d_{i,\mu,v+1} &= d_{i,\mu,v} + \rho H_{\mu,v}(1 - H_{\mu,v})\tau_{\mathrm{P},\mathbf{i},i} \\ &\times \sum_{\zeta=1}^{3}(\mathbf{P}_{\mathrm{P},\mathbf{i}}^{\zeta} - \tilde{\mathbf{P}}_{\mathrm{P},\mathbf{i},v}^{\zeta})g_{\mu,\zeta,v}, \end{aligned} \tag{7.29}$$

$$g_{\mu,\zeta,v+1} = g_{\mu,\zeta,v} + \rho(\mathbf{P}_{\mathrm{P},\mathbf{i}}^{\zeta} - \tilde{\mathbf{P}}_{\mathrm{P},\mathbf{i},v}^{\zeta})H_{\mu,v}, \tag{7.30}$$

$$\begin{aligned} \phi_{\mu,v+1} &= \phi_{\mu,v} + \rho H_{\mu,v}(1 - H_{\mu,v}) \sum_{\zeta=1}^{3}(\mathbf{P}_{\mathrm{P},\mathbf{i}}^{\zeta} \\ &- \tilde{\mathbf{P}}_{\mathrm{P},\mathbf{i},v}^{\zeta})g_{\mu,\zeta,v}, \end{aligned} \tag{7.31}$$

$$\varphi_{\zeta,v+1} = \varphi_{\zeta,v} + \rho(\mathbf{P}_{\mathrm{P},\mathbf{i}}^{\zeta} - \tilde{\mathbf{P}}_{\mathrm{P},\mathbf{i},v}^{\zeta}), \tag{7.32}$$

where ρ indicates learning rate.

If $\left\| P_{\mathrm{err},v+1} - P_{\mathrm{err},v} \right\| < \epsilon$ or $v > v_{\max}$ is satisfied, the training process is ended, where ϵ represents the error threshold and $v_{\max}$ denotes the upper limit of training steps.

Step 3. Position Estimation
Based on the acquired parameters $d_{i,\mu}$, $g_{\mu,\zeta}$, ϕ_{μ} and φ_{ζ}, we substitute time difference $\Delta t_{j,i}$ into (7.24), through which the estimations of $x_{\mathrm{S},j}$ and $y_{\mathrm{S},j}$ can be acquired.

7.4.3 Localization when Unlabelled Data Occupies the Majority

A large number of labelled data is not always feasible, and it is essential to combine the unlabelled data with labelled data. Inspired by this, we adopt semisupervised DRL to estimate the position information of sensor node $j \in \mathcal{I}_{\mathrm{S}}$, i.e., $x_{\mathrm{S},j}$ and $y_{\mathrm{S},j}$. In particular, the semisupervised DRL employs a small set of labelled data along with a large number amount of unlabelled data to train the network (Zhang et al. 2014). Based on the above, we divide the data into two parts, i.e., labelled data and unlabelled data. The labelled data is represented by $\{\mathbf{P}_{\mathrm{P},\mathbf{i}}, \tau_{\mathrm{P},\mathbf{i},i}\}_{\mathbf{i}=1}^{\mathbf{n}}$, which is pre-known to sensor node $j \in \mathcal{I}_{\mathrm{S}}$. The unlabelled data is denoted by $\{\mathbf{P}_{\mathrm{P},\bar{\imath}}, \tau_{\mathrm{P},\bar{\imath},i}\}_{\bar{\imath}=\mathbf{n}+1}^{n}$, which is unknown and requires to be explored by sensor node $j \in \mathcal{I}_{\mathrm{S}}$. With regard to this, a semisupervised DRL localization estimator is developed to seek the optimal solution, as presented in Algorithm 5. Accordingly, its architecture is shown in Fig. 7.6, and the main procedure can be described as follows.

Step 1: Initial Position Estimation
In Sect. 7.4.1, the initial position vector $\hat{\mathbf{P}}_{\mathrm{S},j,1}$ is stochastically selected, because sensor node $j \in \mathcal{I}_{\mathrm{S}}$ does not have any information on position labels. When the initial position is far away from the actual position, the convergence speed of

Algorithm 5: Semisupervised DRL localization estimator

Input: The messages in (7.7), the capacity of memory pool N_{m}, total exploration step N_{e} and probability ϵ

Output: Estimated position of sensor node j

1 Estimate parameters $d^{\#}_{i,\bar{\mu}}$, $g^{\#}_{\bar{\mu},\zeta}$, $\phi^{\#}_{\bar{\mu}}$ and $\varphi^{\#}_{\zeta}$ with labelled data, and then get the initial position vector $\hat{\mathbf{P}}_{\mathrm{S},j,1}$
2 **for** *episode*=$\mathbf{n}+1{:}n$ **do**
3 Sensor node j randomly selects an action execution based on the current position
4 Store tuples $\{\hat{\mathbf{P}}_{\mathrm{S},j,\bar{\iota}}, \mathbf{u}_{\mathrm{S},j,\bar{\iota}}, r_{\bar{\iota}}, \hat{\mathbf{P}}_{\mathrm{S},j,\bar{\iota}+1}\}^{n}_{\bar{\iota}=\mathbf{n}+1}$ in memory pool
5 Train the DQN by combining random samples into memory pool and target $\hat{Q}$-network
6 **while** $\mathcal{H}(\hat{x}_{\mathrm{S},j,k+1}, \hat{y}_{\mathrm{S},j,k+1}) \geq \Delta\mathcal{H}$ *or* $k+1 > K$ **do**
7 **if** *explore* N_p *steps* **then**
8 Target $\hat{Q}$-network replicates the DQN, and updates the parameters of DQN
9 **if** *within probability* $1-\epsilon$ **then**
10 Select the optimal policy $\pi^{*}(\hat{\mathbf{P}}_{\mathrm{S},j,k})$
11 **else**
12 Randomly select action in action space
13 Update $\hat{\mathbf{P}}_{\mathrm{S},j,k+1}$ and $\mathcal{H}(\hat{x}_{\mathrm{S},j,k+1}, \hat{y}_{\mathrm{S},j,k+1})$
14 **return** *Estimated position of sensor node* j, *i.e.*, $\hat{\mathbf{P}}_{S,j,k+1}$

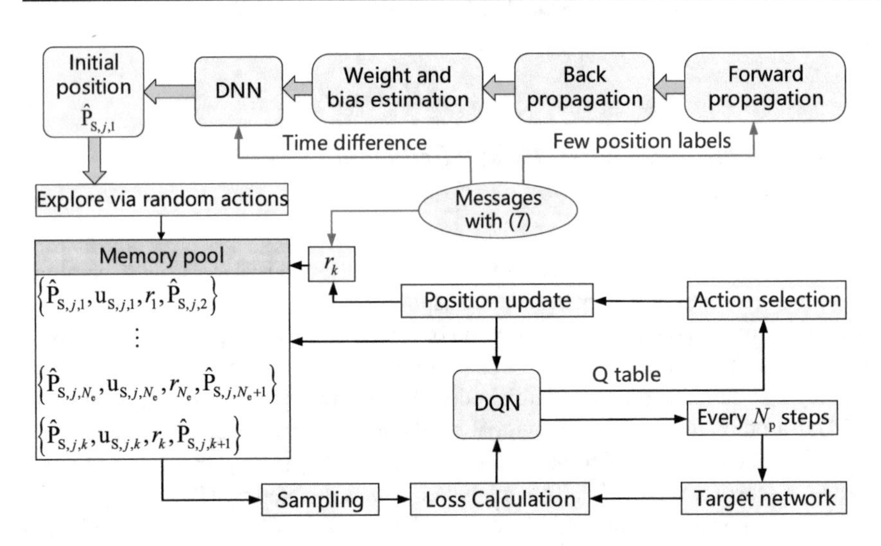

Fig. 7.6 Description of semisupervised DRL localization estimator

localization estimator may be very slow. To solve the above deficiency, we adopt the labelled position data to estimate $\hat{\mathbf{P}}_{\mathrm{S},j,1}$, and thereby, we can obtain

$$\hat{\mathbf{P}}^{\zeta}_{\mathrm{S},j,1} = \sum_{\bar{\mu}=1}^{\bar{\mu}_{\max}} \Gamma^{\#}_{\bar{\mu}} g^{\#}_{\bar{\mu},\zeta} + \varphi^{\#}_{\zeta}, \tag{7.33}$$

where $\Gamma_{\bar{\mu}}^{\#} = \frac{1}{1+\exp(-(\sum_{i=1}^{M} \Delta t_{j,i} d_{i,\bar{\mu}}^{\#} + \phi_{\bar{\mu}}^{\#}))}$. specially, $\hat{\mathbf{P}}_{\mathrm{S},j,1}^{\zeta}$ is the ζ-th element of $\hat{\mathbf{P}}_{\mathrm{S},j,1}$ for $\zeta \in \{1, 2, 3\}$, $\bar{\mu}_{\max}$ indicates the number of hidden units, and $g_{\bar{\mu},\zeta}^{\#}$ denotes the weight from hidden unit $\Gamma_{\bar{\mu}}^{\#}$ to output unit $\hat{\mathbf{P}}_{\mathrm{S},j,1}^{\zeta}$. Besides, $\varphi_{\zeta}^{\#}$ is the trained bias of output unit $\hat{\mathbf{P}}_{\mathrm{S},j,1}^{\zeta}$, and $d_{i,\bar{\mu}}^{\#}$ is the weight from the input unit $\Delta t_{j,i}$ to the hidden unit $\Gamma_{\bar{\mu}}^{\#}$. In addition, $\phi_{\bar{\mu}}^{\#}$ stands for the trained bias of hidden unit $\Gamma_{\bar{\mu}}^{\#}$.

Particularly, the weights and biases can be updated as

$$\begin{aligned} d_{i,\bar{\mu},v+1}^{\#} = d_{i,\bar{\mu},v}^{\#} + \rho H_{\bar{\mu},v}^{\#}(1 - H_{\bar{\mu},v}^{\#})\tau_{\mathrm{P},\mathbf{i},i} \\ \times \textstyle\sum_{\zeta=1}^{3}(\mathbf{P}_{\mathrm{P},\mathbf{i}}^{\zeta} - \tilde{\mathbf{P}}_{\mathrm{P},\mathbf{i},v}^{\zeta\#})g_{\bar{\mu},\zeta,v}^{\#}, \end{aligned} \tag{7.34}$$

$$g_{\bar{\mu},\zeta,v+1}^{\#} = g_{\bar{\mu},\zeta,v}^{\#} + \rho(\mathbf{P}_{\mathrm{P},\mathbf{i}}^{\zeta} - \tilde{\mathbf{P}}_{\mathrm{P},\mathbf{i},v}^{\zeta\#})H_{\bar{\mu},v}^{\#}, \tag{7.35}$$

$$\begin{aligned} \phi_{\bar{\mu},v+1}^{\#} = \phi_{\bar{\mu},v}^{\#} + \rho H_{\bar{\mu},v}^{\#}(1 - H_{\bar{\mu},v}^{\#}) \textstyle\sum_{\zeta=1}^{3}(\mathbf{P}_{\mathrm{P},\mathbf{i}}^{\zeta} \\ - \tilde{\mathbf{P}}_{\mathrm{P},\mathbf{i},v}^{\zeta\#})g_{\bar{\mu},\zeta,v}^{\#}, \end{aligned} \tag{7.36}$$

$$\varphi_{\zeta,v+1}^{\#} = \varphi_{\zeta,v}^{\#} + \rho(\mathbf{P}_{\mathrm{P},\mathbf{i}}^{\zeta} - \tilde{\mathbf{P}}_{\mathrm{P},\mathbf{i},v}^{\zeta\#}), \tag{7.37}$$

where $\tilde{\mathbf{P}}_{\mathrm{P},\mathbf{i},v}^{\zeta\#} = \sum_{\bar{\mu}=1}^{\bar{\mu}_{\max}} H_{\bar{\mu},v}^{\#} g_{\bar{\mu},\zeta,v}^{\#} + \varphi_{\zeta,v}^{\#}$ and $H_{\bar{\mu},v}^{\#} = \frac{1}{1+\exp(-(\sum_{i=1}^{M} \tau_{\mathrm{P},\mathbf{i},i} d_{i,\bar{\mu},v}^{\#} + \phi_{\bar{\mu},v}^{\#}))}$. Note that, $d_{i,\bar{\mu},v}^{\#}$, $g_{\bar{\mu},\zeta,v}^{\#}$, $\phi_{\bar{\mu},v}^{\#}$ and $\varphi_{\zeta,v}^{\#}$ denote the estimations of $d_{i,\bar{\mu}}^{\#}$, $g_{\bar{\mu},\zeta}^{\#}$, $\phi_{\bar{\mu}}^{\#}$ and $\varphi_{\zeta}^{\#}$ at training step v, respectively.

Step 2: Exploring the Environment
It is noted that $\hat{\mathbf{P}}_{\mathrm{S},j,1}$ is a coarse estimation value of $\mathbf{P}_{\mathrm{S},j}$, since the number of labelled data in this scenario is few. For such case, it is necessary to further explore the environment, through which the unlabelled data $\{\mathbf{P}_{\mathrm{P},\bar{i}},\ \tau_{\mathrm{P},\bar{i},i}\}_{\bar{i}=\mathbf{n}+1}^{n}$ can be explored and stacked into the memory pool. Furthermore, sensor node j selects an unexplored random action $\mathbf{u}_{\mathrm{S},j,1}$ based on position vector $\hat{\mathbf{P}}_{\mathrm{S},j,1}$, and then the reward r_1 is calculated. With the updated position, one explores the environment via random selection, whose detailed procedure is similar to Step 1 in Sect. 7.4.1. Repeating this procedure, the unlabelled data can be transformed into labelled data.

Step 3. Training the Entire Network
Next, we employ the explored environment information to train the network. Specificial, the difference between target value $y_{j,k}^{\#}$ and the Q-value $Q(\hat{\mathbf{P}}_{\mathrm{S},j,k}, \mathbf{u}_{\mathrm{S},j,k}; \eta^{\#})$ is used to optimize network parameter $\eta^{\#}$. Consequently, the loss function at the k-th iteration is given as

$$L(\eta^{\#}) = E[(y_{j,k}^{\#} - Q(\hat{\mathbf{P}}_{\mathrm{S},j,k}, \mathbf{u}_{\mathrm{S},j,k}; \eta^{\#}))^2], \tag{7.38}$$

where $y^{\#}_{j,k} = r_k + \ell \max_{\mathbf{u}'_{\mathrm{S},j,k+1}} \hat{Q}(\hat{\mathbf{P}}_{\mathrm{S},j,k+1}, \mathbf{u}'_{\mathrm{S},j,k+1}; \hat{\eta}^{\#})$. In addition, $\hat{Q}(\hat{\mathbf{P}}_{\mathrm{S},j,k+1}, \mathbf{u}'_{\mathrm{S},j,k+1}; \hat{\eta}^{\#})$ and $\hat{\eta}^{\#}$ are the Q-value and network parameter for target $\hat{Q}$-network, respectively.

Therefore, the stochastic gradient descent method is adopted to train the entire network, and thereby, the gradient of loss function is calculated by the first derivative, i.e.,

$$\begin{aligned}\nabla_{\eta^{\#}} L(\eta^{\#}) = E\Big\{&2(y^{\#}_{j,k} - Q(\hat{\mathbf{P}}_{\mathrm{S},j,k}, \mathbf{u}_{\mathrm{S},j,k}; \eta^{\#})) \\ &\times \nabla_{\eta^{\#}} Q(\hat{\mathbf{P}}_{\mathrm{S},j,k}, \mathbf{u}_{\mathrm{S},j,k})\Big\}.\end{aligned} \tag{7.39}$$

Step 4. Position Estimation
With the optimized $\eta^{\#}$, we update the Q table by considering $\hat{\mathbf{P}}_{\mathrm{S},j,k}$ as the input of DQN. Therefore, the optimal strategy $\pi^*(\hat{\mathbf{P}}_{\mathrm{S},j,k})$ can be acquired, where the ϵ-greedy strategy is employed to select action. When position vector $\hat{\mathbf{P}}_{\mathrm{S},j,k+1}$ is obtained, $\mathcal{H}(\hat{x}_{\mathrm{S},j,k+1}, \hat{y}_{\mathrm{S},j,k+1})$ can be calculated. If $\mathcal{H}(\hat{x}_{\mathrm{S},j,k+1}, \hat{y}_{\mathrm{S},j,k+1})$ is smaller than $\Delta\mathcal{H}$, the $\hat{\mathbf{P}}_{\mathrm{S},j,k+1}$ can be regarded as the final position vector of sensor node $j \in \mathcal{I}_{\mathrm{S}}$. Else, repeat the above procedure until $\mathcal{H}(\hat{x}_{\mathrm{S},j,k+1}, \hat{y}_{\mathrm{S},j,k+1}) < \Delta\mathcal{H}$ or $k+1 > K$.

7.4.4 Performance Analysis

(1) Equivalence of the Localization Protocols
In (7.11), the localization optimization problem for sensor node $j \in \mathcal{I}_{\mathrm{S}}$ is constructed. In order to solve this problem, one should know $\Delta t_{j,i}$ and $\tau_{j,i}$. Obviously, $\Delta t_{j,i}$ cannot expose the position privacy, since $\Delta t_{j,i}$ is a measurement value and it does not rely on the position information. In contrast, $\tau_{j,i}$ can expose the position privacy, since it relies on the position information of anchor and sensor nodes, as defined in (7.3). Thereby, it is necessary to hide the position information during the construction of $\tau_{j,i}$. Based on the above, we analyze the equivalence and privacy of travel time $\tau_{j,i}$ for $i \in \mathcal{I}_{\mathrm{A}}$ and $j \in \mathcal{I}_{\mathrm{S}}$. Before processing, we consider a traditional scenario, i.e., the privacy preservation of position information is ignored. Hence, by referring to Ramezani et al. (2013), Mortazavi et al. (2017), and Yan et al. (2020c), the travel time $\breve{\tau}_{j,i}$ at iteration step k is given as

$$\breve{\tau}_{j,i,k} = -\frac{1}{a}\left(\ln\frac{1+\sin\breve{\theta}_{j,k}}{\cos\breve{\theta}_{j,k}} - \ln\frac{1+\sin\breve{\theta}_i}{\cos\breve{\theta}_i}\right), \tag{7.40}$$

with

$$\begin{aligned}\breve{\theta}_{j,k} &= \arctan \frac{z_{S,j} - z_{A,i}}{\sqrt{(\hat{x}_{S,j,k} - x_{A,i})^2 + (\hat{y}_{S,j,k} - y_{A,i})^2}} \\ &- \arctan \frac{a\sqrt{(\hat{x}_{S,j,k} - x_{A,i})^2 + (\hat{y}_{S,j,k} - y_{A,i})^2}}{2b + a(z_{S,j} + z_{A,i})}, \\ \breve{\theta}_i &= \arctan \frac{z_{S,j} - z_{A,i}}{\sqrt{(\hat{x}_{S,j,k} - x_{A,i})^2 + (\hat{y}_{S,j,k} - y_{A,i})^2}} \\ &+ \arctan \frac{a\sqrt{(\hat{x}_{S,j,k} - x_{A,i})^2 + (\hat{y}_{S,j,k} - y_{A,i})^2}}{2b + a(z_{S,j} + z_{A,i})}.\end{aligned} \tag{7.41}$$

By adopting the proposed localization protocol in Sect. 7.3, the travel time $\tau_{j,i}$ at iteration step k is calculated as

$$\tau_{j,i,k} = -\frac{1}{a}\left(\ln \frac{1 + \sin\theta_{j,k}}{\cos\theta_{j,k}} - \ln \frac{1 + \sin\theta_i}{\cos\theta_i}\right), \tag{7.42}$$

where $\theta_{j,k}$ represents the acoustic ray angle at sensor node j and θ_i is the acoustic ray angle at anchor node i. It is noticed that $\theta_{j,k}$ and θ_i can be constructed by the localization protocol in Sect. 7.3.

Theorem 7.1 *By adopting the localization protocol in Sect. 7.3, the travel time $\tau_{j,i,k}$ in (7.42) is equivalent to $\breve{\tau}_{j,i,k}$ in (7.40). Furthermore, the private position information of anchor and sensor nodes in (7.42) is not exposed to the network.*

Proof From (7.14), (7.15), (7.16) and (7.17), $\theta_{j,k}$ and θ_i can be proposed as the following forms, i.e.,

$$\begin{aligned}\theta_{j,k} &= \arctan \frac{z_{S,j} - z_{A,i}}{\sqrt{(\hat{x}_{S,j,k} - x_{A,i})^2 + (\hat{y}_{S,j,k} - y_{A,i})^2}} \\ &- \arctan \frac{a\sqrt{(\hat{x}_{S,j,k} - x_{A,i})^2 + (\hat{y}_{S,j,k} - y_{A,i})^2}}{2b + a(z_{S,j} + z_{A,i})},\end{aligned} \tag{7.43}$$

$$\begin{aligned}\theta_i &= \arctan \frac{z_{S,j} - z_{A,i}}{\sqrt{(\hat{x}_{S,j,k} - x_{A,i})^2 + (\hat{y}_{S,j,k} - y_{A,i})^2}} \\ &+ \arctan \frac{a\sqrt{(\hat{x}_{S,j,k} - x_{A,i})^2 + (\hat{y}_{S,j,k} - y_{A,i})^2}}{2b + a(z_{S,j} + z_{A,i})}.\end{aligned} \tag{7.44}$$

With regard to Case 1 (see Sect. 7.4.1), one has

$$\begin{aligned}
&(\hat{x}_{\mathrm{S},j,k} - x_{\mathrm{A},i})^2 + (\hat{y}_{\mathrm{S},j,k} - y_{\mathrm{A},i})^2 \\
&= (\hat{x}_{\mathrm{S},j,k-1} - x_{\mathrm{A},i})^2 + (\hat{y}_{\mathrm{S},j,k-1} - y_{\mathrm{A},i})^2 \\
&\quad + 2s(\hat{x}_{\mathrm{S},j,1} + k_1 s - x_{\mathrm{A},i}) + s^2, \\
&= (\hat{x}_{\mathrm{S},j,k-1} - x_{\mathrm{A},i})^2 + (\hat{y}_{\mathrm{S},j,k-1} - y_{\mathrm{A},i})^2 \\
&\quad + 2s(\hat{x}_{\mathrm{S},j,1} - x_{\mathrm{A},i}) + s^2 - (\varepsilon_{i,\mathrm{x}} + \varepsilon_{i,\mathrm{y}}) + 2k_1 s^2 \\
&\quad + (\hat{x}_{\mathrm{S},j,1} - x_{\mathrm{A},i})^2 + \varepsilon_{i,\mathrm{x}} + (\hat{y}_{\mathrm{S},j,1} - y_{\mathrm{A},i})^2 + \varepsilon_{i,\mathrm{y}} \\
&\quad - \left((\hat{x}_{\mathrm{S},j,1} - x_{\mathrm{A},i})^2 + (\hat{y}_{\mathrm{S},j,1} - y_{\mathrm{A},i})^2\right), \\
&= \Delta R^2_{j,k-1,i} + \Delta x'_{j,1,i} + 2k_1 s^2 \\
&\quad + \Delta x_{j,1,i} + \Delta y_{j,1,i} - \Delta R^2_{j,1,i}.
\end{aligned} \tag{7.45}$$

Define $F^{\mathrm{C1}}_{j,k,i} = \Delta R^2_{j,k-1,i} + \Delta x'_{j,1,i} + 2k_1 s^2 + \Delta x_{j,1,i} + \Delta y_{j,1,i} - \Delta R^2_{j,1,i}$, and then (7.43) and (7.44) are rearranged as

$$\theta_{j,k} = \arctan \frac{z_{\mathrm{S},j} - z_{\mathrm{A},i}}{\sqrt{F^{\mathrm{C1}}_{j,k,i}}} - \arctan \frac{a\sqrt{F^{\mathrm{C1}}_{j,k,i}}}{2b + a(z_{\mathrm{S},j} + z_{\mathrm{A},i})}, \tag{7.46}$$

$$\theta_i = \arctan \frac{z_{\mathrm{S},j} - z_{\mathrm{A},i}}{\sqrt{F^{\mathrm{C1}}_{j,k,i}}} + \arctan \frac{a\sqrt{F^{\mathrm{C1}}_{j,k,i}}}{2b + a(z_{\mathrm{S},j} + z_{\mathrm{A},i})}. \tag{7.47}$$

By emplying (7.45), the following two conclusions are acquired: 1) $\theta_{j,k} = \breve{\theta}_{j,k}$ and $\theta_i = \breve{\theta}_i$ since $F^{\mathrm{C1}}_{j,k,i} = (\hat{x}_{\mathrm{S},j,k} - x_{\mathrm{A},i})^2 + (\hat{y}_{\mathrm{S},j,k} - y_{\mathrm{A},i})^2$; 2) private positions of anchor and sensor nodes, i.e., $x_{\mathrm{A},i}$, $y_{\mathrm{A},i}$, $x_{\mathrm{S},j}$ and $y_{\mathrm{S},j}$, are not exposed to the network, since $F^{\mathrm{C1}}_{j,k,i}$ is calculated by the elements in (7.4) where the private positions are not released to the network.

Similar to Case 1, one can get

$$\begin{aligned}
&(\hat{x}_{\mathrm{S},j,k} - x_{\mathrm{A},i})^2 + (\hat{y}_{\mathrm{S},j,k} - y_{\mathrm{A},i})^2 \\
&= (\hat{x}_{\mathrm{S},j,k-1} - x_{\mathrm{A},i})^2 + (\hat{y}_{\mathrm{S},j,k-1} - y_{\mathrm{A},i})^2 \\
&\quad - 2s(\hat{x}_{\mathrm{S},j,1} + k_1 s - x_{\mathrm{A},i}) + s^2, \\
&= \Delta R^2_{j,k-1,i} - \Delta x'_{j,1,i} + 2(-k_1 + 1)s^2 \\
&\quad - \Delta x_{j,1,i} - \Delta y_{j,1,i} + \Delta R^2_{j,1,i}.
\end{aligned} \tag{7.48}$$

For clarity, we define $F^{\mathrm{C2}}_{j,k,i} = \Delta R^2_{j,k-1,i} - \Delta x'_{j,1,i} + 2(-k_1+1)s^2 - \Delta x_{j,1,i} - \Delta y_{j,1,i} + \Delta R^2_{j,1,i}$. Hence, the $\theta_{j,k}$ and θ_i in Case 2 can be denoted as

$$\theta_{j,k} = \arctan \frac{z_{\mathrm{S},j} - z_{\mathrm{A},i}}{\sqrt{F^{\mathrm{C2}}_{j,k,i}}} - \arctan \frac{a\sqrt{F^{\mathrm{C2}}_{j,k,i}}}{2b + a(z_{\mathrm{S},j} + z_{\mathrm{A},i})}, \tag{7.49}$$

$$\theta_i = \arctan \frac{z_{\mathrm{S},j} - z_{\mathrm{A},i}}{\sqrt{F^{\mathrm{C2}}_{j,k,i}}} + \arctan \frac{a\sqrt{F^{\mathrm{C2}}_{j,k,i}}}{2b + a(z_{\mathrm{S},j} + z_{\mathrm{A},i})}. \tag{7.50}$$

Clearly, we can conclude that $\theta_{j,k} = \breve{\theta}_{j,k}$ and $\theta_i = \breve{\theta}_i$. In addition, private positions of anchor and sensor nodes are not released to the network, because the private positions in $F^{\mathrm{C2}}_{j,k,i}$, i.e., $x_{\mathrm{A},i}$, $y_{\mathrm{A},i}$, $x_{\mathrm{S},j}$ and $y_{\mathrm{S},j}$, are not exposed to the network.

For Case 3 and Case 4, one can acquire

$$\begin{aligned}&(\hat{x}_{\mathrm{S},j,k} - x_{\mathrm{A},i})^2 + (\hat{y}_{\mathrm{S},j,k} - y_{\mathrm{A},i})^2\\ &= \Delta R^2_{j,k-1,i} + \Delta y'_{j,1,i} + 2k_2 w^2\\ &\quad + \Delta x_{j,1,i} + \Delta y_{j,1,i} - \Delta R^2_{j,1,i},\end{aligned} \tag{7.51}$$

$$\begin{aligned}&(\hat{x}_{\mathrm{S},j,k} - x_{\mathrm{A},i})^2 + (\hat{y}_{\mathrm{S},j,k} - y_{\mathrm{A},i})^2\\ &= \Delta R^2_{j,k-1,i} - \Delta y'_{j,1,i} + 2(-k_2+1) w^2\\ &\quad - \Delta x_{j,1,i} - \Delta y_{j,1,i} + \Delta R^2_{j,1,i},\end{aligned} \tag{7.52}$$

and thereby, we can also conclude that $\theta_{j,k} = \breve{\theta}_{j,k}$ and $\theta_i = \breve{\theta}_i$.

Consequently, we can acquire that $\breve{\tau}_{j,i,k} = \tau_{j,i,k}$, and more importantly, the private positions of anchors and sensors in our localization protocol are not exposed to the network. □

(2) CRLB

We calculate the CRLB for the proposed localization problem. In particular, CRLB is a lower bound to the error variance of parameter estimator, which is regarded as an important tool in the performance evaluation of parameter estimation. For sensor node j, the log-likelihood function is regarded as $\ln \Phi(\hat{\bar{\mathbf{P}}}_{\mathrm{S},j,k})$, where $\bar{\mathbf{P}}_{\mathrm{S},j}$ is the ground truth. Accordingly, the FIM can be expressed as

$$\begin{aligned}\Upsilon(\hat{\bar{\mathbf{P}}}_{\mathrm{S},j,k}) &= E\{[\nabla_{\hat{\bar{\mathbf{P}}}_{\mathrm{S},j,k}} \ln \Phi(\hat{\bar{\mathbf{P}}}_{\mathrm{S},j,k})]\\ &\quad \times [\nabla_{\hat{\bar{\mathbf{P}}}_{\mathrm{S},j,k}} \ln \Phi(\hat{\bar{\mathbf{P}}}_{\mathrm{S},j,k})]^{\mathrm{T}}\}|_{\hat{\bar{\mathbf{P}}}_{\mathrm{S},j,k} = \bar{\mathbf{P}}_{\mathrm{S},j}}\\ &= -E\{\nabla_{\hat{\bar{\mathbf{P}}}_{\mathrm{S},j,k}} \nabla^{\mathrm{T}}_{\hat{\bar{\mathbf{P}}}_{\mathrm{S},j,k}} \ln \Phi(\hat{\bar{\mathbf{P}}}_{\mathrm{S},j,k})\}|_{\hat{\bar{\mathbf{P}}}_{\mathrm{S},j,k} = \bar{\mathbf{P}}_{\mathrm{S},j}}.\end{aligned} \tag{7.53}$$

Then the log-likelihood function is defined as

$$\ln \Phi(\hat{\bar{\mathbf{P}}}_{\mathrm{S},j,k}) = \frac{1}{\sigma_{\mathrm{mea}}^2} \sum_{i=1}^{M} [\Delta t_{j,i} - \tau_{j,i,k}]^2. \quad (7.54)$$

By directly calculating the first derivative of position, we obtain the following elements from (7.54), i.e.,

$$\begin{aligned} \Upsilon_{1,1} &= \frac{2}{\sigma_{\mathrm{mea}}^2} \sum_{i=1}^{M} [-\frac{\partial \tau_{j,i,k}}{\partial \hat{x}_{\mathrm{S},j,k}}]^2, \\ \Upsilon_{1,2} &= \Upsilon_{2,1} \\ &= \frac{2}{\sigma_{\mathrm{mea}}^2} \sum_{i=1}^{M} [\frac{\partial \tau_{j,i,k}}{\partial \hat{x}_{\mathrm{S},j,k}} \frac{\partial \tau_{j,i,k}}{\partial \hat{y}_{\mathrm{S},j,k}}], \\ \Upsilon_{2,2} &= \frac{2}{\sigma_{\mathrm{mea}}^2} \sum_{i=1}^{M} [-\frac{\partial \tau_{j,i,k}}{\partial \hat{y}_{\mathrm{S},j,k}}]^2, \end{aligned} \quad (7.55)$$

where $\Upsilon_{p,q}$ denotes the (p, q)-th element of Υ and $p, q \in \{1, 2\}$.

Soon afterwards, the CRLB is given as

$$\begin{aligned} \mathrm{CRLB} &= [\Upsilon^{-1}(\hat{\bar{\mathbf{P}}}_{\mathrm{S},j,k})]_{1,1} + [\Upsilon^{-1}(\hat{\bar{\mathbf{P}}}_{\mathrm{S},j,k})]_{2,2} \\ &= \mathrm{tr}\{\Upsilon^{-1}(\hat{\bar{\mathbf{P}}}_{\mathrm{S},j,k})\}|_{\hat{\bar{\mathbf{P}}}_{\mathrm{S},j,k}=\bar{\mathbf{P}}_{\mathrm{S},j}}. \end{aligned} \quad (7.56)$$

From (7.56), the localization error satisfies the relationship of

$$\mathbf{E}\{\|\hat{\bar{\mathbf{P}}}_{\mathrm{S},j,k} - \bar{\mathbf{P}}_{\mathrm{S},j}\|^2\} \geq \mathrm{tr}\{\Upsilon^{-1}(\hat{\bar{\mathbf{P}}}_{\mathrm{S},j,k})\}|_{\hat{\bar{\mathbf{P}}}_{\mathrm{S},j,k}=\bar{\mathbf{P}}_{\mathrm{S},j}}. \quad (7.57)$$

(3) Privacy Preservation Analysis

We now analyze the privacy of our localization solution, and therefore, the following theorems are developed.

Theorem 7.2 *When the nodes in USNs do not collude, sensor node $j \in \mathcal{I}_{\mathrm{S}}$ and anchor nodes can achieve Level-1 Privacy.*

Proof The proof is to prove that: (1) sensor node $j \in \mathcal{I}_{\mathrm{S}}$ can estimate its position $\mathbf{P}_{\mathrm{S},j}$; (2) sensor node $j \in \mathcal{I}_{\mathrm{S}}$ cannot estimate the position of anchor node $i \in \mathcal{I}_{\mathrm{A}}$, i.e., $\mathbf{P}_{\mathrm{A},i}$; 3) anchor node i cannot estimate the positions of sensor node $j \in \mathcal{I}_{\mathrm{S}}$ and the other anchor nodes.

Argument 1 is valid by adopting the privacy-preserving localization protocol. In particular, sensor node j obtains $\mathcal{B}_{i,j}$ and $\mathcal{D}$ (corresponding to Steps 3-4), through which it calculates time difference and reward function by (7.8) and (7.13). Based on the above, localization estimators can be utilized to seek its position $\mathbf{P}_{\mathrm{S},j}$ (corresponding to Sects. 7.4.1, 7.4.2, and 7.4.3).

Argument 2 is correct. From (7.5) it is obvious that the calculation of $\mathbf{P}_{\mathrm{A},i}$ needs the value of $\boldsymbol{\gamma}_{j,i}$ or the values of $\varepsilon_{i,\mathrm{x}}$ and $\varepsilon_{i,\mathrm{y}}$. For independent sensor node j, it cannot obtain the above values through the legitimate information.

Argument 3 is also correct. It is noted that the calculation of $\mathbf{P}_{S,j}$ by anchor node i needs $t_{s,j}$ and $t^{s}_{i,j}$. For anchor node i, the received message from sensor node j is $\mathcal{C}_{j,i}$ for $i \in \mathcal{I}_{A}$, where the values of $t_{s,j}$ and $t^{s}_{i,j}$ are not included. Thereby, the calculation of $\mathbf{P}_{S,j}$ cannot be achieved since the time difference in (7.8) is lacked. Meanwhile, one obtains that anchor node i cannot estimate $\mathbf{P}_{A,i^{\#}}$ for $i^{\#} \in \mathcal{I}_{A}/\{i\}$.

□

Theorem 7.3 *When sensor node j colludes with anchor nodes, sensor node j and anchor nodes can both achieve Level-2 Privacy. In addition, if sensor node j does not involve in collusion, sensor node j can achieve Level-3 Privacy while the anchor nodes can achieve Level-2 Privacy.*

Proof To begin with, one can know that sensor node j can estimate its position $\mathbf{P}_{S,j}$ in whether or not the collusion is occurred. To prove the rest of the theorem, we consider the following two scenarios: (1) sensor node j colludes with anchor nodes; (2) sensor node j does not involve in collusion, however the collusion occurs between anchor nodes.

For scenario 1, sensor node j shares its message to anchor nodes, including $t_{s,j}$, $t^{s}_{i,j}$, $\mathcal{B}_{i,j}$, $\mathcal{C}_{j,i^{\#}}$ and $\mathcal{D}$. Meanwhile, anchor node i shares its collected message to sensor node j and the other anchor nodes, including $\mathcal{A}^{j}_{i}$, $\mathcal{B}_{i,j}$, $\mathcal{C}_{j,i}$ and $\mathcal{D}$. In above context, sensor node j and the other anchor nodes can figure out the position of anchor node i, since $\mathbf{P}_{A,i} = \mathrm{L}_{j,i} + \boldsymbol{\gamma}_{j,i}$ where $\mathrm{L}_{j,i}$ and $\boldsymbol{\gamma}_{j,i}$ are included in $\mathcal{B}_{i,j}$ and $\mathcal{A}^{j}_{i}$, respectively. Besides that, anchor node i can figure out the position of sensor node j, since the time difference is calculated by (7.8). Combining the above conclusions, one knows that sensor node j and anchor nodes can both achieve Level-2 Privacy.

For scenario 2, anchor node i shares its collected message to the other anchor nodes, which includes $\mathcal{A}^{j}_{i}$, $\mathcal{B}_{i,j}$, $\mathcal{C}_{j,i}$ and $\mathcal{D}$. In view of above, anchor node i cannot figure out the position of sensor node j, since the calculation of $\mathbf{P}_{S,j}$ needs the values of $t_{s,j}$ and $t^{s}_{i,j}$. Nevertheless, $t_{s,j}$ and $t^{s}_{i,j}$ are not revealed to anchor node i. In addition, sensor node j cannot figure out the position of anchor node i, since $\boldsymbol{\gamma}_{j,i}$ is transmitted between anchor nodes and anchor nodes do not share $\boldsymbol{\gamma}_{j,i}$ to sensor node j. Besides, $\mathbf{P}_{A,i}$ can be estimated by other anchors involved in collusion because $\mathrm{L}_{j,i}$ and $\boldsymbol{\gamma}_{j,i}$ are included in $\mathcal{B}_{i,j}$ and $\mathcal{A}^{j}_{i}$, respectively. Accordingly, sensor node j can achieve Level-3 Privacy, and anchor nodes can achieve Level-2 Privacy. □

(4) Global Optimum Analysis

As mentioned above, DRL-based localization estimators in this chapter can obtain the global optimal solution. In order to prove this conclusion, we classify our localization estimators into two types: (1) DQN based estimator, i.e., unsupervised estimator (corresponding to Sect. 7.4.1) and semisupervised estimator (corresponding to Sect. 7.4.3); (2) DNN based estimator, i.e., supervised estimator (corresponding to Sect. 7.4.2). In order to support the proof, the following lemma is developed.

Lemma 7.1 (Melo (2001)) *The random process in the form of*

$$\Delta Q(\hat{\mathbf{P}}_{S,j,k+1}) = (1-\alpha_k)\Delta Q(\hat{\mathbf{P}}_{S,j,k}) + \alpha_k F_k(\hat{\mathbf{P}}_{S,j,k}) \tag{7.58}$$

converges to zero with probability 1 if

1. $\alpha_k \in (0, 1)$, $\sum_k \alpha_k = \infty$ and $\sum_k \alpha_k^2 < \infty$;
2. $\left\| \mathbf{E}\left[F_k(\hat{\mathbf{P}}_{S,j,k})|\mathcal{F}_k\right]\right\|_W \leq \ell \left\|\Delta Q(\hat{\mathbf{P}}_{S,j,k})\right\|_W$ for $\ell \in (0, 1]$;
3. $\mathbf{var}\left[F_k(\hat{\mathbf{P}}_{S,j,k})|\mathcal{F}_k\right] \leq V(1+\|\Delta Q\|_W^2)$ for $V \in \mathcal{R}^+$.

Note that, $\mathcal{F}_k = \{\Delta Q(\hat{\mathbf{P}}_{S,j,k}), \Delta Q(\hat{\mathbf{P}}_{S,j,k-1}), \ldots, F_{k-1}, \ldots, \alpha_{k-1}, \ldots\}$ is the past at step k. F_{k-1} and α_{k-1} are allowed to depend on the past insofar as the above conditions remain valid. In addition, $\|\cdot\|_W$ represents some weighted maximum norm. $\mathbf{var}[\cdot]$ is the calculation of variance value.

Based on the above, the following theorem is developed.

Theorem 7.4 *Given finite tuple* $\{\hat{\mathbf{P}}_{S,j,k}, \mathbf{u}_{S,j,k}, r_k, \hat{\mathbf{P}}_{S,j,k+1}\}_{k=1}^{N_e}$, *Q-function* $Q(\hat{\mathbf{P}}_{S,j,k}, \mathbf{u}_{S,j,k}; \eta)$ *as updated by (7.21), and learning rate* $\alpha_k \in (0, 1)$ *satisfying* $\sum_k \alpha_k = \infty$ *and* $\sum_k \alpha_k^2 < \infty$, *subsequently,* $Q(\hat{\mathbf{P}}_{S,j,k}, \mathbf{u}_{S,j,k}; \eta)$ *converges to the optimal value* $Q^*(\hat{\mathbf{P}}_{S,j,k}, \mathbf{u}_{S,j,k}; \eta)$ *with probability 1.*

Proof Subtracting (7.19) from both sides of (7.21), we can acquire

$$\begin{aligned} &Q(\hat{\mathbf{P}}_{S,j,k}, \mathbf{u}_{S,j,k}; \eta) - Q^*(\hat{\mathbf{P}}_{S,j,k}, \mathbf{u}_{S,j,k}; \eta) \\ &= (1-\alpha_k)(Q(\hat{\mathbf{P}}_{S,j,k}, \mathbf{u}_{S,j,k}; \eta) - Q^*(\hat{\mathbf{P}}_{S,j,k}, \mathbf{u}_{S,j,k}; \eta)) \\ &\quad + \alpha_k(r_k + \ell \max Q(\hat{\mathbf{P}}_{S,j,k+1}, \mathbf{u}_{S,j,k+1}; \eta) \\ &\quad - Q^*(\hat{\mathbf{P}}_{S,j,k}, \mathbf{u}_{S,j,k}; \eta)). \end{aligned} \tag{7.59}$$

Define $\Delta Q(\hat{\mathbf{P}}_{S,j,k}, \mathbf{u}_{S,j,k}; \eta) = Q(\hat{\mathbf{P}}_{S,j,k}, \mathbf{u}_{S,j,k}; \eta) - Q^*(\hat{\mathbf{P}}_{S,j,k}, \mathbf{u}_{S,j,k}; \eta)$ and $F_k(\hat{\mathbf{P}}_{S,j,k}, \mathbf{u}_{S,j,k}) = r_k + \ell \max Q(\hat{\mathbf{P}}_{S,j,k+1}, \mathbf{u}_{S,j,k+1}; \eta) - Q^*(\hat{\mathbf{P}}_{S,j,k}, \mathbf{u}_{S,j,k}; \eta)$. Then, one has

$$\begin{aligned} &\mathbf{E}\left\{F_k(\hat{\mathbf{P}}_{S,j,k}, \mathbf{u}_{S,j,k})|\mathcal{F}_k\right\} \\ &= \sum \mathcal{P}_u(\hat{\mathbf{P}}_{S,j,k}, \hat{\mathbf{P}}_{S,j,k+1})(r_k \\ &\quad + \ell \max Q(\hat{\mathbf{P}}_{S,j,k+1}, \mathbf{u}_{S,j,k+1}; \eta)) - Q^*(\hat{\mathbf{P}}_{S,j,k}, \mathbf{u}_{S,j,k}; \eta) \\ &= (\mathbf{G}Q)(\hat{\mathbf{P}}_{S,j,k}, \mathbf{u}_{S,j,k}; \eta) - (\mathbf{G}Q^*)(\hat{\mathbf{P}}_{S,j,k}, \mathbf{u}_{S,j,k}; \eta), \end{aligned} \tag{7.60}$$

where $\mathcal{P}_u(\hat{\mathbf{P}}_{S,j,k}, \hat{\mathbf{P}}_{S,j,k+1})$ is transition probability, and the optimal Q-function is a fixed point of contraction operator $\mathbf{G}$.

Referring to Melo (2001) and noticing with the contraction operator in sup norm, Eq. (7.60) can be rearranged as

$$\left\| \mathbf{E}(F_k(\hat{\mathbf{P}}_{S,j,k}, \mathbf{u}_{S,j,k})|\mathcal{F}_k) \right\|_\infty \leq \ell \left\| Q - Q^* \right\|_\infty = \ell \left\| \Delta Q \right\|_\infty, \tag{7.61}$$

which yields the second condition.

In order to acquire the third condition, one has

$$\begin{aligned} &\mathbf{var}(F_k(\hat{\mathbf{P}}_{S,j,k}, \mathbf{u}_{S,j,k})|\mathcal{F}_k) \\ &= \mathbf{var}(r_k(\hat{\mathbf{P}}_{S,j,k}, \mathbf{u}_{S,j,k}, \widetilde{\mathbf{P}}(\hat{\mathbf{P}}_{S,j,k}, \mathbf{u}_{S,j,k})) \\ &\quad + \ell \max Q(\hat{\mathbf{P}}_{S,j,k+1}, \mathbf{u}_{S,j,k+1}; \eta)|\mathcal{F}_k), \end{aligned} \tag{7.62}$$

where $\widetilde{\mathbf{P}}(\hat{\mathbf{P}}_{S,j,k}, \mathbf{u}_{S,j,k})$ is a random position.

It is noticed that $r_k(\hat{\mathbf{P}}_{S,j,k}, \mathbf{u}_{S,j,k}, \widetilde{\mathbf{P}}(\hat{\mathbf{P}}_{S,j,k}, \mathbf{u}_{S,j,k}))$ is bounded, and hence, Eq. (7.62) can be rearranged as

$$\mathbf{var}(F_k(\hat{\mathbf{P}}_{S,j,k}, \mathbf{u}_{S,j,k})|\mathcal{F}_k) \leq V(1 + \parallel \Delta Q \parallel_W^2), \tag{7.63}$$

which yields the third condition.

According to Lemma 7.1, (7.61) and (7.63), we know that ΔQ converges to zero with probability 1, such that $Q(\hat{\mathbf{P}}_{S,j,k}, \mathbf{u}_{S,j,k}; \eta)$ converges to $Q^*(\hat{\mathbf{P}}_{S,j,k}, \mathbf{u}_{S,j,k}; \eta)$ with probability 1. □

Remark 7.2 The main difference between unsupervised and semisupervised estimators is the selection of initial position. In view of this, we can conclude that the global optimum of semisupervised estimator (corresponding to Sect. 7.4.3) can also be guaranteed. The detailed proof is similar to the one in Theorem 7.4, and it is omitted due to page limitation.

Next, we study the global optimum of supervised estimator. Referring to Du et al. (2019), the global optimum of general supervised estimator can be provided.

Lemma 7.2 (Du et al. (2019)) *Given labelled input data and DNN, if the number of hidden units is large enough, global optimum can be obtained where the training error linearly converges to* 0.

By utilizing Lemma 7.2, one can easily acquire that the global optimum of our supervised estimator in Sect. 7.4.2 can also be guaranteed. In particular, the labelled input data in Sect. 7.4.2 is $\{\mathbf{P}_{P,\mathbf{i}}, \tau_{P,\mathbf{i},i}\}_{\mathbf{i}=1}^{\mathbf{n}}$, and the DNN is described in Fig. 7.5.

(5) Computation Complexity

Of note, the calculation is mainly operated on sensor nodes. According to Hunger and Report (2017); Xiao et al. (2020), the computational complexities for sensor node j in the above three scenarios are proposed.

It is worth mentioned that the computation complexity of the privacy-preserving localization protocol is $\mathcal{O}(3M)$. Furthermore, the computation complexities for **S1** in *Step* 1 − 3 are $\mathcal{O}(M(52N_e - 2o_1) + 3N_e)$, $\mathcal{O}(9G_1\breve{\mu}_{\max,1\#} + 20G_1\breve{\mu}_{\max,2\#} + 3G_1\breve{\mu}_{\max,1\#}\breve{\mu}_{\max,2\#})$ and $\mathcal{O}(M(52k - 2o_2) + 3k + (k-1)(3G_1\ \breve{\mu}_{\max,1\#} + G_1\breve{\mu}_{\max,1\#}\breve{\mu}_{\max,2\#} + 5G_1\breve{\mu}_{\max,2\#}))$, respectively. Note that, o_1 is the operation number for the exploration actions of $[s,\ 0]^{\mathrm{T}}$ and $[0,\ w]^{\mathrm{T}}$. Besides, o_2 is the operation number for the localization actions of $[s,\ 0]^{\mathrm{T}}$ and $[0,\ w]^{\mathrm{T}}$. $\breve{\mu}_{\max,1\#}$ and $\breve{\mu}_{\max,2\#}$ are the numbers of neurons for the two hidden layers in DNN, respectively. Moreover, the unsupervised estimator samples G_1 experiences per time.

The computation complexities for **S2** is $\mathcal{O}(4G_2M\breve{\mu}_{\max,3\#} + 4G_2\breve{\mu}_{\max,3\#}\breve{\mu}_{\max,4\#} + 15G_2\breve{\mu}_{\max,4\#})$, where $\breve{\mu}_{\max,3\#}$ and $\breve{\mu}_{\max,4\#}$ denote the numbers of neurons for the two hidden layers in DNN, respectively. Note that, the supervised estimator samples G_2 experiences per time.

The computation complexities for **S3** in *Step* 1 − 4 are $\mathcal{O}(4G_3M\breve{\mu}_{\max,5\#} + 4G_3\breve{\mu}_{\max,5\#}\breve{\mu}_{\max,6\#} + 15G_3\breve{\mu}_{\max,6\#})$, $\mathcal{O}(M(52n - 52\mathbf{n} - 2o_3) + 3n - 3\mathbf{n})$, $\mathcal{O}(9G_4\breve{\mu}_{\max,7\#} + 3G_4\breve{\mu}_{\max,7\#}\breve{\mu}_{\max,8\#} + 20G_4\breve{\mu}_{\max,8\#})$ and $\mathcal{O}(M(52k - 2o_4) + 3k)$, respectively. $\breve{\mu}_{\max,5\#}$ and $\breve{\mu}_{\max,6\#}$ denote the numbers of neurons for the two hidden layers in DNN that estimate the initial position. Moreover, o_3 denotes the operation number for the exploration actions of $[s,\ 0]^{\mathrm{T}}$ and $[0,\ w]^{\mathrm{T}}$. Besides, $\breve{\mu}_{\max,7\#}$ and $\breve{\mu}_{\max,8\#}$ are the numbers of neurons for the two hidden layers in DNN, respectively. o_4 is the operation number for the localization actions of $[s,\ 0]^{\mathrm{T}}$ and $[0,\ w]^{\mathrm{T}}$. The semisupervised estimator samples G_3 and G_4 experiences per time for the DNN and DQN, respectively.

(6) Communication Complexity

Similar to Shi and Wu (2018); Wang et al. (2018), the communication complexity is studied by counting the transmitted and received scalars for each node. Note that, the communication is conducted in *Step* 1 − 4 for the localization protocol. For anchor 1, the numbers of transmitted scalars in *Step* 1 − 4 are 0, 3, 5 and 7, respectively; the numbers of received scalars in *Step* 1 − 4 are 1, 3, 0 and 5, respectively. For anchor $h_\# \in \mathcal{I}_A/\{1\}$, the numbers of transmitted scalars in *Step* 1 − 4 are 0, 3, 11 and 0, respectively; the numbers of received scalars in *Step* 1 − 4 are 1, 3, 5 and 0, respectively. For sensor node j, the numbers of transmitted scalars in *Step* 1 − 4 are M, 0, $5M - 5$ and 5, respectively; the numbers of received scalars in *Step* 1 − 4 are 0, 0, $11M - 6$ and 7, respectively. Totally, anchor 1 transmits 15 scalars and receives 9 scalars, while anchor $h_\# \in \mathcal{I}_A/\{1\}$ transmits 14 scalars and receives 9 scalars. In addition that, sensor node $j \in \mathcal{I}_S$ transmits $6M$ scalers and receives $11M + 1$ scalars.

7.5 Simulation Results

In this section, simulation results are conducted on MATLAB 2016a. Particularly, four anchor nodes are deployed to localize three sensor nodes. Some parameters are given as follows: $a = 0.017\ \mathrm{s}^{-1}$, $b = 1473$ m/s, $s = 1.3$m, $w = 0.5$ m, $c = 1$ m,

$v_{\max} = 1000$, $\sigma_{\text{mea}}^2 = 0.00003$, $\epsilon = 0.001$, $N_{\text{m}} = 10^7$, $N_{\text{e}} = 50000$, $\Delta\mathcal{H} = 0.0005$ s and $K = 600$.

A. Equivalence of the Localization Protocol
Different from the traditional localization protocols, i.e., Ramezani et al. (2013), Mortazavi et al. (2017), and Yan et al. (2020c), the proposed localization protocol in this chapter can not only preserve the position privacy, but also ensure the equivalence of travel time. In order to verify this conclusion, we take the travel times $\tau_{j,i}$ and $\breve{\tau}_{j,i}$ as the example, where $i \in \{1, 2, 3, 4\}$ and $j \in \{1, 2, 3\}$. It is worth mentioning that the calculation of $\tau_{j,i}$ does not require the private position information of anchor nodes, i.e., $x_{\text{A},i}$ and $y_{\text{A},i}$. However, the calculation of $\breve{\tau}_{j,i}$ requires the private position information of anchor nodes, as presented in Ramezani et al. (2013), Mortazavi et al. (2017), and Yan et al. (2020c). The travel times $\tau_{j,i}$ and $\breve{\tau}_{j,i}$ are shown in Fig. 7.7a, through which we know the travel times obtained by our solution and the traditional localization protocols (Ramezani et al. 2013; Mortazavi et al. 2017; Yan et al. 2020c) are completely the same. This result verifies the effectiveness of Theorem 7.1.

In addition, a special scenario is considered, i.e., the stratification effect is ignored in the localization protocol. We apply the algorithm in Cheng et al. (2008), and Zhao et al. (2020) to calculate the travel time, where the acoustic speed is fixed as 1500 m/s. The travel time is shown on the top of Fig. 7.7b, whose errors are presented on the bottom of Fig. 7.7b. It is clear that the stratification effect can influent the calculation accuracy. Thereby, the consideration of stratification effect is necessary for USNs.

B. Effectiveness of the Localization Estimator
An unsupervised DRL-based localization estimator is presented in Sect. 7.4.1 to acquire $x_{\text{S},j}$ and $y_{\text{S},j}$. To be specific, a DNN with two hidden layers, each has seven neurons, is used in the DQN. Based on this, the explored and estimated trajectories for sensor node 1 are shown in Fig. 7.8a. Similarly, the trajectories for sensor nodes 2 and 3 are presented in Fig. 7.8b, c, respectively. Clearly, the explored trajectories are stochastically selected, however, the estimated trajectories have a close trend to the actual position point. Accordingly, the final estimated position information of sensor nodes is presented in Fig. 7.8d, where the depth information is accurately measured by depth units. To show more clearly, the localization error of sensor node $j \in \mathcal{I}_{\text{S}}$ at iteration step k is defined as $\text{error}_{j,k} = \sqrt{(\hat{x}_{\text{S},j,k} - x_{\text{S},j})^2 + (\hat{y}_{\text{S},j,k} - y_{\text{S},j})^2}$. Thus, these localization errors are depicted in Fig. 7.8e, and the total reward for each sensor node is shown in Fig. 7.8f. From Fig. 7.8d–f, one knows the position information of sensor nodes can be correctly estimated, since the localization errors approximately convergent to zeros while the reward can reach to the maximum value. Meanwhile, the localization errors for each sensor node are very close to the CRLB.

In Sect. 7.4.2, a supervised DRL-based localization estimator is designed for sensor nodes, where the DNN consists two hidden layers and each hidden layer

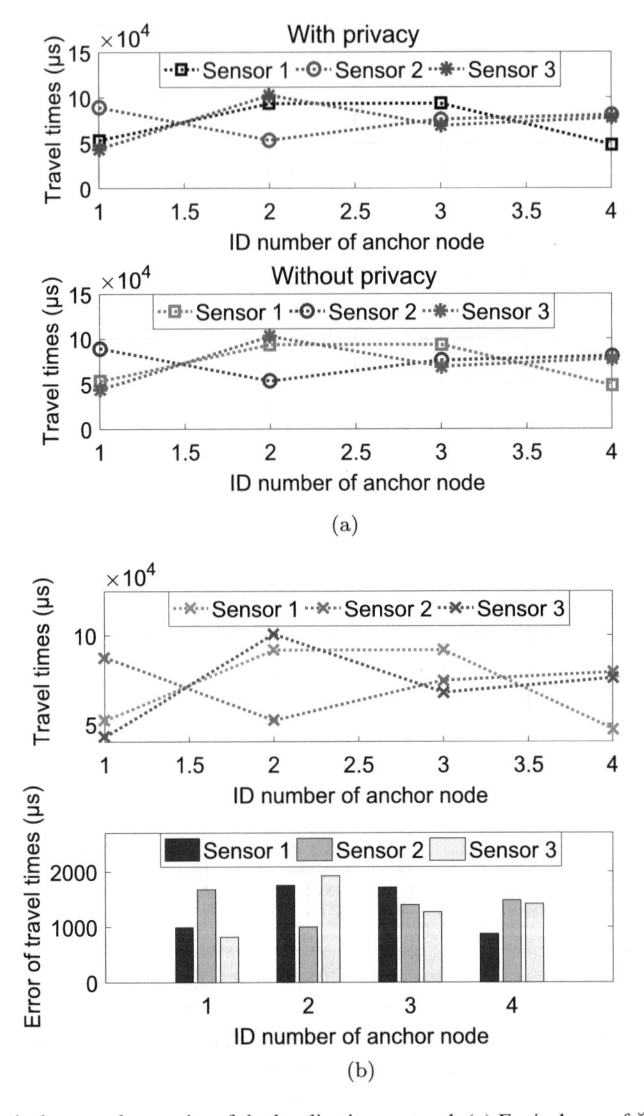

Fig. 7.7 Equivalence and necessity of the localization protocol. (**a**) Equivalence of $\breve{\tau}_{j,i}$ and $\tau_{j,i}$. (**b**) Travel time in a special scenario

has seven neurons. According to the number of labelled data, the following two cases are considered, i.e., Case 1: $\mathbf{n} = 35000$; Case 2: $\mathbf{n} = 350000$. Based on this, Fig. 7.9a shows the final position estimation of sensor nodes, and meanwhile, the localization errors are depicted on the top of Fig. 7.9b. Clearly, the localization accuracy in Case 2 is higher than the one in Case 1. The reason associated with this phenomenon is that the number of labelled data in Case 2 is larger that the one in Case 1. In order to further verify this conclusion, we increase the number of labelled

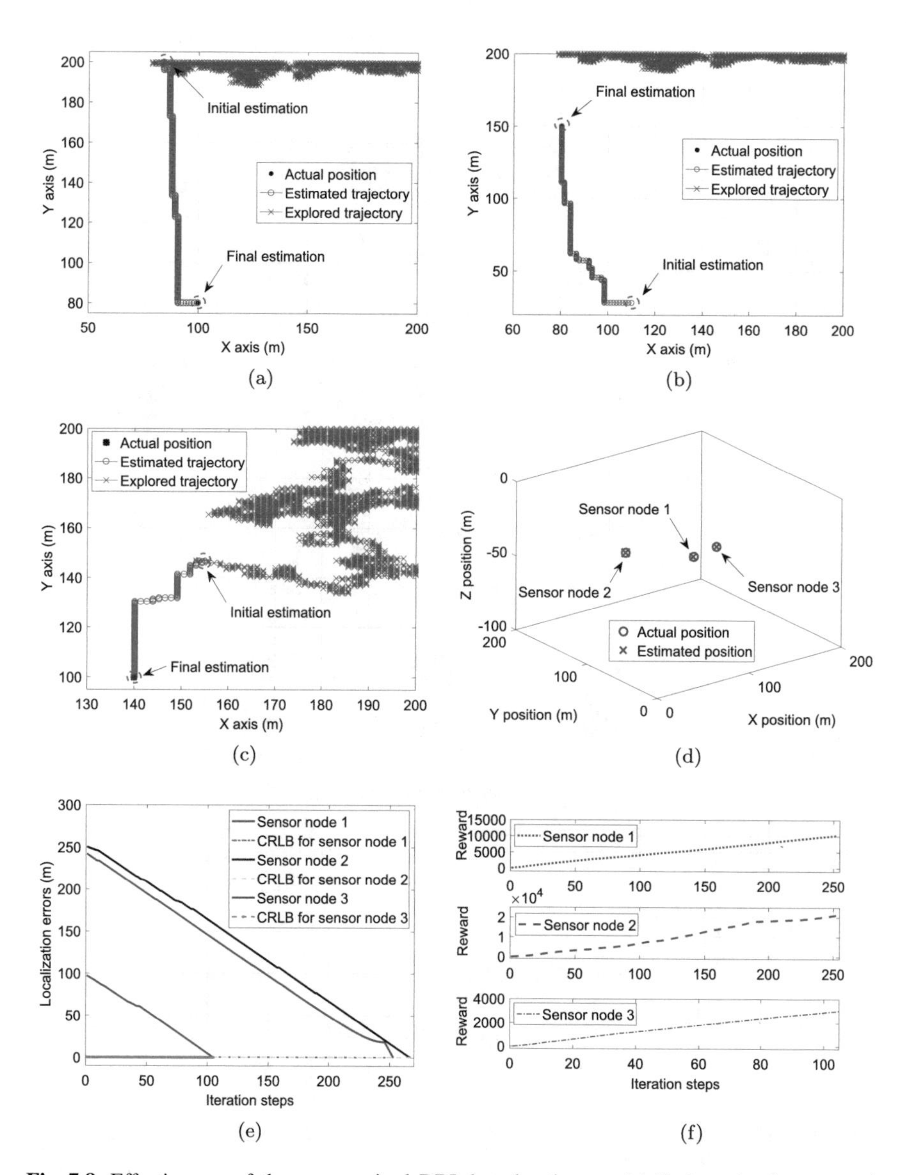

Fig. 7.8 Effectiveness of the unsupervised DRL-based estimator. (**a**) Trajectories for sensor 1. (**b**) Trajectories for sensor 2. (**c**) Trajectories for sensor 3. (**d**) Positions of sensor nodes. (**e**) Localization errors for sensor nodes. (**f**) Reward for each sensor node

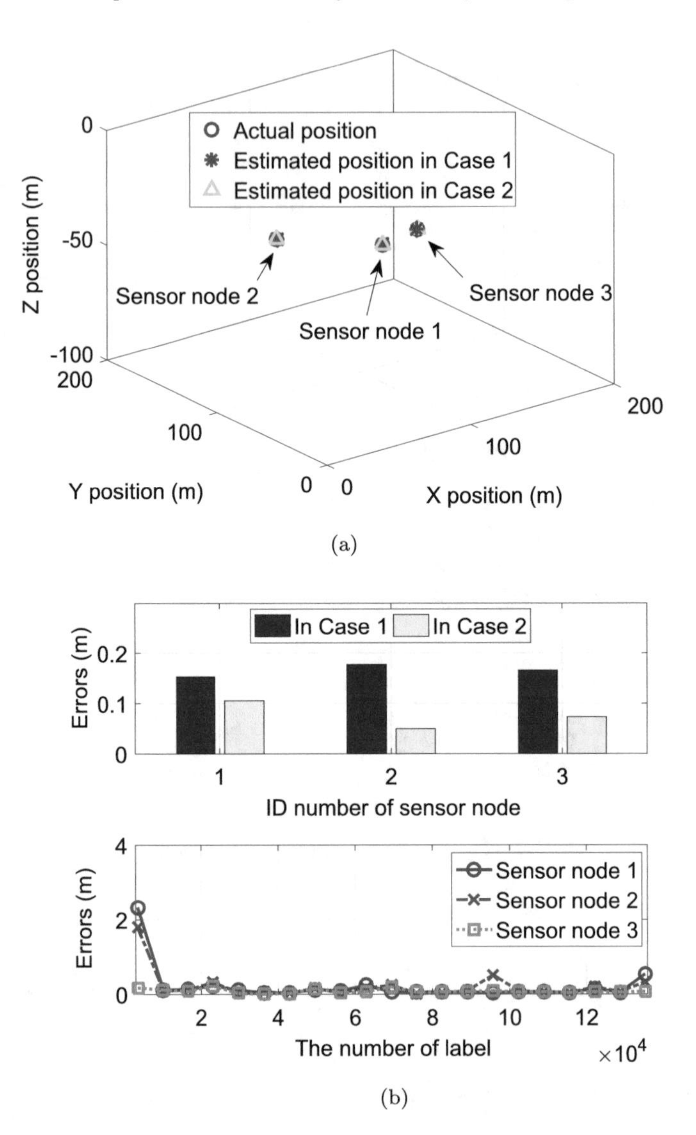

Fig. 7.9 Effectiveness of the supervised DRL-based estimator. (**a**) Positions of sensor nodes. (**b**) Localization errors for sensor nodes

data, and the localization errors are presented on the bottom of Fig. 7.9b. Obviously, the number of labelled data plays a critical role on localization accuracy.

Beyond that, we propose a semisupervised DRL-based localization estimator in Sect. 7.4.3. In order to verify the effectiveness of semisupervised DRL-based localization estimator, the exploration and training trajectories of sensor nodes are represented in Fig. 7.10a–c. The final estimated position information of sensor nodes is depicted in Fig. 7.10d, and the localization errors are presented in Fig. 7.10e. Correspondingly, the total reward is presented in Fig. 7.10f. From Fig. 7.10a–f, we know the localization of sensor nodes can be achieved.

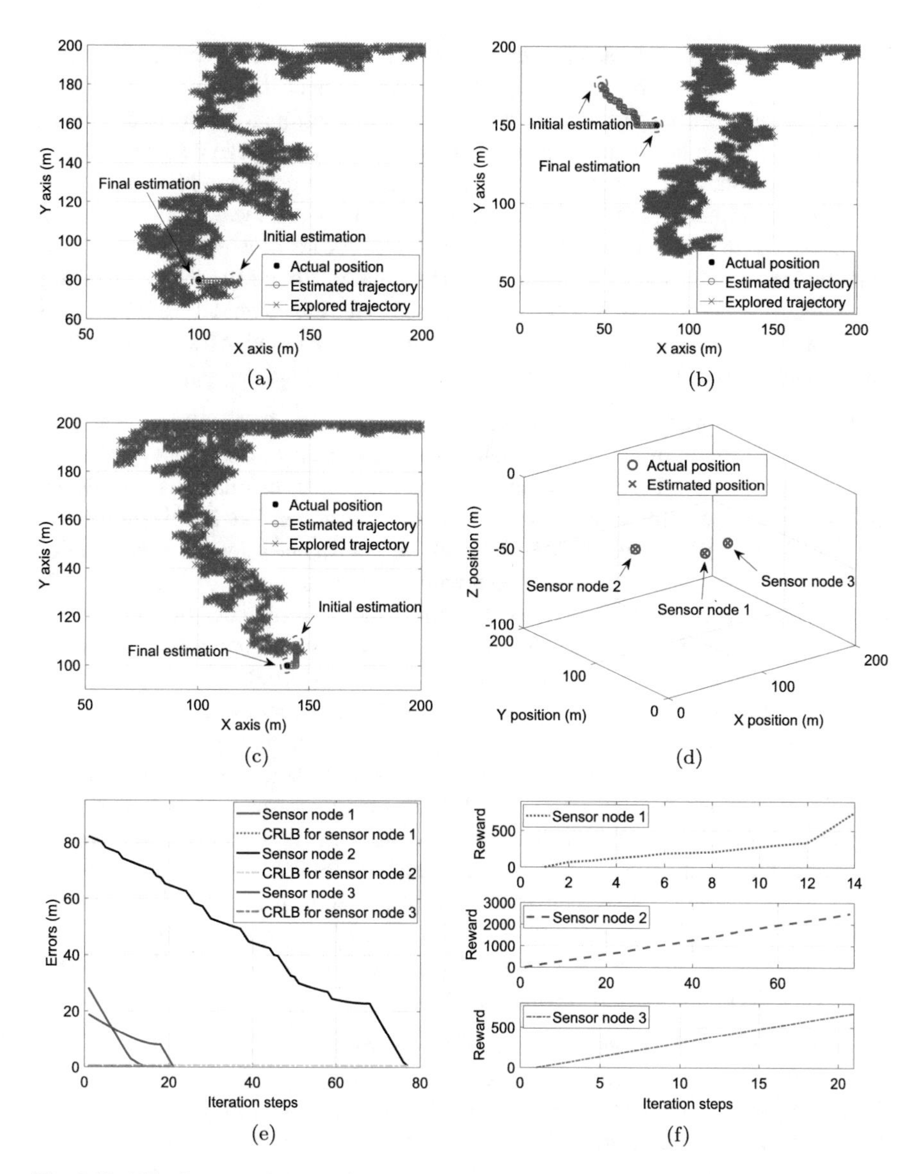

Fig. 7.10 Effectiveness of the semisupervised DRL-based estimator. (**a**) Trajectories for sensor 1. (**b**) Trajectories for sensor 2. (**c**) Trajectories for sensor 3. (**d**) Positions of sensor nodes. (**e**) Localization errors for sensor nodes. (**f**) Reward for each sensor node

C. Comparison of the Three Localization Estimators

As mentioned in Li et al. (2020) and Zhang et al. (2014), the unsupervised estimator is based on a well-exploration mechanism, and it does not require position labels. However, the convergence speed of unsupervised localization estimator may be slow when the initial position is far away from the actual position. Alternatively, the convergence speed of semisupervised localization estimator can be very fast, because the initial position vector is deterministically selected by DNN. On the other hand, the convergence speed of supervised localization estimator is depended on the number of position labels, and hence, the following two cases can be occurred, i.e., (1) when the number of position labels is small, the convergence speed of supervised localization estimator is faster that the one in unsupervised localization estimator; (2) otherwise, the convergence speed of supervised localization estimator is slower that the one in unsupervised localization estimator. To verify this conclusion, Fig. 7.11a, b are presented. It is clear that, these results are coincided with the above conclusion. In addition, some other related results can be found in Yan et al. (2020a) and Yan et al. (2021).

The localization accuracy of supervised estimator is also depended on the number of position labels. Figure 7.11c is given to show the localization errors for Case 1. It is noticed that the localization accuracy of supervised estimator is not always higher that the one with unsupervised localization estimator. The localization errors for Case 2 are presented in Fig. 7.11d. From Fig. 7.11d, one knows the localization accuracy of supervised localization estimator is always higher that the one with unsupervised estimator. Accordingly, we know the performances of convergence speed and localization accuracy for semisupervised estimator are better that the ones with unsupervised estimator. Meanwhile, the performances of convergence speed and localization accuracy for supervised estimator are depended on the number of position labels. However, a large number amount of labelled data is not always feasible. For example, when the labelled data is not correct or the number of labelled data is small, the localization performance may be greatly degraded. In order to verify this conclusion, Gauss noises are added to the labelled data, and the localization errors are presented in Fig. 7.11e. Besides, we can also reduce the number of available data labels from 46500 to 3300, through which the localization errors are presented in Fig. 7.11f. These results reflect that semisupervised localization estimator is an appreciate choice when the available position labels do not occupy the majority.

D. Advantage of DRL-Based Localization Estimator

We assume that the localization procedure is disturbed by periodic noise, and thus the noise can be given as $\Psi = 0.05\cos(0.38\hat{x}_{S,j} - 0.05)\sin(0.7\hat{y}_{S,j} + 0.05)$. Then, the localization optimization problem can be rearranged as $(x^*_{S,j}, y^*_{S,j}) = \text{argmin}_{(\hat{x}_{S,j}, \hat{y}_{S,j})}\left[\mathcal{H}(\hat{x}_{S,j}, \hat{y}_{S,j}) + \Psi\right]$, where the global optimum is $(x^*_{S,j}, y^*_{S,j}) = (100, 80)$. By using least squares-based estimator, e.g., Liu et al. (2016), Mortazavi et al. (2017), and Yan et al. (2020c), the estimated position is given as (91.69, 88.88) with cost 59093, as presented in Fig. 7.12a–c. Meanwhile, the estimated positions with unsupervised, supervised and semisupervised estimators are calculated as (99.8, 80) with cost 13005, (99.94, 80.01) with cost 1281.5, and (99.9, 79.9) with cost 3250.3, respectively. From Fig. 7.12a–c, it is noticed that the DRL-based localization estimators can obtain global optimal solution, while the least squares-based strategy can easily fall into local optimum.

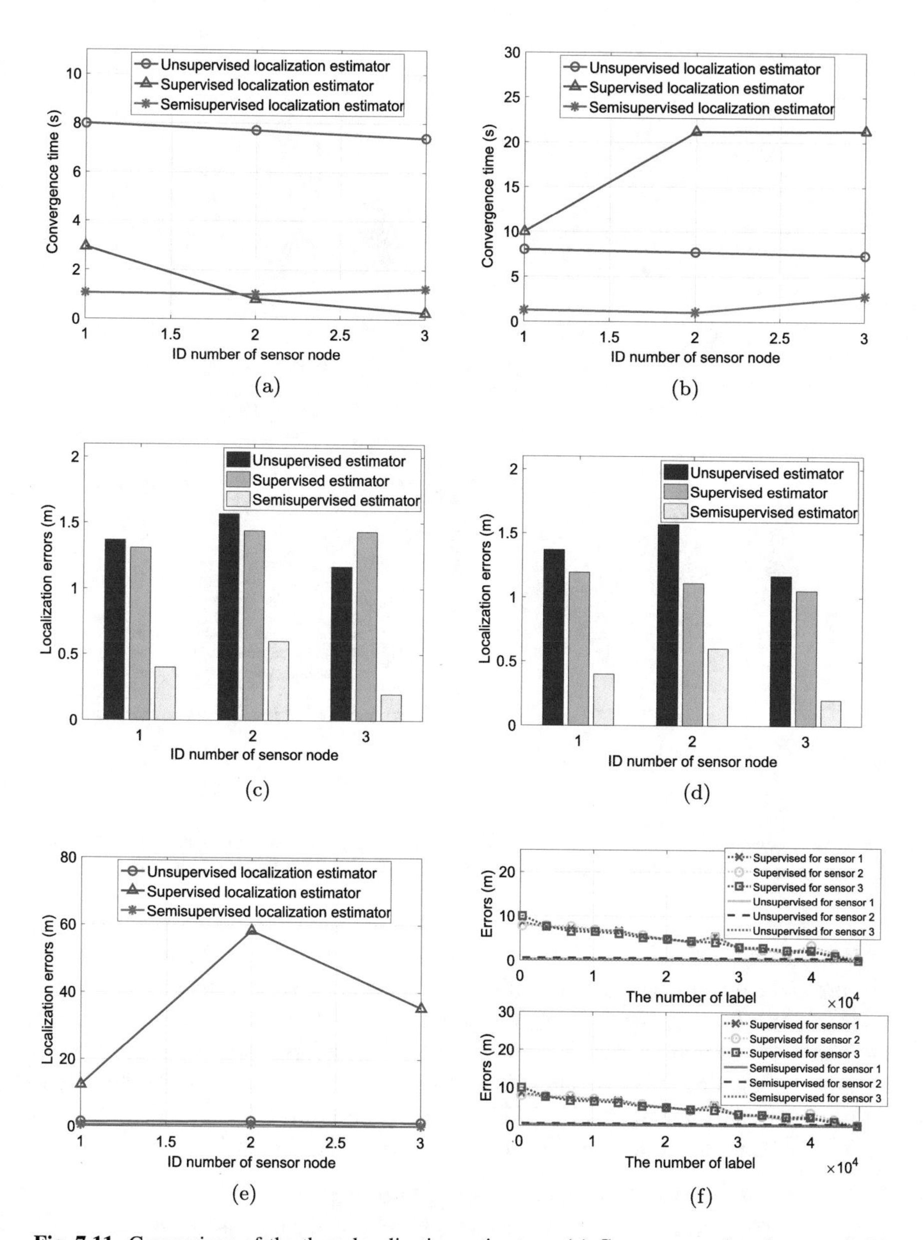

Fig. 7.11 Comparison of the three localization estimators. (**a**) Convergence time for case 1. (**b**) Convergence time for case 2. (**c**) Localization error for case 1. (**d**) Localization error for case 2. (**e**) Error when labelled data is wrong. (**f**) Error when the number is small

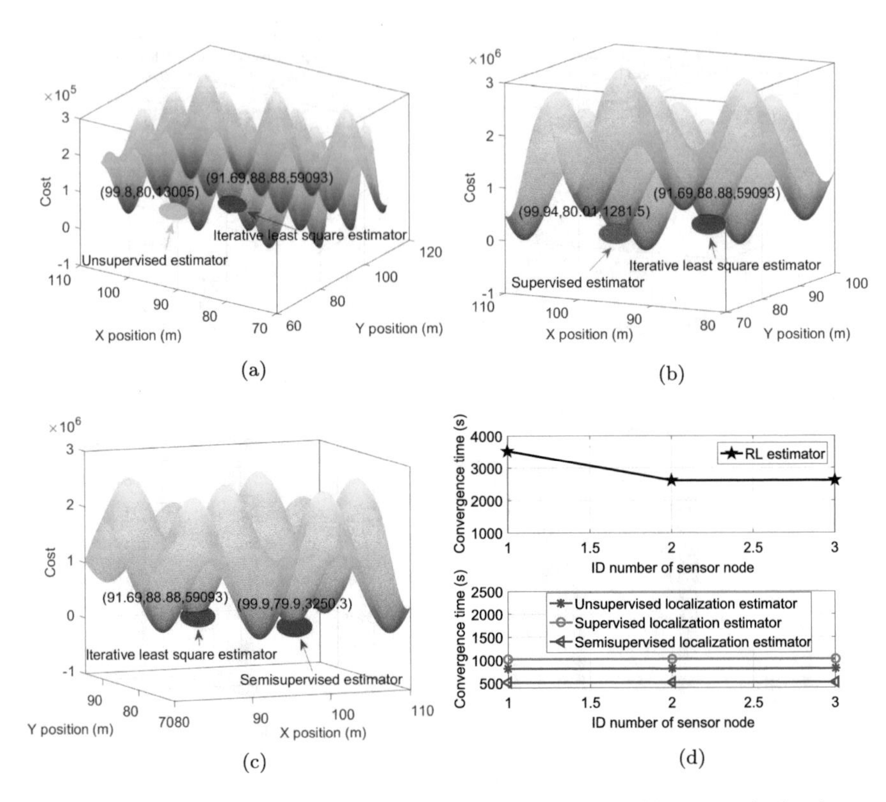

(a)

(b)

(c)

(d)

Fig. 7.12 Advantage of the DRL-based localization estimator. (**a**) With unsupervised estimator. (**b**) With supervised estimator. (**c**) With semisupervised estimator. (**d**) With RL estimator

It is noted that, RL localization estimator, e.g., Yan et al. (2020b), needs to explore and gain knowledge of an entire environment. The above operation requires a lot of time to seek the best policy. Alternatively, the DRL based localization estimator can overcome this deficiency, since DRL adopts deep neural networks to train the learning process. Thus, the iteration processes by using RL and DRL estimators are presented in Fig. 7.12d. Clearly, the convergence speeds of DRL based estimators are faster than the one with RL based estimator.

E. Level Evaluation for the Location Privacy Preservation

We evaluate the location privacy preservation level. Specially, assume there are M^* colluding anchor nodes, we study the privacy preservation levels of the localization system in two cases: (1) sensor node $j \in \mathcal{I}_S$ colludes with anchor nodes; (2) sensor node $j \in \mathcal{I}_S$ does not collude with anchor nodes. Based on this, the preservation levels for Case 1 and Case 2 are depicted by Fig. 7.13a, b, respectively. The above results are coincided with the conclusions in Theorem 7.2 and Theorem 7.3. In order to show more clearly, we label sensor node j as '0', while the anchor nodes are sequently labeled as '1', '2', '3' and '4'. On this basis, Table 7.1 is presented

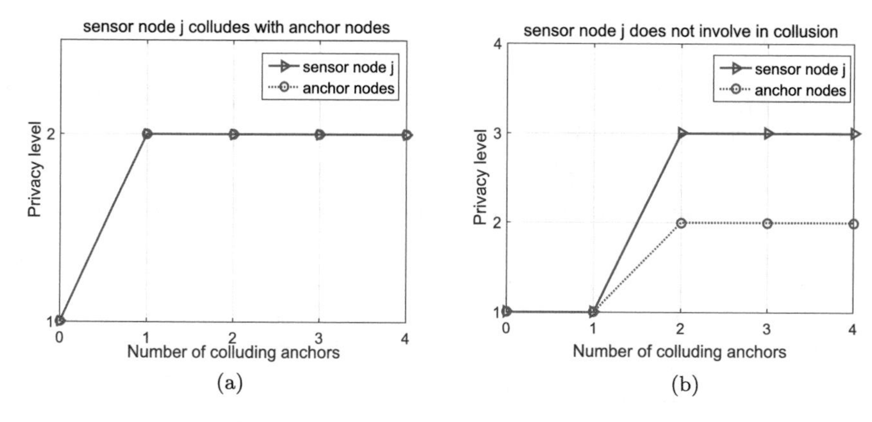

Fig. 7.13 Privacy preservation levels for the localization system. (**a**) Case 1. (**b**) Case 2

Table 7.1 Location privacy ability for each node in Case 2

Colluding nodes	0	1	2	3	4
(1, 2)	√	×	×	√	√
(1, 3)	√	×	√	×	√
(1, 4)	√	×	√	√	×
(2, 3)	√	√	×	×	√
(2, 4)	√	√	×	√	×
(3, 4)	√	√	√	×	×
(1, 2, 3)	√	×	×	×	√
(1, 2, 4)	√	×	×	√	×
(1, 3, 4)	√	×	√	×	×
(2, 3, 4)	√	√	×	×	×
(1, 2, 3, 4)	√	×	×	×	×

to show the location privacy ability for each node in Case 2, where symbol '×' denotes the position privacy is exposed and symbol '√' denotes the position privacy is preserved. From Table 7.1, one knows the position privacy may be exposed in the presence of colluding nodes, and it is necessary to reduce the occurrence of node collusion in some special situations.

7.6 Conclusion

This chapter investigates a DRL-based privacy-preserving localization issue for USNs. With consideration of stratification effect, a privacy-preserving localization protocol is developed to hide private position information in inhomogeneous underwater medium. Based on this, the unsupervised, supervised and semisupervised scenarios are considered, respectively. Correspondingly, three DRL-based localization estimators are developed to localize sensor nodes. Finally, simulation results are conducted.

References

Ameer P, Jacob L (2010) Localization using ray tracing for underwater acoustic sensor networks. IEEE Commun Lett 14(10):930–932

Carroll P, Mahmood K, Zhou S, Zhou H, Xu X, Cui J (2014) On-demand asynchronous localization for underwater sensor networks. IEEE Trans Signal Process 62(13):3337–3348

Cheng X, Shu H, Liang Q, Du D (2008) Silent positioning in underwater acoustic sensor networks. IEEE Trans Veh Technol 57(3):1756–1766

Du S, Poczós B, Zhai X, Singh A (2019) Gradient descent provably optimizes over-parameterized neural networks. In: Proceedings IEEE, ICLR, New Orleans, pp 1–19

Gong Z, Li C, Jiang F, Zheng J (2020) AUV-aided localization of underwater acoustic devices based on doppler shift measurements. IEEE Trans Wireless Commun 19(4):2226–2239

Han G, Jiang J, Sun N, Shu L (2015) Secure communication for underwater acoustic sensor networks. IEEE Commun Mag 53(8):54–60

Hunger R, Report T (2017) Floating point operations in matrix-vector calculus. Technische Universit ät München, München

Im J, Jeon S, Lee M (2020) Practical privacy-preserving face authentication for smartphones secure against malicious clients. IEEE Trans. Inf. Forensics Secur. 15(1):2386–2401

Jia T, Ho K, Wang H, Shen X (2019) Effect of sensor motion on time delay and doppler shift localization: analysis and solution. IEEE Trans Signal Process 67(22):5881–5895

Jiang S (2019) On securing underwater acoustic networks: a survey. IEEE Commun Surv. Tutorials 21(1):729–752

Jondhale S, Shubair R, Labade R, Lloret J, Gunjal P (2020) Application of supervised learning approach for target localization in wireless sensor network. Adv Intell Syst Comput 1132(1):493–519

Li H, He Y, Cheng X, Zhu H, Sun L (2015) Security and privacy in localization for underwater sensor networks. IEEE Commun Mag 53(11):56–62

Li Y, Hu X, Zhuang Y, Gao Z, Zhang P, El-Sheimy N (2020) Deep reinforcement learning (DRL): another perspective for unsupervised wireless localization. IEEE Internet Things J. 7(7):6279–6287

Liu J, Wang Z, Cui J, Zhou S, Yang B (2016) A joint time synchronization and localization design for mobile underwater sensor networks. IEEE Trans. Mob. Comput. 15(3):530–543

Liu Y, Fang G, Chen H, Xie L, Fan R, Su X (2018) Error analysis of a distributed node positioning algorithm in underwater acoustic sensor networks. In: Proceedings of IEEE WCSP, Hangzhou, pp 1–6

Luo H, Wu K, Ruby R, Liang Y, Guo Z, Ni L (2018) Software-defined architectures and technologies for underwater wireless sensor networks: a survey. IEEE Commun Surv Tutorials 20(4):2855–2888

Luong N, Hoang D, Gong S, Niyato D, Wang P, Liang Y, Kim D (2019) Applications of deep reinforcement learning in communications and networking: a survey. IEEE Commun Surv Tutorials 21(4):3133–3174

Melo F (2001) Convergence of Q-learning: a simple proof. Institute of Systems and Robotics, Tech. Rep., pp 1–4

Meng W, Zheng Q, Yang L, Li P, Pan G (2020) Qualitative measurements of policy discrepancy for return-based deep Q-network. IEEE Trans Neural Networks Learn Syst 31(10):4374–4380

Mnih V, Kavukcuoglu K, Silver D et al (2015) Human-level control through deep reinforcement learning. Nature 518(7540):529–533

Mortazavi E, Javidan R, Dehghani M, Kavoosi V (2017) A robust method for underwater wireless sensor joint localization and synchronization. Ocean Eng 137(1):276–286

Ramezani H, Jamali-Rad H, Leus G (2013) Target localization and tracking for an isogradient sound speed profile. IEEE Trans Signal Process 61(6):1434–1446

Saeed N, Alouini M, Al-Naffouri T (2020) Accurate 3-D localization of selected smart objects in optical internet of underwater things. IEEE Internet Things J 7(2):937–947

Shi X, Wu J (2018) To hide private position information in localization using time difference of arrival. IEEE Trans Signal Process 66(18):4946–4956

Shu T, Chen Y, Yang J (2015) Protecting multi-lateral localization privacy in pervasive environments. IEEE/ACM Trans Netw 23(5):1688–1701

Soares C, Gomes J, Ferreira B, Costeira J (2017) LocDyn: robust distributed localization for mobile underwater networks. IEEE J Oceanic Eng 42(4):1063–1074

Shu T, Chen Y, Yang J, Williams A (2014) Multi-lateral privacy-preserving localization in pervasive environments. In: Proceedings of IEEE INFOCOM, Toronto, pp 2319–2327

Wang G, He J, Shi X, Pan J, Shen S (2018) Analyzing and evaluating efficient privacy-preserving localization for pervasive computing. IEEE Internet Things J 5(4):2993–3007

Xiao L, Sheng G, Wan X, Su W, Cheng P (2019) Learning-based PHY-layer authentication for underwater sensor networks. IEEE Commun Lett 23(1):60–63

Xiao Y, Xiao L, Lu X, Zhang H, Yu S, Poor H (2020) Deep reinforcement learning based user profile perturbation for privacy aware recommendation. IEEE Internet Things J 8:4560–4568. https://doi.org/10.1109/JIOT.2020.3027586

Yang Z, Javinen K (2018) The death and rebirth of privacy-preserving WiFi fingerprint localization with paillier encryption. In: Proceedings of IEEE INFOCOM, Honolulu, pp 1223–1231

Yan J, Meng Y, Yang X, Luo X, Guan X (2020a) Privacy-preserving localization for underwater sensor networks via deep reinforcement learning. IEEE Trans Inform Foren Secur 16(1):1880–1895

Yan J, Gong Y, Chen C, Luo X, Guan X (2020b) AUV-aided localization for internet of underwater things: a reinforcement learning-based method. IEEE Internet Things J. 7(10):9728–9746

Yan J, Guo D, Luo X, Guan X (2020c) AUV-aided localization for underwater acoustic sensor networks with current field estimation. IEEE Trans Veh Technol 69(8):8855–8870

Yan J, Meng Y, Luo X, Guan X (2021) To hide private position information in localization for internet of underwater things. IEEE Internet Things J https://doi.org/10.1109/JIOT.2021.3068298

Zhang Q, Yin Y, Zhan D, Peng J (2014) A novel serial multimodal biometrics framework based on semisupervised learning techniques. IEEE Trans Inf Forensics Secur 9(10):1681–1694

Zhao H, Yan J, Luo X, Guan X (2020) Privacy preserving solution for the asynchronous localization of underwater sensor networks. IEEE/CAA J Autom Sin 7(6):1511–1527

Zheng C, Sun D, Cai L, Li X (2018) Mobile node localization in underwater wireless networks. IEEE Access 6(1):17232–17244

Chapter 8
Future Research Directions

We have addressed several new canonical localization schemes for USNs, including asynchronous localization with mobility prediction, consensus-based UKF localization, RL-based localization in weak communication channel, privacy preserving asynchronous localization, privacy preserving asynchronous localization with malicious attacks, and DRL-based privacy preserving localization. In this chapter, we present several research directions that depict future investigation on underwater localization, including the network architecture, communication protocol and optimization estimator.

8.1 Space-Air-Ground-Sea Network Architecture

In order to satisfy the requirement of the future underwater localization, an integrated space-air-ground-sea network architecture is required, as shown in Fig. 8.1. The space-air-ground-sea network is an infrastructure based on the group network, while it is supplemented by the space, air and sea networks. Particularly, the components of such system can be presented as follows.

- **Satellite Network.** Satellite network can provide communication services to achieve full coverage of remote areas such as the polar regions and open sea. Based on this, it can realize real-time data processing, and other marine functions without relying on the ground data centers.
- **Air Network.** Air network includes the high-altitude platforms (HAPs) and unmanned aerial vehicles (UAVs). Its communication delay is shorter than the one of satellite network, while its coverage range is larger than the one of ground network. Of note, the deployment of air network is very convenient and flexible, hence, it can provide low-latency communication and timely services for maritime hot spots.

J. Yan et al., *Localization in Underwater Sensor Networks*, Wireless Networks,
https://doi.org/10.1007/978-981-16-4831-1_8

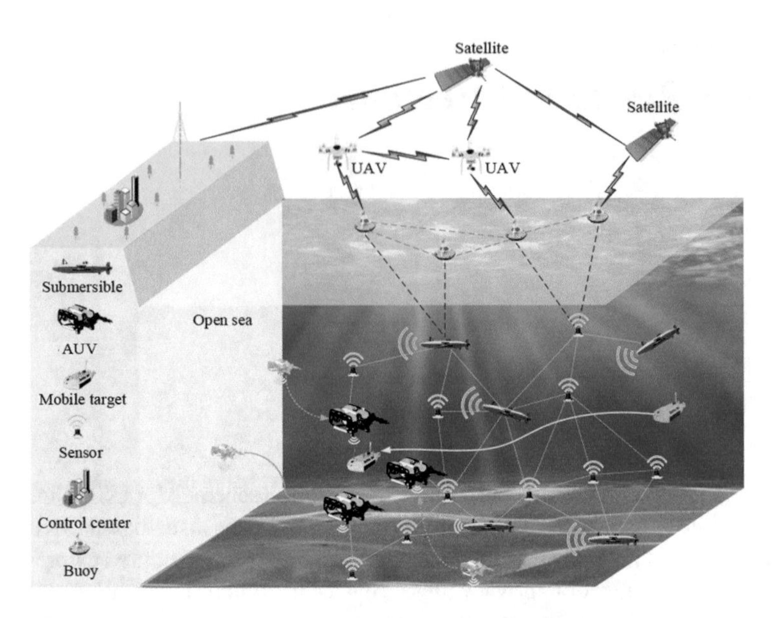

Fig. 8.1 Illustration of the integrated space-air-ground-sea network

- **Ground Network**. By acoustic and electromagnetic communication links, the ground network on coast (or in a vessel on water offshore) collects all available data and then makes mission decision. Based on this, the mission decision is transmitted to underwater nodes via the aid of buoys.
- **Underwater Network**. Underwater network is composed of static sensors and mobile vehicles, which are deployed to perform cooperative monitoring tasks. The deployment of underwater network can enhance the monitoring capacity for various applications such as intrusion surveillance, marine resource protection, navigation, geographic mapping and petroleum exploration. However, its communication and computing ability is still limited due to the weak communication property.

8.2 Intergradation Design of Localization Protocol

At present, the sensing, communication and control systems for USNs are usually independent to each other. Specifically, the sensing system mainly focuses on how to use the active (or passive) sonar to receive the target radiated noise and scattered echo. At the same time, the beamforming, azimuth estimation and other signal processing technologies are employed to determine the target shape and position information. With the consideration of the characteristics of underwater acoustic channel such as multipath interference, Doppler frequency shift, narrow bandwidth and high energy consumption, the communication system mainly focuses on how to design the information forwarding strategy, through which the signal of source node can be transmitted to the remote receiving transducer. In addition, the control system mainly focuses on how to use the feedforward information of the control center and the feedback information of the underwater nodes, such that a high-performance controller can be designed to ensure the stable control of the underwater nodes, e.g., the flocking control and formation control.

Clearly, the sensing and communication systems have a certain overlap in the working principle and signal processing. If they are designed and used separately, they will compete with each other in bandwidth utilization and energy consumption. In addition, the communication and control systems have strong complementarity in function. If they do not interact with each other, the control command of the submersible will face incomplete information loss. On the other hand, the communication topology and resource allocation of the sensor will lack effective feedback optimization mechanism. Therefore, it is urgent to integrate the underwater sensing, communication networking and cooperative control system, so as to realize the all-round and all-weather real-time localization of underwater mobile targets. The research results of this idea can not only make up for the shortcomings of the existing static sensor network, but also provide the necessary theoretical basis and technical support for underwater localization application. In order to explain more clearly, an example of the sensing-communication-control intergradation design of underwater localization protocol is depicted in Fig. 8.2.

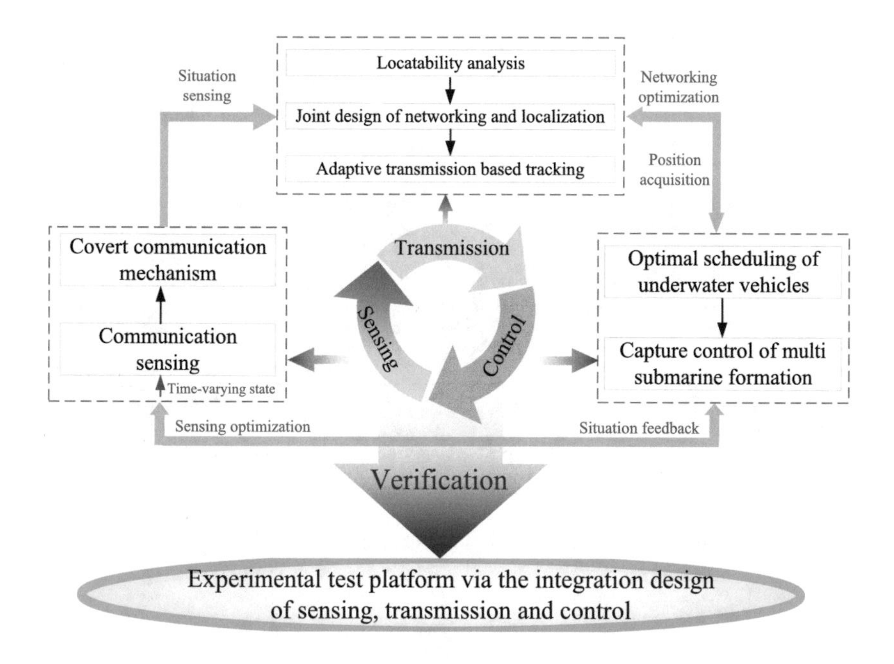

Fig. 8.2 Example of the sensing-communication-control intergradation design

8.3 Learning-Based Optimization Estimator

Most of the existing underwater localization approaches focus on localization accuracy and they ignore the influence of underwater acoustic communication channel. Specifically, it is assumed that the underwater communication quality is ideal within a certain range of the transmitter and invalid otherwise. However, underwater communication channel suffers from strong shadow and multipath fading. Besides that, USNs are usually deployed in harsh environment. Thus, the ignoring of attack detection and privacy preservation can make localization system easily vulnerable to failure.

In our previous work, we have incorporated the privacy-preserving strategy into the localization of USNs. However, more practical scenarios should be considered in the future localization algorithm. If the complex constraints are considered, the optimization problem will be very hard to solve. A feasible way is to adopt the DRL strategy to solve the optimization problem, and we believe the localization performance will benefit a lot from DRL. With consideration of the weak communication channel and intrusion attack, an example of the DRL-based localization procedure is depicted in Fig. 8.3.

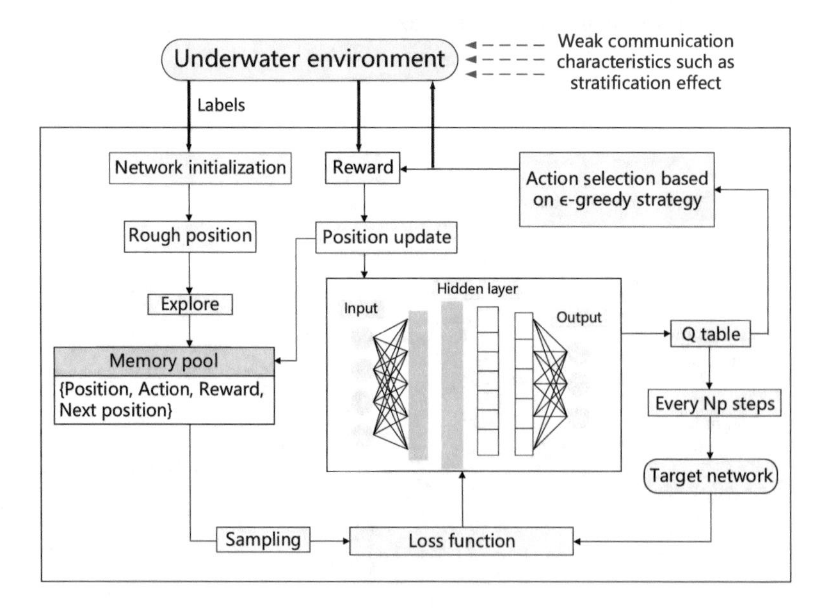

Fig. 8.3 Example of the DRL-based localization procedure